ESO ASTROPHYSICS SYMPOSIA
European Southern Observatory

Series Editor: Jacqueline Bergeron

Physics and Astronomy | ONLINE LIBRARY

http://www.springer.de/phys/

Springer-Verlag Berlin Heidelberg GmbH

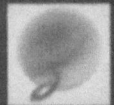
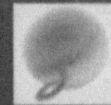

L. Kaper E.P.J. van den Heuvel
P.A. Woudt (Eds.)

Black Holes in Binaries and Galactic Nuclei: Diagnostics, Demography and Formation

Proceedings of the ESO Workshop
Held at Garching, Germany,
6-8 September 1999,
in Honour of Riccardo Giacconi

 Springer

Volume Editors

Lex Kaper
Edward P.J. van den Heuvel
Astronomical Institute "Anton Pannekoek"
University of Amsterdam
Kruislaan 403
1098 SJ Amsterdam, The Netherlands

Patrick A. Woudt
Department of Astronomy
University of Cape Town
Private Bag
Rondebosch 7700, Republic of South Africa

Series Editor

Jacqueline Bergeron
European Southern Observatory
Karl-Schwarzschild-Strasse 2
85748 Garching, Germany

Library of Congress Cataloging-in-Publication Data applied for.

Die Deutsche Bibliothek - CIP-Einheitsaufnahme

Black holes in binaries and galactic nuclei : diagnostics, demography
and formation ; proceedings of the ESO workshop, held at Garching,
Germany, 6 - 8 September 1999, in honour of Riccardo Giacconi / L.
Kaper ... (ed.). - Berlin ; Heidelberg ; New York ; Barcelona ; Hong
Kong ; London ; Milan ; Paris ; Singapore ; Tokyo : Springer, 2001
  (ESO astrophysics symposia)
  (Physics and astronomy online library)

ISBN 978-3-662-30799-1   ISBN 978-3-540-44562-3 (eBook)
DOI 10.1007/978-3-540-44562-3

© Springer-Verlag Berlin Heidelberg 2001
Originally published by Springer-Verlag Berlin Heidelberg New York 2001.
Softcover reprint of the hardcover 1st edition 2001

Typesetting: Camera-ready by the authors/editors
Cover design: Erich Kirchner, Heidelberg

Printed on acid-free paper   SPIN: 10720995   55/3141/du - 5 4 3 2 1 0

Preface

In recent decades astrophysicists have become convinced that black holes do exist in nature, even though the existence of such objects might pose problems for fundamental physics, for example in relation to quantum mechanics. This conviction has resulted from the discovery of objects with large masses, small dimensions and very strong gravitational fields, both in binary systems (double stars) and in the nuclei of galaxies. The dimensions of these objects are so small and their masses, derived from their gravitational effects, so large that, assuming general relativity to be correct, they can only be black holes.

Black holes in binary systems betray their presence by the copious emission of X-rays, powered by the accretion of matter captured from the companion star. We now know at least a dozen such objects in our galaxy and the Large Magellanic Cloud, with masses ranging from three to more than ten times the mass of the Sun.

Another class of black holes is found in the nuclei of galaxies: supermassive black holes, with masses ranging from a million to several billion times the mass of the Sun. The presence of black holes in the centers of active galaxies and quasars has been suspected for a long time, but that supermassive black holes also lurk in the nuclei of ordinary galaxies like our own is a recent big surprise.

Riccardo Giacconi played a crucial role in the discovery of the first black-hole X-ray binary Cygnus X-1 and in the recognition that this object is most likely a black hole. It therefore seemed appropriate to honour him, on the occasion of his retirement as Director General of the European Southern Observatory, with an international workshop on the astrophysics of black holes. The aim of this workshop was to bring together astrophysicists working on the rather separate fields of stellar-mass and supermassive black holes, who study similar objects, but on different physical scales.

The workshop was held at the ESO Headquarters in Garching from 6 to 8 September 1999 and was attended by more than 100 astrophysicists from all over the world. The Scientific Organizing Committee was composed of R. Bacon, R. Bender, R.D. Blandford, P.T. de Zeeuw, L. Kaper, G. Monnet, M.J. Rees, A. Renzini, D.O. Richstone, R. Sunyaev, J. Trümper, Y. Tanaka, and E.P.J. van den Heuvel (Chair). The programme consisted of invited review talks, contributed talks and poster papers. The local organization was in the hands of L. Kaper (Chair), C. Stoffer, and P.A. Woudt.

These proceedings are aimed at providing an up-to-date overview of the observational evidence for the existence of black holes in binaries and galactic nuclei. Special attention is given to the formation, the physical properties and the environments (such as disks, jets, and accretion tori), and the demographics of stellar-mass and supermassive black holes. The important recent evidence on stellar-mass black-hole formation provided by gamma-ray bursts is included.

We would very much like to thank Christina Stoffer (who was in charge of all the local arrangements that had to be made) and Britt Sjöberg for their help in organizing and running the workshop. Pamela Bristow played a major role in preparing the proceedings for publication. Ed Janssen designed the workshop poster. We are also very grateful for the generous financial support received from the European Southern Observatory and from the Dutch research organization NWO, through Spinoza grant SPI 78-327.

Finally, we would like to thank all the workshop participants for the stimulating atmosphere fostered during the workshop and their contributions to the proceedings.

Garching, September 2000   *Lex Kaper*
                                                   Ed van den Heuvel
                                                    Patrick Woudt

Riccardo Giacconi, ESO Director General 1993–1999

Contents

Part 1. SETTING THE STAGE: BLACK HOLES, PAST AND FUTURE PERSPECTIVES

Part 2. BLACK-HOLE DIAGNOSTICS: DYNAMICAL EVIDENCE

Part 3. BLACK-HOLE PHENOMENOLOGY: VARIABILITY, JETS, DISKS AND ACCRETION TORI

Part 4. BLACK-HOLE DEMOGRAPHY

Part 5. BLACK-HOLE FORMATION

List of Participants

Name	Institution
ANDERS, Stephan	MPI für extraterrestrische Physik, Garching `anders@mpe.mpg.de`
BACON, Roland	CRAL – Observatoire de Lyon `bacon@obs.univ-lyon1.fr`
BEGELMAN, Mitchell	University of Colorado, JILA `mitch@jila.colorado.edu`
BELLONI, Tomaso	Osservatorio Astronimico di Brera `belloni@merate.mi.astro.it`
BÖHRINGER, Hans	MPI für extraterrestrische Physik, Garching `hxb@mpe.mpg.de`
BROCKSOPP, Catherine	Open University `cb@star.cpes.susx.ac.uk`
CELOTTI, Annalisa	S.I.S.S.A., Trieste `celotti@sissa.it`
CESARSKY, Catherine	ESO, Garching `ccesarsk@eso.org`
CHARLES, Phil	Oxford University, Dept. of Astropysics `pac@astro.ox.ac.uk`
DAVIES, Melvyn	Leicester University, Dept. of Physics & Astronomy `mbd@star.le.ac.uk`
DE ZEEUW, Tim	Leiden Observatory `tim@strw.leidenuniv.nl`
DORAN, Rosa	Astronomy Observatory of Lisbon `rdoran@oal.ul.pt`
ECKART, Andreas	MPI für extraterrestrische Physik, Garching `eckart@mpe.mpg.de`

EMSELLEM, Eric — Centre de Recherche Astronomique de Lyon
emsellem@obs.univ-lyon1.fr

ESIN, Ann — CALTECH
aidle@tapir.caltech.edu

FABIAN, Andrew — Institute of Astronomy,
University of Cambridge
acf@ast.cam.ac.uk

FALCKE, Heino — MPI für Radioastronomie, Bonn
hfalcke@mpifr-bonn.mpg.de

FENDER, Rob — University of Amsterdam
rpf@astro.uva.nl

FOELLMI, Cedric — Université de Montréal
foellmi@astro.umontreal.ca

FREITAG, Marc — Observatoire de Genève
marc.freitag@obs.unige.ch

FROMERTH, Michael — University of Arizona, Physics Dept.
fromerth@physics.arizona.edu

FRYER, Chris — University of California, Santa Cruz,
Lick Observatory
cfryer@ucolick.org

FUCHS, Yael — C.E.A., Saclay – Service d'Astrophysique
yfuchs@discovery.saclay.cea.fr

GEBHARDT, Karl — University of California, Santa Cruz,
Lick Observatory
gebhardt@ucolick.org

GERHARD, Ortwin — University of Basel, Astronomical Institute
gerhard@astro.unibas.ch

GHEZ, Andrea — University of California, Los Angeles
ghez@astro.ucla.edu

GIACCONI, Riccardo — Associated Universities, Inc., Washington
giacconi@AUI.EDU

GRACIA CALVO, Juan José — Landessternwarte Heidelberg
jgracia@lsw.uni-heidelberg.de

HAEHNELT, Martin — MPI für Astrophysik, Garching
haehnelt@mpa-garching.mpg.de

HARLAFTIS, Emilios — National Observatory of Athens
ehh@astro.noa.gr

HEGER, Alexander — University of California, Santa Cruz, Lick Observatory — alex@ucolick.org

HUTHOFF, Fredrik — University of Amsterdam, Astron. Inst. — huthoff@astro.uva.nl

IKHSANOV, Nazar — Central Astronomical Observatory Pulkovo, St. Petersburg — ikhsanov@gao.spb.su

ISRAELIAN, Garik — Instituto de Astrofisica de Canarias — gil@iac.es

JANIUK, Agnieszka — N. Copernicus Astronomical Center, Warsaw — agnes@camk.edu.pl

KALOGERA, Vicky — Harvard Smithsonian Center for Astrophysics — vkalogera@cfa.harvard.edu

KAPER, Lex — University of Amsterdam — lexk@eso.org, lexk@astro.uva.nl

KAUFFMANN, Guinevere — MPI für Astrophysik, Garching — gamk@mpa-garching.mpg.de

KING, Andrew — Leicester University, Dept. of Physics & Astronomy — ark@star.le.ac.uk

KOCH MIRAMOND, Lydie — C.E.A., Saclay – Service d'Astrophysique — lkoch@discovery.saclay.cea.fr

KOMOSSA, Stefanie — MPI für extraterrestrische Physik, Garching — skomossa@xray.mpe.mpg.de

LANGER, Norbert — Universität Potsdam — ntl@astro.physik.uni-potsdam.de

LASOTA, Jean-Pierre — Institut d'Astrophysique de Paris — lasota@iap.fr

LEEBER, Dawn — New Mexico State University, Dept. of Astronomy — dleeber@nmsu.edu

MACFADYEN, Andrew — University of California, Santa Cruz, Dept. of Astronomy — andrew@ucolick.org

MARCONI, Alessandro — Osservatorio Astrofisico di Arcetri — marconi@arcetri.astro.it

McALLISTER, Jo — University of Edinburgh, Institute for Astronomy jrm@roe.ac.uk

MEIER, David — Jet Propulsion Lab., Pasadena dlm@cena.jpl.nasa.gov

MEURS, Evert — Dunsink Observatory, Dublin ejam@dunsink.dias.ie

MEYER, Friedrich — MPI für Astrophysik, Garching frm@mpa-garching.mpg.de

MEYER-HOFMEISTER, Emmi — MPI für Astrophysik, Garching emm@mpa-garching.mpg.de

MIRABEL, Félix — CEA Saclay mirabel@discovery.saclay.cea.fr

MONNET, Guy — ESO, Garching gmonnet@eso.org

MÜLLER, Horst — Landessternwarte Heidelberg hmueller@lsw.uni-heidelberg.de

MURRAY, James — Leicester University, Dept. of Physics & Astronomy jmu@star.le.ac.uk

NARAYAN, Ramesh — Harvard College Observatory, Cambridge rnarayan@cfa.harvard.edu

NELEMANS, Gijs — University of Amsterdam gijsn@astro.uva.nl

OGILVIE, Gordon — MPI für Astrophysik, Garching gordon@mpa-garching.mpg.de

OLLING, Rob P. — Rutgers University, Dept. of Physics & Astronomy olling@astro.rutgers.edu

OROSZ, Jerome — University of Utrecht, Faculty of Physics and Astronomy J.A.Orosz@astro.uu.nl

PEITZ, Jochen — Harvard Smithsonian Center for Astrophysics jpeitz@cfa.harvard.edu

PELLEGRINI, Silvia — Università di Bologna, Dip. Astronomia pellegrini@astbo3.bo.astro.it

PHINNEY, E.S. — Caltech, Theoretical Astrophysics esp@tapir.caltech.edu

POOLEY, Guy   University of Cambridge, Cavendish Laboratory `ggp1@cam.ac.uk`

PORTEGIES ZWART, Simon   Boston University / MIT `spz@komodo.bu.edu`

POTTSCHMIDT, Katja   Universität Tübingen, Institut für Astronomie & Astrophysik `katja@astro.uni-tuebingen.de`

RASIO, Frederic   MIT, Cambridge `rasio@mit.edu`

REES, Martin   Institute of Astronomy, University of Cambridge `mjr@mail.ast.cam.ac.uk`

REIPRICH, Thomas H.   MPI für extraterrestrische Physik, Garching `reiprich@mpe.mpg.de`

RENZINI, Alvio   ESO, Garching `arenzini@eso.org`

RIX, Hans-Walter   MPI for Astronomy, Heidelberg `rix@mpia-hd.mpg.de`

RUFFINI, Remo   Institute for Cosmology & Relativistic Astrophysics, Rome `Ruffini@icra.it`

RUSZKOWSKI, Mateusz   Institute of Astronomy, University of Cambridge `ruszkows@ast.cam.ac.uk`

SCHREIER, Ethan   STScI, Baltimore `schreier@stsci.edu`

SILLANPÄÄ, Aimo   Turku University `aimosill@astro.utu.fi`

SRINIVASAN, Ganesan   Raman Research Institute, Bangalore `srini@rri.ernet.in`

SUNYAEV, Rashid   MPI für Astrophysik, Garching `sunyaev@mpa-garching.mpg.de`

SZUSZKIEWICZ, Ewa   Torun Centre for Astronomy `esz@astri.uni.torun.pl`

TANAKA, Yasuo   Institute of Space and Aeronautical Science – ISAS `ytanaka@xray.mpe.mpg.de`

THATTE, Niranjan   MPI für extraterrestrische Physik, Garching `thatte@mpe.mpg.de`

TRAMS, Norman   ESA-ESTEC (SAG)
ntrams@astro.estec.esa.nl

TRÜMPER, Joachim   MPI für extraterrestrische Physik, Garching
jtrumper@mpe-garching.mpg.de

TSURU, Takeshi   Kyoto University, Dept. of Physics
tsuru@cr.scphys.kyoto-u.ac.jp

TSURUTA, Sachiko   Montana State University
uphst@gemini.oscs.montana.edu

TÜRLER, Marc   ISDC – INTEGRAL Science Data Centre, Versoix
marc.turler@obs.unige.ch

VAN DEN HEUVEL, Edwin   University of Amsterdam
edvdh@astro.uva.nl

VAN DER MAREL, Roeland   STScI, Baltimore
marel@stsci.edu

VAN KERKWIJK, Marten   Astronomical Institute, Utrecht
M.H.vanKerkwijk@astro.uu.nl

VAN PARADIJS, Johannes   University of Amsterdam
jvp@astro.uva.nl

VENNES, Stéphane   The Australian National University, Canberrra
vennes@maths.anu.edu.au

VERBUNT, Frank   Institute of Astronomy, Utrecht
verbunt@phys.uu.nl

WELLSTEIN, Stephan   Universität Potsdam
stephan@astro.physik.uni-potsdam.de

WHITE, Simon D.M.   MPI für Astrophysik, Garching
swhite@mpa-garching.mpg.de

WILMS, Jörn   Universität Tübingen,
Institut für Astronomie & Astrophysik
wilms@astro.uni-tuebingen.de

WISOTZKI, Lutz   Hamburger Sternwarte
lwisotzki@hs.uni-hamburg.de

WOOSLEY, Stan   Univ. of California, Santa Cruz, Astronomy Dept.
woosley@ucolick.org

WOUDT, Patrick   ESO, Garching
pwoudt@eso.org

YONEHARA, Atsunori   Kyoto University, Dept. of Physics
yonehara@kusastro.kyoto-u.ac.jp

ZANOTTI, Olindo   S.I.S.S.A., Trieste
zanotti@sissa.it

ZENSUS, Anton   MPI für Radioastronomie, Bonn
azensus@mpifr-bonn.mpg.de

Part 1

SETTING THE STAGE: BLACK HOLES, PAST AND FUTURE PERSPECTIVES

Black Hole Research Past and Future

Riccardo Giacconi[1,2]

[1] Associated Universities, Inc., Suite 730, 1400 16th Street, NW,
Washington, DC 20036
[2] The Johns Hopkins University, The Bloomberg Center,
Charles and 34th Street, Baltimore, MD 21218-2686

1   Introduction

I am truly happy to be back here at ESO for a short visit and see so many
colleagues and friends with whom we have shared the last few years of won-
derful enterprises. I am truly touched by the honor you have wanted to do me
by organizing this symposium. The only fly in the ointment for me has been
that Lex Kaper had convinced me to give a talk before I realized what was
happening. Between the recent physical move to the U.S. with my papers in
transit, my continual conceptual move to the red from X-rays to visible to
millimeter astronomy and the excitement of the *Chandra* launch, I was lucky
to find some viewgraphs to share with you. I was sustained in the thought (I
believe due to Pascal) that one should not underestimate the pleasure that
people have in listening to something they already know.

The idea of the existence of black holes has an ancient and honorable history with which I am sure you are all familiar. Very briefly, starting from the
observation made in 1795 by Laplace that (as a consequence of Newtonian
gravity and corpuscular theory of light) light could not escape an object of
sufficiently large mass and small enough radius, we move to the Schwarzschild
1916 general relativistic solution of Einstein equations for the gravitational
field surrounding a spherical mass which gave us the concept of a Schwarzschild radius. The Chandrasekhar 1930 formulation of an upper limit to the
mass of a completely degenerate configuration made the formation of black
holes the inevitable fate of the evolution of massive stars as was realized by
Eddington in 1935. Yet in the early thirties both Eddington and Landau re-
coiled from accepting the reality of such a state of matter and it was only
in 1939 that Oppenheimer and Snyder gave the first rigorous demonstration
of the collapse of a black hole. The subject languished up to the late 50s,
when Wheeler and collaborators began a serious investigation of the problem of collapse and in 1968 Wheeler coined the name of "black hole." The
work by Kerr and Newman in the early 60s provided a complete solution
of the external gravitational and electromagnetic fields of a stationary black
hole. The discovery of quasars in 1963, pulsars in 1968 and even earlier, of
compact X-ray sources in 1962 helped give new impetus to these theoretical
studies. Observations of the binary X-ray source Cyg X-1 in the early 70s

provided the first plausible evidence that black holes may actually exist in the Universe.

CEN X-3

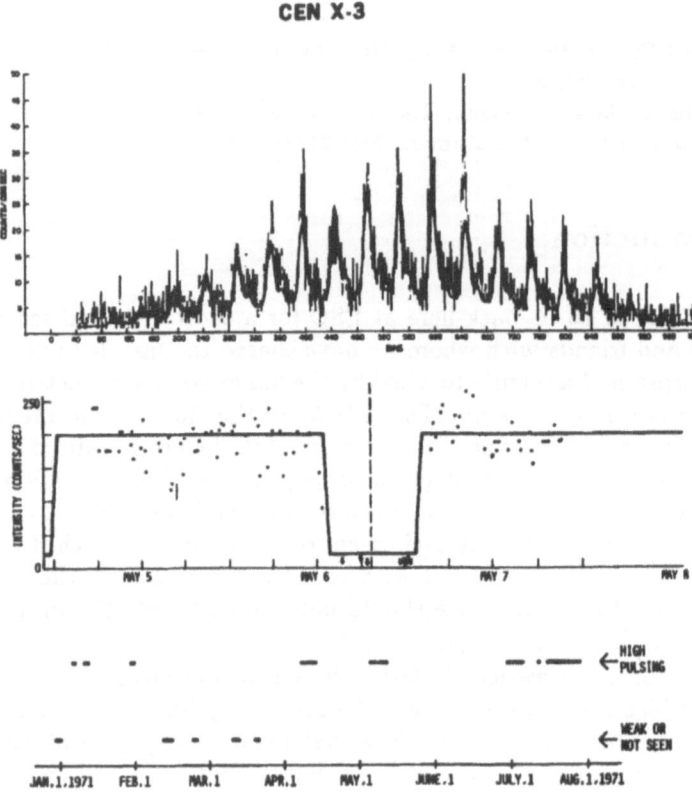

Fig. 1. Regular pulsations and an eclipse in the X-ray emission of Cen X-3 (Schreier et al. 1971).

2   Uhuru

The connection between my own work and black-hole research dates from more than 30 years ago in the golden age of discoveries of the early 60s and came to fruition with the launch of the orbiting X-ray observatory Uhuru on December 12, 1970 which made possible the discovery of X-ray sources in binary systems and gave us a powerful new tool to study the physical processes occurring in stars near the end point of stellar evolution.

Zeldovich and Novikov had suggested as early as 1964 that condensed stars could be found as X-ray sources accreting matter from binary companions. Such an idea received some support from the fact that Sco X-1, the

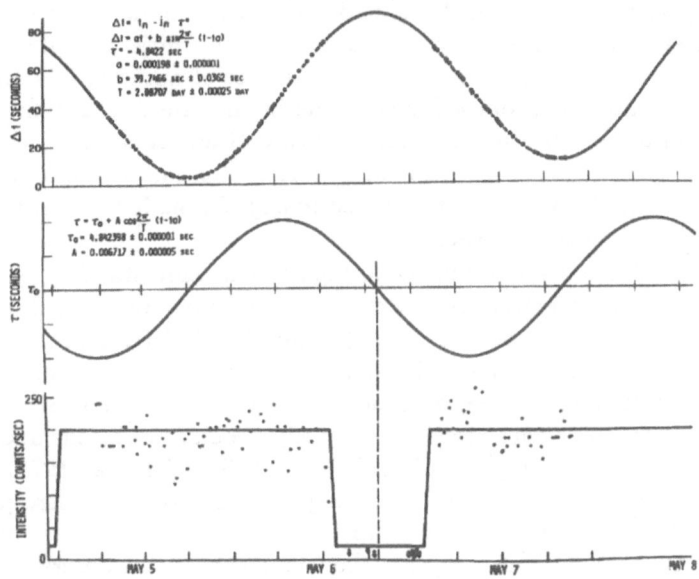

Fig. 2. The Doppler shift of the pulsation frequency of Cen X-3 with zero phase coinciding with X-ray eclipse.

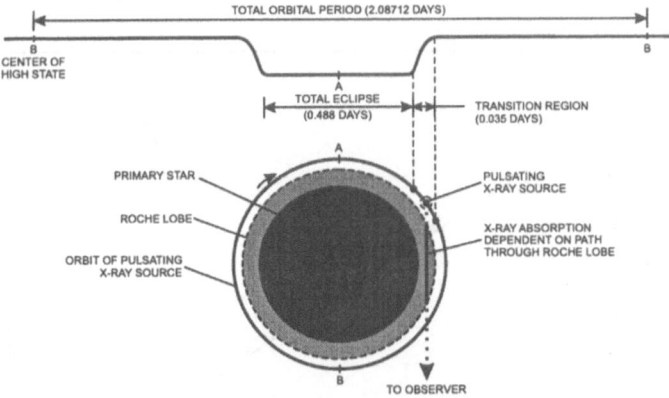

Fig. 3. A sketch of the system with the X-ray pulsar orbiting the primary star.

first X-ray star discovered in 1962, had a spectrum similar to that of an old
nova, systems which are known to consist of close mass exchange binaries
containing a white dwarf companion. A number of authors, including Bur-
bidge and Shklovsky, proposed this model in the late 60s, but the absence
of any definite evidence on the binary nature of any X-ray source rendered
for a time such models somewhat unpopular. Alternate explanations based
on some analogy to pulsars were favored to explain the X-ray emissions from
sources other than supernovae.

Then in 1971 a dramatic breakthrough in our understanding of compact
x-ray sources was achieved as a result of observations by Uhuru. Two peri-
odically pulsating X-ray sources, Cen X-3 and Her X-1 (Giacconi et al. 1971,
Tananbaum et al. 1972), were discovered. The observation of eclipses and a
Doppler variation in their period conclusively established their binary nature
on the basis of X-ray data alone. Very briefly, regular pulsations and eclipses
were observed in the X-ray emission of Cen X-3 (Schreier et al. 1972) (Fig. 1).
A Doppler shift of the pulsation frequency was observed with zero phase coinciding with zero phase of the eclipse (Fig. 2), leading to the picture in
Fig. 3. Over a period of many years the compact star was observed to speed
up rather than slow down its pulsation frequency (Fig. 4). This excludes rotational energy and leaves only gravitational infall as the source of energy for
the X-ray emission (Fig. 5).

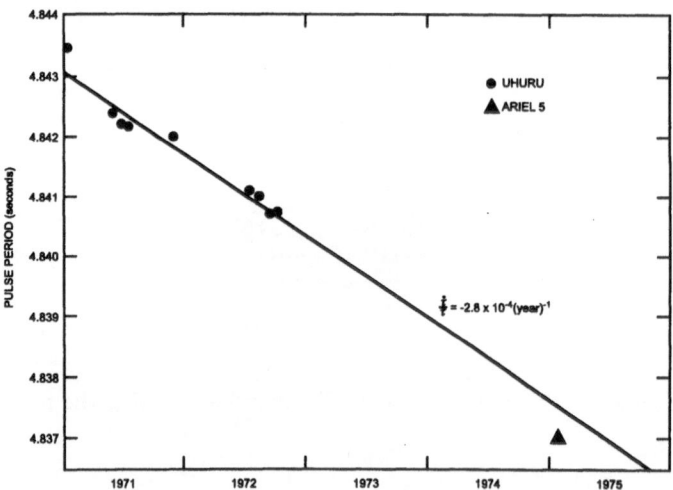

Fig. 4. Over a period of many years the X-ray pulsar Cen X-3 was observed to
speed up rather than slow down.

If I may be pardoned for quoting myself, in 1973 (Giacconi 1974) I wrote:
"Today, as a result of combined radio, optical and X-ray observations, we can

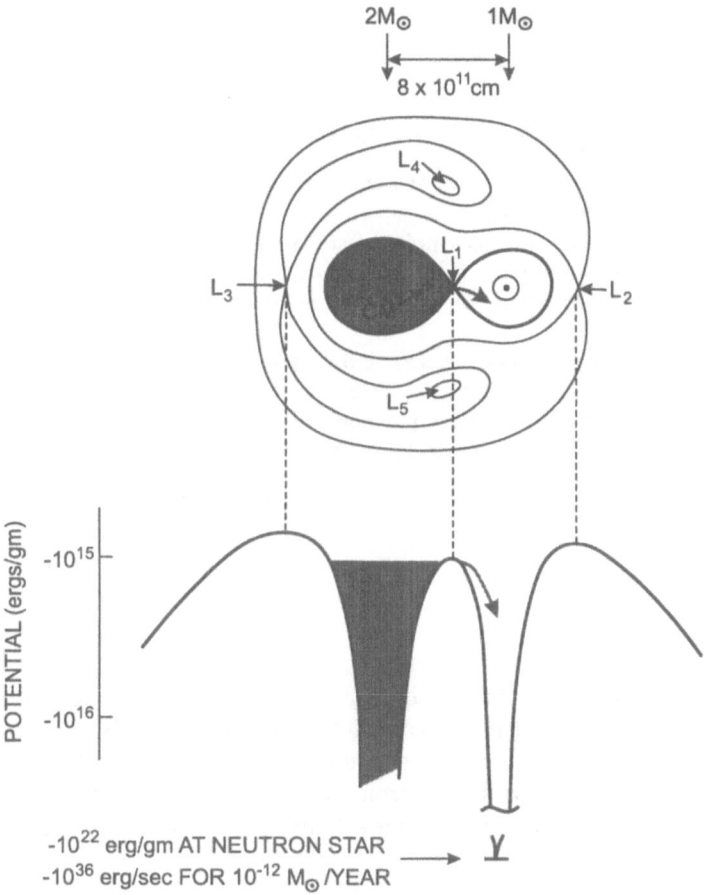

Fig. 5. Roche-lobe overflow results in accretion of matter onto the compact star.

make a fairly convincing case that all compact X-ray sources not associated with supernova remnants are associated with mass transfer binaries containing a collapsed star. Perhaps, most significant of all, in the case of Cyg X-1, X-ray astronomy has furnished the strongest evidence yet for the existence of a new class of objects, black holes." The significance for general relativity of the discovery of black holes had been pointed out by Ruffini and Wheeler in their famous article of 1971 (Ruffini and Wheeler 1971). Seen from the perspective of more than two decades, I think that the discovery of binary X-ray systems had a profound influence on our thinking about astronomical systems in general, well beyond the immediate interests of relativistic astrophysicists. In Table 1, I summarize some of the points I consider had the greatest impact. Perhaps one of the most significant is to make plausible the role of gravitational infall from an accreting disk onto a supermassive central black hole in providing the stupendous energy source required to power active galactic nuclei.

Table 1. Consequences of the Discovery of Binary X-ray Systems

- Existence of binary systems containing a neutron star or a black hole
- Existence of black holes of stellar mass
- Measurement of mass, radius, moment of inertia and equation of state for neutron stars (density 10^{15} g cm^{-3})
- A new source of energy: infall of accreting material in a strong gravitational field. For celestial objects 100 times more effective per nucleon than fission
- A model (now generally accepted) of the nuclei of active galaxies and quasars

3   Cyg X-1

Returning to the theme of this conference I would like to tell you a little about the role that the study of Cyg X-1 played in all of this. Uhuru was one of the first satellites from which we could obtain data within 24 hours. Herb Gursky and I were the senior astronomers of the group. We decided that together with the young astronomers (who included Harvey Tannenbaum, Ethan Schreier and Ed Kellogg), we would look at each piece of data as it arrived. Since we had complete control of the satellite we were anxious to have information on new and unexpected discoveries which we could follow up almost in real time.

One aspect of the data that we were concerned with had to do with timing analysis. We had learned of the discovery of pulsars by Hewish too late in the program to convince NASA of the need to increase the time resolution of the Uhuru detectors to better than 100 milliseconds. Still we were quite aware of the possibility of rapid time variations. We reasoned that looking at the counting rate of each source modulated by the triangular response of the detectors as they swept by, we could observe large fluctuations in emission provided they represented many standard deviations. We invited Minoru Oda, who was visiting a nearby institution, MIT, to work with us on this aspect of the data. We were delighted when he found a funny looking set of data in correspondence to Cyg X-1.

To investigate the apparent large and rapid fluctuations, we decided to spend more time on the source. We could do this by slowing down the spin of the satellite so that each object would linger in the field of view of the detectors for a longer time. We could extend the observations from 10 seconds to more than 100 seconds. It was this technique, first used for Cyg X-1, which became crucial in the studying of the regularly pulsating sources such as Her X-1 and Cen X-3. But to return to Cyg X-1, Cygnus had been observed in some of the earliest surveys resulting in accurate locations for Cyg X-1, Cyg X-2 and Cyg X-3. As a result the optical counterpart for Cyg X-2 was found. No candidate object could be identified for Cyg X-3 or Cyg X-1. Energy spectra for the sources were also measured. The spectrum of Cyg X-1, covering the range from 1 to 80 keV had been measured in balloons and rockets. It appeared to be similar to spectrum of Crab nebula with a flat power low spectral shape $E^{-\alpha}$ with $\alpha = 0.7$. It was puzzling that no radio emissions could be found to a limit of 1/500 that of Crab although the X-ray intensity and spectrum were so similar. Since we only knew of Sco X-1 like sources and supernovae, this was considered evidence of a different type of source.

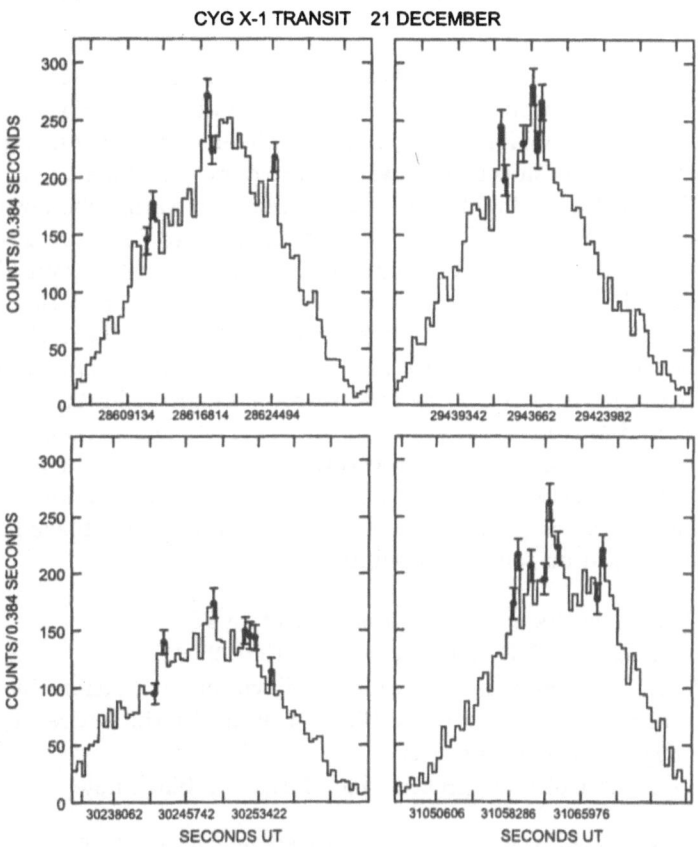

Fig. 6. Variations in X-ray flux of Cyg X-1 as reported by Oda et al. (1971).

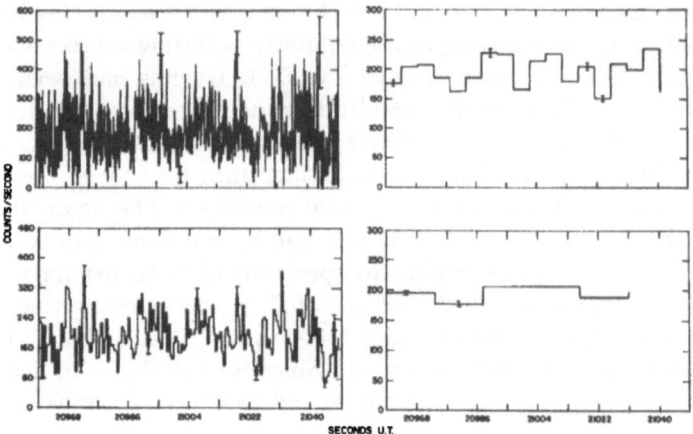

Fig. 7. The variations in Cyg X-1's X-ray flux on different timescales.

How truly different was not revealed to us until Uhuru detected the existence of X-ray pulsations from Cyg X-1 (Oda et al. 1971) (Fig. 6). Figure 7 shows the chaotic behavior of Cyg X-1 on different time scales. Even faster time variability than shown here was reported by scientists at MIT, NRL and GSFC. The GSFC flight in particular reported variations down to the one millisecond scale. This finding alone compels us to consider an emitting region of 10^9 cm or less. The location obtained from an MIT rocket flight and from Uhuru led to the discovery of a radio source. The correlation between variability of the radio and X-ray emission established by Uhuru clinched the identification (Fig. 8). The precise radio location led to the optical identification of Cyg X-1 with the 5.6 day spectroscopic binary HDE 226868 (Fig. 9) by Webster and Murdin (1972) and Bolton (1972). The central object of the system is a B0 supergiant and conservative mass estimates for the primary lead to a mass in excess of several $M_\odot$ for the unseen companion.

If the companion is the compact X-ray source then it is a black hole. This conclusion is based on four main points:

1. HDE 226868 is the optical counter part of Cyg X-1.
2. The mass of the HDE 226868 is greater than 20 $M_\odot$.
3. The X-ray emitting region is compact. Since the fluctuations in emission are so large and close to the Eddington limit for the entire object, the entire object must be compact.
4. A compact star of mass greater than 3 $M_\odot$ is a black hole.

Time does not permit me to discuss each point in detail. Suffice it to say that this interpretation has withstood the scrutiny of 20 years. The situation was summarized by E.E. Salpeter who said: "A black hole in Cyg X-1 is the most conservative hypothesis." Is Cyg X-1 the only black hole in the

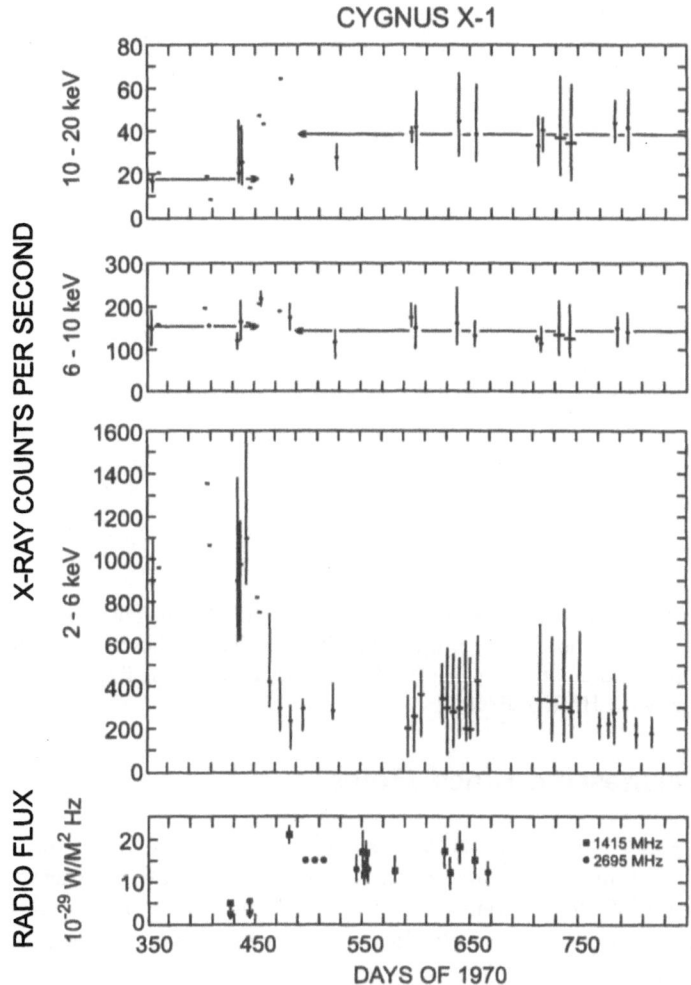

Fig. 8. Correlation between the variability of the radio- (bottom panel) and X-ray emission of Cyg X-1.

galaxy and is it a black hole? Recent compilations cite nine "confirmed black holes" of which Cyg X-1 is still the best candidate. All of these sources are considered black holes on the basis of the evidence provided by the optically determined mass function. As to the question of whether such compact stars of mass greater than the 3 $M_\odot$ allowed by fundamental physics (a limit found by Rhoades and Ruffini 1973) really correspond to the singularities described by Wheeler, there is still no answer. This is due to the fact that as yet no specific and relativistic effects which are unique to black holes and could give the signature of the singularity have been observed in a galactic X-ray source.

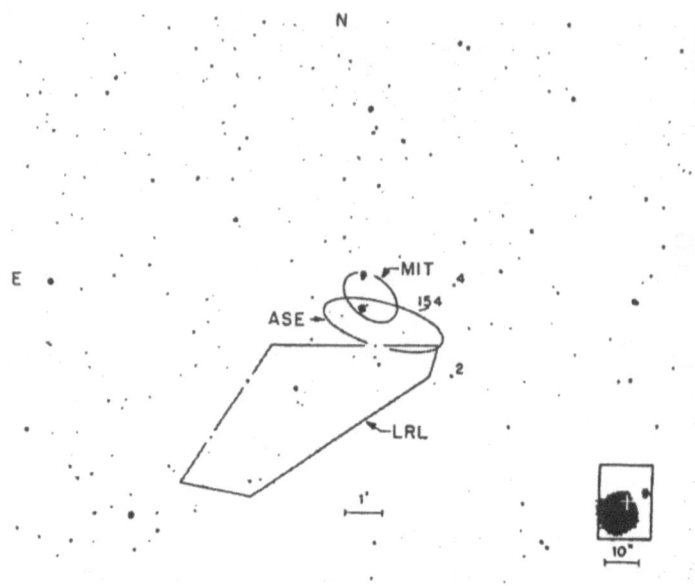

Fig. 9. X-ray location of Cygnus X-1. HDE 226868 is the bright star in the overlap between the MIT and AS & E error boxes. The insert shows the radio location which was reduced to an uncertainty of less than 1" after the figure was drawn and is coincident with HDE 226868.

4   Supermassive Black Holes

In recent years new impetus has been given to the study of black holes by the advent of a number of new observational tools and techniques which give us confidence of the existence of not only stellar-mass black holes but also that of supermassive black holes in the center of active galaxies, as proposed by Hills in 1975.

The higher angular resolution offered by the *Hubble Space Telescope* permits us to study the velocity dispersion of stars and the rotation curves of stars and gas ever closer to the central engine. Continuing in his classical studies initiated in the late 70s, Sargent has been able to demonstrate that M87 contains a central dark mass of 3×10^9 $M_\odot$ in a region too compact to be a central cluster of stars. Genzel and Eckart have demonstrated from direct measurement of proper motions in the star cluster at the center of our own galaxy the existence of a $3 \times 10^6 M_\odot$ object with a density as large as $10^{12} M_\odot pc^{-3}$! In NGC 4258 Miyoshi and his collaborators (Miyoshi et al 1995) have used maser emission from water molecules to determine the position, radial velocities, radial accelerations and proper motions of the maser spots. The analysis leads to a consistent picture of a nearly edge-on warped annulus extending from 0.13 pc to 0.26 pc from the nucleus (Herenstein et al. 1999). The enclosed mass results from the measured rotation curve to be

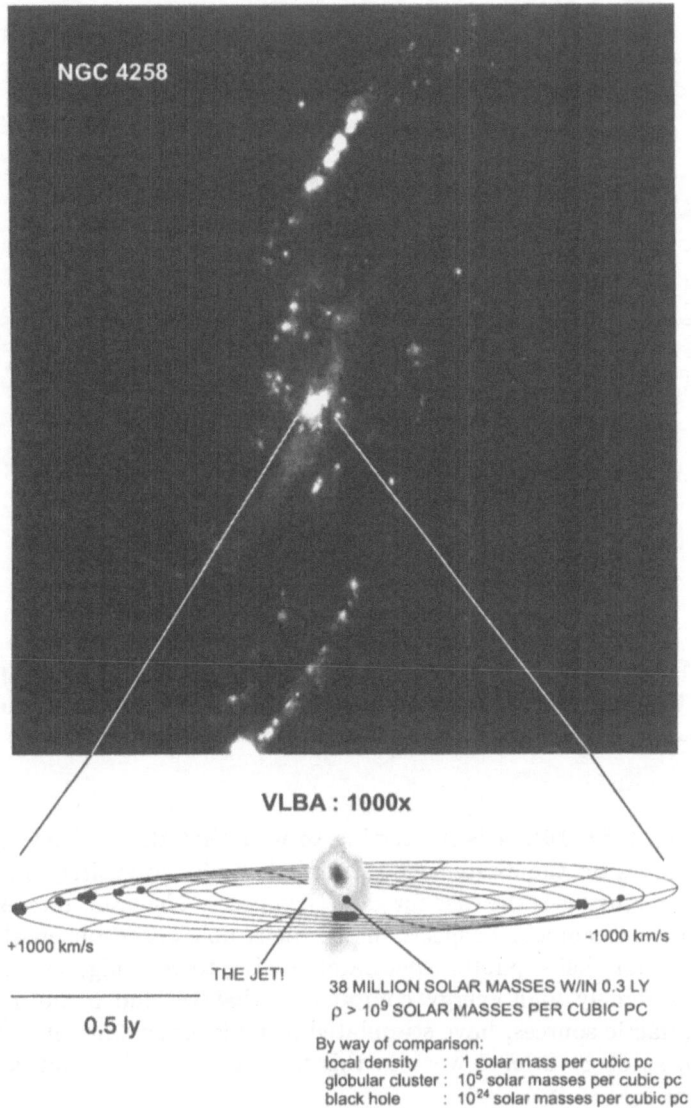

Fig. 10. Evidence for a supermassive black hole in the nucleus of NGC 4258 (Herenstein et al. 1999).

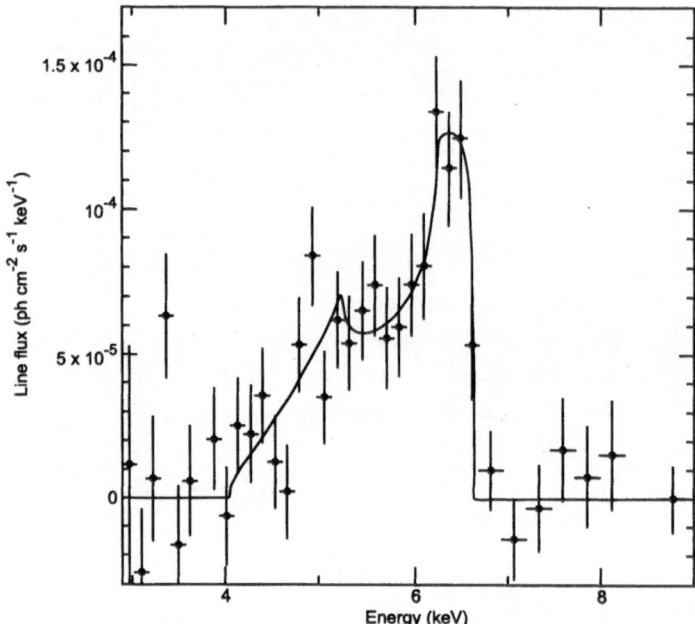

Fig. 11. The Fe Kα line in MCG-6-30-15 shows a relatively narrow blue wing boosted in intensity by the radial Doppler shift and a broad red wing shaped by the combination of gravitational redshift and transverse Doppler shift.

$3.6 \times 10^7\ M_\odot$ (Fig. 10). It is interesting to note that the evidence given by these techniques for the existence of black holes is based entirely on dynamical arguments just as is the case for galactic sources. The great interest in the complex phenomenological aspects of the black hole accretion on all scales, namely accretion disks, photon spectra extending to very high energies, and jets with extremely high energy components that we find in both galactic and extragalactic sources, have stimulated a tremendous amount of theoretical as well as observational work in this field of which this conference is a consequence.

The most intriguing new aspects of the study, however, still deal with the search for unique signatures of the relativistic effects due to the emission occurring near the event horizon. Here the most promising observations have been provided by the ASCA Japanese satellite. Tanaka and his collaborators (Tanaka et al. 1995) have revealed extremely broadened iron Kα emission lines with Doppler width as high as $c/3$ in some cases. These lines are thought to arise from fluorescence of relatively cool (10^6 K or less) optically thick gas exposed to hard X-rays produced in an optically thin corona. As discussed by Fabian, et al. in 1989, if the fluorescing gas forms the inner part of an accretion disk orbiting a black hole, then the line profile should typically display a relatively narrow blue wing boosted in intensity by the radial Doppler shift and broad red wing shaped by the combination of gravitational redshift and transverse Doppler shift. The best studied case, MCG-6-30-15, shows exactly these features in its high intensity state (Fig. 11). Although some doubts on the precise interpretations of the data still are advanced, there is no doubt that such investigations will give us a powerful tool to study the inner regions of the accretion disk.

The study of such features and of reverberation phenomena from radiation produced in flares or transient events in the inner region of the accretion disks of AGNs will constitute one of the main lines of research with the current generation of X-ray missions such as *Chandra*, XMM and ASTRO-E. Definitive results may require an even greater combination of sensitivity and spectral resolution such as will be achieved in the Constellation X and XEUS missions currently under study. But if recent history is a guide, we can expect a rich harvest of results to come also from radio, optical and gamma-ray observations.

I am confident that in the next few decades the astrophysical study of these bizarre and fascinating objects where we are testing the limit of applicability of all known physics will yield abundant and rich results.

References

1. Bolton, C. T.(1972) Nature **235**, 271
2. Fabian, A. C., Rees, M. J., Stella L., White, N. E. (1989), MNRAS, **238**, 729
3. Giacconi, R., Gursky, H., Kellogg, E., Schreier, E., Tananbaum, H.(1971) ApJ **167**, L67
4. Giacconi, R. (1974) Proc. IAU Symp 64, Warsaw, D., Reidel Publishing Co., C. de Witt - Morette ed., 147-180
5. Herenstein, J. R., Moran, J. M., Greenhill, L. J., Diamond, P. J., Moue, M., Nakai, N., Miyoshi, M., Henkel, C., Riess, A. (1999) Nature **400**, 539
6. Hills, J.G. (1975) Nature **254**, 295
7. Miyoshi, M., et al. (1995) Nature **373**, 127
8. Oda, M., Gorenstein, P., Gursky, H., Kellogg, E., Schreier, E., Tananbaum, H., Giacconi, R. (1971) ApJL **166**, L1
9. Rhoades, C., Ruffini, R. (1974) Phys. Rev. Lett., **32**, 324
10. Ruffini, R., Wheeler, J. A.(1971) Physics Today **24**, 30-41
11. Schreier, E., Levinson, R., Gursky, H., Kellogg, E., Tananbaum, H., Giacconi, R. (1972) ApJ **172**, L79
12. Tanaka,Y., Nandra, K., Fabian, A. C., Moue, H., Otaui, C., Dotaui, T., Hayashida, K., Iwasawa, K., Kli, T., Kunieda, H., Makino, F., Matsuoka, M. (1995) Nature **375**, 659
13. Tananbaum, H., Gursky, H., Kellogg, E., Levinson, R., Schreier, E., Giacconi, R.(1972) ApJ **174**, L143
14. Webster, L., Murdin, P. (1972) Nature **235**, 37

Overview: Black Holes in the Universe

Mitchell C. Begelman

JILA, University of Colorado, Boulder CO 80309-0440, USA

Abstract. New observational techniques and theoretical ideas are enabling us to find black holes and measure their masses with increasing precision, and we may soon be able to measure black hole spins. We are now in a position to study the formation and demography of black holes, and to develop a deeper understanding of how they interact with their environments.

1 Introduction

As the nature of Cygnus X-1 was becoming apparent in the early 1970's and for a few years thereafter, black holes were regarded as fanciful by many astronomers. A speaker with the presumption to use a title like mine might have hedged his bets by putting one or two question marks at the end. The more conservative among us still retained the hope that the mass of the X-ray emitting star in Cyg X-1 could be squeezed below 3 solar masses. And the prescient speculations about black holes in AGN — by Salpeter, Zel'dovich, Lynden-Bell, and others — were by no means compelling to everyone.

Now, the evidence for supermassive black holes, not only in AGN but in the nuclei of most normal galaxies, is even more solid than the (also compelling) evidence for black-hole stellar remnants. We have clearly moved beyond the stage of producing evidence for the existence of black holes and into an era of studying their formation, demographics, and the ways in which they interact with their surroundings.

We have made significant progress, but are nowhere near the denouement of this field of research. We still grapple with major uncertainties about the phenomenology of black hole accretion and the nature of relativistic jets. At the more fundamental level, we face questions such as: How did supermassive black holes form and what role do they play in galaxy formation and evolution? Are the compact dark masses we detect really the black holes predicted by general relativity? Do stellar mass black holes signal their creation by producing gamma ray bursts? Fortunately, we are able to study black hole phenomena over 8 orders of magnitude in mass and size, so we have a real chance to perform comparative as well as intensive studies. We have some idea of how black-hole phenomenology should scale with mass; one of our principal goals should be to test these scalings. A workshop such as this one, in which black holes in X-ray binaries are discussed side-by-side with their supermassive counterparts, is an ideal venue for such comparisons.

2   Black-Hole Demography

There are two well-established populations of black holes. The black hole remnants of massive stars, with masses in the range 3 to $\sim 30\ M_\odot$, have been detected in a handful of X-ray binaries and suspected in 1–2 dozen more. Presumably, there are somewhere between $\sim 10^6$ and $\sim 10^8$ stellar-mass black holes in the Galaxy, with the uncertainty arising from our lack of knowledge about the high end of the initial mass function and factors (such as mass threshold, rotation, etc.) that determine whether a star becomes a black hole or a neutron star. Hopefully this situation will improve soon, given advances in computational power and renewed interest — fueled by the possible connection to gamma ray bursts — in the formation of black holes through supernova explosions.

Observationally, little is known about the mass distribution of stellar-remnant holes. Until we find a way to detect isolated stellar-mass holes, we will have to content ourselves with working backwards from the observed distribution of hole masses in binaries. This would require us to learn how to correct for the biases inherent in binarity.

Demographic data are more complete for supermassive black holes ($\sim 10^6$ to $\sim 10^9\ M_\odot$) in galactic nuclei. The evidence suggests that a black hole is found at the center of virtually every galaxy [50]. There are indications that the mass of the hole is correlated roughly linearly with the mass of the galactic bulge [28], although the statistics are still poor and selection effects incompletely understood. A linear correlation might reflect the history of black hole growth either by accretion or by hierarchical merging of smaller "seed" black holes. Alternatively, the growing hole could react back on the host galaxy, controlling its mass by regulating either the galaxy's stellar dynamics [30] or the gas inflow rate [9]. The distribution of black hole spins could provide one crucial discriminant of various growth scenarios; the abundance and properties of supermassive black-hole binaries could provide another.

Are there other, as yet undiscovered, populations of black holes? One cannot rule out the creation of "mini" black holes in the early Universe, although there are constraints on the frequency of gamma-ray flashes expected from the Hawking evaporation process. Likewise, there is still considerable scope for an unknown population of medium mass holes ($10^2 - 10^6\ M_\odot$).

2.1   Measuring Black Holes

Progress in characterizing black hole demography has been made possible, in part, by a recent explosion in techniques for measuring the properties of black holes. The three "classical" techniques — mass functions for X-ray binaries, velocity dispersions of stars and rotation curves of both stars and gas in galactic nuclei [24] — are becoming ever more refined, as they yield additional black-hole candidates. But the last few years have seen the appearance of at least four "new" techniques. Two of these — measurement of stellar proper motions in the Galactic Center and Keplerian rotation traced by water masers — probe the region from about 10^3 to more than 10^5 Schwarzschild radii. While these new methods are greatly strengthening the case for massive dark objects, they are not testing the general relativistic description of black holes per se. But the other two methods — broad X-ray emission lines and "disko-seismology" — are potentially probing the regions at 10 Schwarzschild radii or less. These techniques offer the exciting promise of probing gravity in the "strong field" limit.

- *Stellar Proper Motions in the Galactic Center.* Direct measurements of proper motions [15][16] have moved the Galactic Center up to second place behind NGC 4258 (see below) as the most compelling detection of a "massive dark object" in a galactic nucleus. Comparing a sample of proper motions with a large number of radial velocities spanning projected distances $0.01 - 0.3$ pc from SgrA*, Genzel and collaborators established 1) that the velocity dispersion in the central star cluster is isotropic; and 2) that the velocity dispersion steadily increases toward the putative Galactic Center in a manner consistent with a 2.6 million $M_\odot$ point mass. Ghez and collaborators (this volume) report further important progress in tracking the orbits of stars close to the black hole. The Galactic Center gives us the opportunity to test theoretical ideas about the evolution of galactic nuclear star clusters in the presence of a black hole. The presence of a large number of bright blue stars, despite little corroborating evidence of recent *in situ* star formation, has led to speculation about the role of stellar coalescence. Among the older, red stars it should be possible to test ideas about mass segregation and the formation of stellar cusps.

- *H_2O Megamasers.* It would seem that only a thin Keplerian disk could give a cleaner signal of a massive dark object than the stellar velocity dispersions mapped out in the Galactic Center. And that is precisely what has been found in the center of the nearby LINER galaxy NGC 4258, using a most unlikely diagnostic — maser emission from water molecules [31][19]. Precise measurements of positions, radial velocities, radial accelerations and proper motions of the maser spots lead to an amazingly consistent (indeed, overdetermined!) fit to a nearly edge-on, warped annulus extending from 0.13 pc to 0.26 pc from the nucleus. The rotation curve is fit by a Keplerian $r^{-1/2}$ law to within 0.3%, giving our most accurate black hole mass to date: 3.6×10^7 $M_\odot$. Is NGC 4258 a fluke, or does it herald a new standard of evidence in the search for black holes? The fact that the disk must be nearly edge-on in order to observe the maser activity would make these systems somewhat rare. If the masers are pumped by X-rays emitted close to the accreting black hole [39], then a warp is also necessary in order for the opaque disk to receive enough incident radiation. Given these conditions, a standard thin accretion disk with enough mass flow to power the X-rays can account nicely for the scale and intensity of the maser emission [40]. Indeed, water maser emission has been detected from nearly 20 AGN. Although a simple kinematic signature may not be ubiquitous, at least one or two sources besides NGC 4258 show evidence for a thin disk, notably the archetypal type 2 Seyfert galaxy NGC 1068 [18]. The central mass is about $10^7 \, M_\odot$, implying that this luminous AGN is radiating at close to its theoretical maximum, the Eddington limit.

- *Broad X-ray Lines.* X-ray observations of extremely broad iron lines by the *ASCA* satellite may be providing the first direct evidence for disk-like flow close to the event horizon of a black hole [33][14][53]. These lines are thought to arise from fluorescence of relatively cool, optically thick gas exposed to hard X-rays produced in an optically thin corona. If the fluorescing gas forms the inner part of an accretion disk orbiting a black hole, then the line profile should typically display a relatively narrow blue wing boosted in intensity by the radial Doppler shift, and a broad red wing shaped by the combination of gravitational redshift and transverse Doppler shift. The best studied case, MCG–6-30-15, shows exactly these features in its high-intensity state. Its rapid variability [21] indicates that the line is produced close in. Inferences about the disk geometry and black hole spin have been drawn from correlations between the iron line profile and the X-ray continuum flux [21], but the interpretation is non-unique [48]. Efforts are now underway to develop the technique of X-ray reverberation mapping, which will become feasible with the next generation of X-ray satellites [49][55].

- *Diskoseismology.* While rapid variability has been useful as a qualitative indicator of compactness, its role as a quantitative probe has seldom lived up to expectations. But the *stable* QPO frequencies discovered in two "microquasars" (67 Hz in GRS 1915+105 [32] and 300 Hz in GRO J1655-40 [47]) may provide a relatively clean probe of the black hole's mass and spin [43]. "G−mode" oscillations of thin, Keplerian disks can exist only where their frequencies are smaller than the local epicyclic frequency for a nearly circular orbit. Due to the effects of general relativity, the epicyclic frequency has a maximum near the inner edge of an accretion disk around a black hole, and vanishes at the inner edge itself. This means that g−modes can be *trapped* in a narrow band near the inner edge of the disk [41][42]. Although the characteristics of the stable QPOs are consistent with this interpretation, other mechanisms are possible [13][29]. But in all cases, the observed frequency can be used to constrain both the mass of the black hole and its Kerr spin parameter (a/m). If the mass were measured independently using observations of the binary, one could then determine, for the first time, how fast a black hole is spinning.

3   Are Black Holes Fussy Eaters?

The luminous manifestations of black holes — in X-ray binaries and AGN — are all due to the accretion of gas. Yet observations of the Galactic Center, powerful radio galaxies, and weak radio sources in the nuclei of many ellipticals seem to demand that accretion does not always lead to the efficient output of luminosity. This can be accomplished if the accretion flow advects most of the liberated gravitational binding energy through the event horizon instead of radiating it away [20][46][37][1]. It could also occur — with rather different results — if most of the matter at large distances is turned away before reaching the vicinity of the black hole.

"Advection-dominated" accretion can occur at both high and low values of the mass inflow rate. At high accretion rates radiation is produced within the flow but is trapped by the high optical depth [3][5]. At low accretion rates, the gas may be unable to cool and internal thermal (and/or magnetic) energy is then swept into the black hole. The existence of the latter type of flow requires both that ion heating exceed electron heating and that the transfer of energy from ions to electrons be sufficiently slow that $T_i \gg T_e$ is maintained close to the black hole. These requirements are controversial but no one has yet shown that they are impossible [4][7][45].

Models of "advection-dominated accretion flows" (ADAFs) have given satisfactory fits to various observations [36], but they are beset by a well-known flaw in their physical self-consistency. All ADAF models predict that most of the gas in the flow should have a *positive Bernoulli function* [37] [38][35], which means that the gas effectively is not bound to the black hole and could escape if a route is available. This is a fundamental consequence of viscous transport. In the course of transporting angular momentum outward, viscosity unavoidably transports energy as well, depositing it in the flow. This is the reason that the power dissipated locally in a thin Keplerian accretion disk is three times the local rate at which binding energy is released (D. Lynden-Bell and K. Thorne, cited in [44]).

To address this problem, Blandford and Begelman [10] generalized self-similar ADAF models to include the possibility of outflows which reduce the Bernoulli function by carrying away mass, angular momentum, and energy ("adiabatic inflow-outflow solutions", or ADIOS). Two-dimensional numerical hydrodynamic models (without magnetic fields and with a simple stress prescription) have shown not mass loss, per se, but rather a vigorous circulation driven by convection [52]. The net accretion rate, a tiny fraction of the circulating mass flux, releases just enough gravitational binding energy to keep the circulation going.

The observational ramifications of the ADIOS model can be substantial, since the flux of matter reaching the black hole can be much smaller than that supplied at large radii. This means that: 1) Given a fixed mass supply at large r, the radiative efficiency of the overall flow can be even smaller than that predicted by ADAF models. Two-temperature flow can persist for lower values of the viscous stress parameter α. 2) There is a generic relationship between inflow and outflow (or mass supply and circulation) that is based on energetic considerations. 3) If energy transport leads to a wind rather than closed circulation (as could happen, for example, if the flow supports a net magnetic flux), such an outflow could have important dynamical effects on the environment as well as producing direct observational signatures. For example, it could play a key role in producing a jet or emission-line clouds, or it could be responsible for the outflows associated with broad absorption-line (BAL) QSOs. 4) Neutron stars could also have ADIOS, weakening the argument that radiatively inefficient accretion can be used to demonstrate the existence of event horizons [34].

4   Jets

Perhaps no aspect of black hole phenomenology is as astonishing as the creation of highly relativistic jets. It is clear from observations that jets are accelerated and collimated very close to the black hole [22]. In AGN they seem to reach bulk Lorentz factors of at least a few, and, in some cases, retain a highly relativistic velocity and tight collimation out to enormous distances from the black hole [6]. The relativistic jets known to emerge from X-ray binaries seem to have lower velocities. Beyond these remarkable facts, we know relatively little about how and why jets are created.

- What is the launching site and the power source? Spinning magnetized disks can propel jets under certain circumstances [27][8][11]. So can a spinning black hole threaded by a magnetic field supported by currents in external gas [12]. Despite uncertainties about its effectiveness (see below), the Blandford-Znajek mechanism is appealing for its potential to produce highly relativistic flows with little baryon contamination. It is essential to the "spin hypothesis", which seeks to relate the presence or absence of fast jets to the spin of the black hole. It is quite plausible that both mechanisms operate simultaneously, but as yet there is little understanding of how the two flow regions would interact and manifest themselves observationally.
- Where does the matter in the jet come from? Do jets produce their own matter through $\gamma\gamma$-pair production, or is baryonic matter entrained from the disk? In either case the flow would start out strongly dominated by electromagnetic stresses, in which case it could reach high Lorentz factors.
- What determines the maximum bulk Lorentz factor of the jet? Mass loading, mentioned above, is one factor. But the terminal Lorentz factor can also be affected by radiation drag, entrainment of additional material from the jet's surroundings, and dissipation of some of the magnetic energy into radiation (in which case it becomes unavailable for accelerating the jet).

- What determines whether jets form at all? Jets are found in only about 10% of AGN and a small subclass of black-hole X-ray binaries. Is the determining factor the spin of the hole, the magnetic field structure of the accretion flow, quenching of the incipient jet in most cases, e.g., by radiation or gas dynamical drag? Or are other factors involved? The association of powerful radio sources with elliptical galaxies is presumably telling us something, but is it a reflection of the spin history of merging black holes [54] or something about the accretion flow [51][46]?

The variability of jets probably contains clues to their physical properties that have not yet been deciphered. Superluminal motion, gamma-ray flares, and intraday radio variability not only provide strong indications of relativistic flow speed, but also mark disturbances in the jet, the natures of which are not yet understood. (In at least some cases, intraday radio variability may be due to interstellar scintillation [23], but that in itself is interesting since it implies that the source is extremely compact and has a brightness temperature approaching 10^{14} K.) Are these disturbances strong shocks oriented perpendicularly to the flow, weak, oblique shocks, or relatively diffuse regions of enhanced dissipation (e.g., due to reconnection)? Are they triggered by local obstacles, the spontaneous growth of instabilities along the jet, or fluctuations at the jet source?

A lot more work needs to be done on the Blandford-Znajek effect, to determine whether it really is important in producing jets and, if so, how the jets are actually created. Some authors have recently questioned the importance of the BZ effect in systems with thin accretion disks, on the grounds that the magnetic flux threading the black hole is much smaller than has been supposed [17][26]. At the same time, there has been renewed interest in the possible electromagnetic extraction of energy from within the innermost stable orbit of the accretion disk, which could lead to a higher accretion efficiency and also might contribute to the formation of jets [25][2].

5   Concluding Remarks

We have reached the point at which the characterization of black hole demography is a realistic goal. If we are able to measure spins as well as masses, we will have important new clues to the ways in which black holes form, and possibly to the mechanisms that produce jets. At the same time, we must address such fundamental problems as the radiative efficiency of black hole accretion, and whether black holes are in fact very "fussy" about how much they eat. We may learn a great deal if we can determine observationally how the properties of accretion flows and jets scale with black hole mass. This workshop will be a success if it moves us closer to any of these goals.

Acknowledgments. My work on black-hole astrophysics is sponsored by the National Science Foundation through grants AST-9529175 and AST-9876887.

I also acknowledge valuable input from Roger Blandford, Mike Nowak, Martin Rees, and Chris Reynolds.

References

1. Abramowicz, M. A., Chen, X., Kato, S., Lasota, J.-P., Regev, O. 1995, ApJ, 438, L37
2. Agol, E., Krolik, J. 1999, ApJ, in press (astro-ph/9908049)
3. Begelman, M. C. 1979, MNRAS, 187, 237
4. Begelman, M. C., Chiueh, T. 1988, ApJ, 332, 872
5. Begelman, M. C., Meier, D. L. 1982, ApJ, 253, 873
6. Biretta, J. A., Sparks, W. B., Macchetto, F. 1999, ApJ, 520, 621
7. Bisnovatyi-Kogan, G. S., Lovelace, R. V. E. 1997, ApJ, 486, L43
8. Blandford, R. D. 1976, MNRAS, 176, 465
9. Blandford, R. D. 1999, in Galaxy Dynamics ed. D. Merritt, M. Valluri, J. Sellwood, ASP conf. series, in press (astro-ph/9906025)
10. Blandford, R. D., Begelman, M. C. 1999, MNRAS, 303, L1
11. Blandford, R. D., Payne, D. G. 1982, MNRAS, 199, 883
12. Blandford, R. D., Znajek, R. L. 1977, MNRAS, 179, 433
13. Cui, W., Zhang, S. N., Chen, W. 1998, ApJ, 492, L53
14. Fabian, A. C., Nandra, K., Reynolds, C. S., Brandt, W. N., Otani, C., Tanaka, Y., Inoue, H., Iwasawa, K. 1995, MNRAS, 277, L11
15. Genzel, R., Eckart, A., Ott, T., Eisenhauer, F. 1997, MNRAS, 291, 219
16. Ghez, A. M., Klein, B. L., Morris, M., Becklin, E. E. 1998, ApJ, 509, 678
17. Ghosh, P., Abramowicz, M. A. 1997, MNRAS, 292, 887
18. Greenhill, L. J., Gwinn, C. R., Antonucci, R., Barvainis, R. 1996, ApJ, 472, L21
19. Greenhill, L. J., Jiang, D. R., Moran, J. M., Reid, M. J., Lo, K. Y., Claussen, M. J. 1995, ApJ, 440, 619
20. Ichimaru, S. 1977, ApJ, 214, 840
21. Iwasawa, K., Fabian, A. C., Reynolds, C. S., Nandra, K., Otani, C., Inoue, H., Hayashida, K., Brandt, W. N., Dotani, T., Kunieda, H., Matsuoka, M., Tanaka, Y. 1996, MNRAS, 282, 1038
22. Junor, W., Biretta, J. A. 1995, AJ, 109, 500
23. Kedziora-Chudczer, L., Jauncey, D. L., Wieringa, M. H., Walker, M. A., Nicolson, G. D., Reynolds, J. E., Tzioumis, A. K. 1997, ApJ, 490, L9
24. Kormendy, J., Richstone, D. 1995, ARA&A, 33, 581
25. Krolik, J. H. 1999, ApJ, 515, L73
26. Livio, M., Ogilvie, G. I., Pringle, J. E. 1999, ApJ, 512, 100
27. Lovelace, R. V. E. 1976, Nature, 262, 649
28. Magorrian, J., Tremaine, S., Richstone, D., Bender, R., Bower, G., Dressler, A., Faber, S. M., Gebhardt, K., Green, R., Grillmair, C., Kormendy, J., Lauer, T. R. 1998, AJ, 115, 2285
29. Marković, D., Lamb, F. K. 1998, ApJ, 507, 316
30. Merritt, D. 1998, CommAp, 19, 1
31. Miyoshi, M., Moran, J., Herrnstein, J., Greenhill, L., Nakai, N., Diamond, P., Inoue, M. 1995, Nature, 373, 127
32. Morgan, E., Remillard, R., Greiner, J. 1997, ApJ, 482, 993

33. Mushotzky, R. F., Fabian, A. C., Iwasawa, K., Kunieda, H., Matsuoka, M., Nandra, K., Tanaka, Y. 1995, MNRAS, 272, L9
34. Narayan, R., Garcia, M. R., McClintock, J. E. 1997, ApJ, 478, L79
35. Narayan, R., Kato, S., Honma, F. 1997, ApJ, 476, 49
36. Narayan, R., Mahadevan, R., Quataert, E. 1998, in The Theory of Black Hole Accretion Disks, ed. M. A. Abramowicz, G. Bjornsson, J. E. Pringle (Cambridge: Cambridge Univ. Press), 148
37. Narayan, R., Yi, I. 1994, ApJ, 428, L13
38. Narayan, R., Yi, I. 1995, ApJ, 444, 231
39. Neufeld, D. A., Maloney, P. R., Conger, S. 1994, ApJ, 436, L127
40. Neufeld, D. A., Maloney, P. R. 1995, ApJ, 447, L17
41. Nowak, M. A., Wagoner, R. V. 1992, ApJ, 393, 697
42. Nowak, M. A., Wagoner, R. V. 1993, ApJ, 418, 187
43. Nowak, M. A., Wagoner, R. V., Begelman, M. C., Lehr, D. E. 1997, ApJ, 477, L91
44. Pringle, J. E., Rees, M. J. 1972, A&A, 21, 1
45. Quataert, E., Gruzinov, A. 1999, ApJ, 520, 248
46. Rees, M. J., Begelman, M. C., Blandford, R. D., Phinney, E. S. 1982, Nature, 295, 17
47. Remillard, R. A., Morgan, E. H., McClintock, J. E., Bailyn, C. D., Orosz, J. A., Greiner, J. 1998, in Proc. 18th Texas Symp. on Relativistic Astrophysics, ed. A. Olinto, J. Frieman, D. Schramm (Singapore: World Scientific), 750
48. Reynolds, C. S., Begelman, M. C. 1997, ApJ, 488, 109
49. Reynolds, C. S., Young, A. J., Begelman, M. C., Fabian, A. C. 1999, ApJ, 514, 164
50. Richstone, D., Ajhar, E. A., Bender, R., Bower, G., Dressler, A., Faber, S. M., Filippenko, A. V., Gebhardt, K., Green, R., Ho, L. C., Kormendy, J.,Lauer, T. R., Magorrian, J., Tremaine, S. 1998, Nature, 395, 14
51. Sparke, L. S., Shu, F. H. 1980, ApJ, 241, L65
52. Stone, J. M., Pringle, J. E., Begelman, M. C. 1999, MNRAS, in press (astro-ph/9908185)
53. Tanaka, Y., et al. 1995, Nature, 375, 659
54. Wilson, A. S., Colbert, E. J. M. 1995, ApJ, 438, 62
55. Young, A. J., Reynolds, C. S. 1999, ApJ, in press (astro-ph/9910168)

Part 2

BLACK-HOLE DIAGNOSTICS: DYNAMICAL EVIDENCE

Black Holes in X-Ray Binaries

Phil Charles

Dept of Physics & Astronomy, University of Southampton

Abstract. The high mass X-ray binary, Cyg X-1, was for many years the strongest candidate for a stellar-mass black-hole in the Galaxy. It very likely is a black-hole, but it is the "soft X-ray transients", a sub-class of low-mass X-ray binaries, that have revolutionised our ability to accurately measure compact object masses. This is accomplished via optical and IR photometry and spectroscopy of the mass-losing star during the extended X-ray quiescent periods, exploiting essentially classical binary astronomy techniques. I will summarise the optical and IR properties of these objects and review some of the more recent results.

1 Introduction

With Riccardo Giacconi's involvement in the 1962 discovery of Cyg X-1 (with the earliest cosmic X-ray sounding rockets; Gursky et al. 1963), and its subsequent X-ray location (Giacconi et al. 1967) this is a fitting opportunity to review our knowledge of black holes in X-ray binaries. Indeed, this meeting brings together contributions covering a wide range of both observational and theoretical aspects of this field. These include optical (Harlaftis; Israelian), radio (Mirabel; Fender) and X-ray (Tanaka; Belloni) observations, complemented by theoretical studies of accretion flows and jets (Sunyaev; Blandford; King: Narayan), why the black-hole X-ray binaries are transient (Lasota), massive star evolution (Langer; Woosley; Kaper; Wellstein) and binary evolution (Verbunt; Kalogera; Fryer; Rasio). In this review I will therefore concentrate on optical and IR observational properties of black-hole X-ray binaries, with particular emphasis on the X-ray transients where much detailed physical knowledge of the binary parameters has been gleaned in the last decade.

Table 1 contains a list of those X-ray binaries for which a mass function has been determined (and hence accurate mass constraints are available) and are suspected of harbouring black holes. Note that this contains only 3 high-mass X-ray binaries (HMXBs), all of which are bright, "steady" X-ray sources, but a much greater number of low-mass X-ray binaries (LMXBs) virtually all of which are transient in nature. This point is addressed by Lasota (these proceedings).

It is instructive to note that, while for many years considered to be the best black-hole candidate (BHC) in our Galaxy, the uncertainties associated with the compact object mass in the high-mass X-ray binary (HMXB) Cyg X-1 are still considerable. This is of course due to the large and systematic uncertainties in the mass of its supergiant primary HDE226868 and the difficulty in constraining the binary inclination (it does not eclipse). This has

Table 1. Optical/IR Properties of Stellar-Mass Black-Hole Candidates

Source	Outbursts	P (hrs)	Sp. Type	E_{B-V}	V (quiesc)	K	vsini (km/s)	K_2 (km/s)
(a) LMXBs								
J0422+32	1992	5.1	M4-5V	0.3	22	16.2	90	378
A0620-00	1917,75	7.8	K5V	0.35	18.3	6	83	433
GS2000+25	1988	8.3	K5V	1.5	21.5	17	86	518
GRS1124-68	1991	10.4	K0-4V	0.29	20.5	16.9	106	399
H1705-25	1977	12.5	K	0.5	21.5	-	≤79	448
GX339-4	"steady"	14.8	?	1.1	20.5	-	-	-
Cen X-4	1969,79	15.1	K7IV	0.1	18.4	15.0	45	146
V404 Cyg	1938,56,89	155.3	K0IV	1	18.4	12.5	39	208.5
4U1543-47	1971,83,92	27.0	A2V	0.5	16.6	-	-	124
J1655-40	1994+	62.9	F3-6IV	1.2	17.2	-	-	228
(b) HMXBs								
LMC X-3	"steady"	40.8	B3Ve	0.1	17.5	-	130	235
LMC X-1	"steady"	101.3	O7-9III	0.37	14.5	-	~150	68
Cyg X-1	"steady"	134.4	O9.7Iab	1.06	8.9	-	155	75.5

most recently been addressed by Herrero et al. (1995) whose high-resolution optical spectroscopy and detailed atmospheric modelling give a mass range of 12–19 $M_\odot$ for the OBI primary. With an inclination range of 28–67 degrees, the compact object mass is therefore constrained to the range 4–15 $M_\odot$. HDE226868 has also been studied by Lasala et al. (1998), who have derived a much improved orbital ephemeris, and Bałucińska-Church et al. (2000), who used this new ephemeris to study the distribution of X-ray dips as a function of orbital phase (there is a strong orbital modulation).

In spite of this large uncertainty in compact object mass (a feature shared by the LMC HMXBs LMC X-1 and X-3), Cyg X-1 remains a very strong BHC, and its X-ray properties are still invoked as black-hole characteristics. However, many of these (e.g. bimodal spectrum, ultrasoft high state and fast flickering) can be produced in certain circumstances by neutron star systems and so these properties should be used as a guide only. Nevertheless, there is one property, the high-energy power-law tail, that has never been seen in neutron-star binaries and can therefore be cited as a powerful black-hole diagnostic (see Tanaka, these proceedings).

2 Soft X-Ray Transients

The remainder of the sources in Table 1 are very different from Cyg X-1. Almost all are X-ray transients whose X-ray spectrum during outburst often

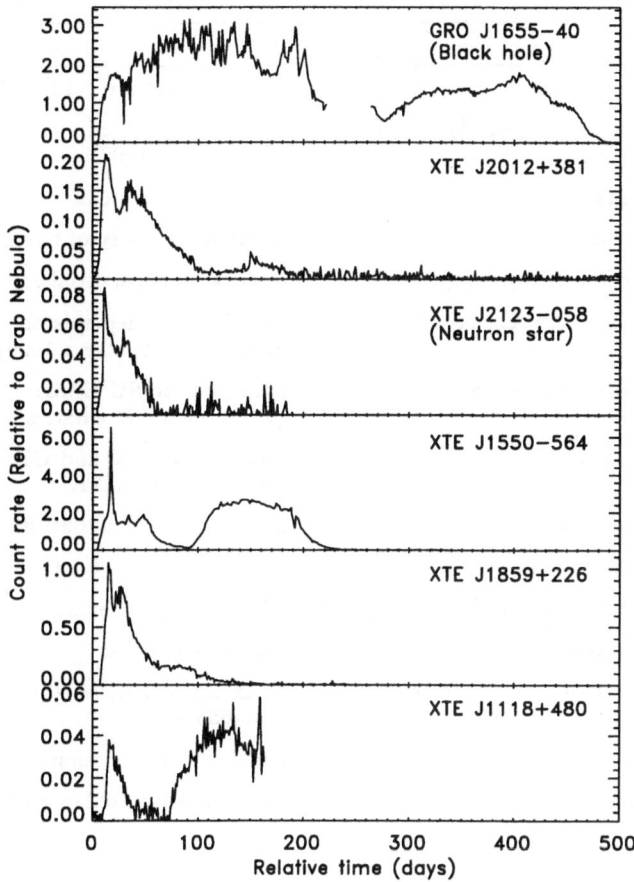

Fig. 1. Selection of SXT lightcurves from the RXTE ASM WWW database.

exhibits an "ultra-soft" component, hence the name "soft X-ray transient" (SXT; although they are also sometimes referred to as "X-ray novae"). The prototype system is A0620-00 (Nova Mon 1975), which at its peak was the brightest X-ray source ever detected. Its optical counterpart also brightened dramatically, from V∼17 in quiescence to ∼11 (see Kuulkers 1998 for a re-appraisal of the A0620-00 outburst data). This large brightness range is a key SXT property and is a result of the X-ray irradiated disc light dominating that of the intrinsically faint, low-mass companion star. These properties have been reviewed by Charles (1998a) and are summarised in Table 2.

The transient outbursts have a characteristic fast rise, exponential decay form (on timescales of ∼30 days; but see the range of light curves displayed in RXTE ASM data in fig 1), although Shahbaz et al (1998) show that it can

Table 2. Observational Properties of Soft X-ray Transients

Property	Notes
Transient outbursts	Typical peak $L_X \sim 10^{37-39} \text{erg s}^{-1}$
Fast rise, exp.decay ($\sim$30d) light curves	Recur every 10–50y (Chen et al. 1997)
Secondary max.	$\sim$60–90d after peak
Mini-outbursts	occur within $\sim$1y of main outburst
Hard power law X-ray compt	often with substantial soft excess
Large amplitude optical brightening	$\sim$ 5–7 mags (depends on P; Shahbaz & Kuulkers 1998)
Superhumps during outburst	analogous to SU UMa's (O'Donoghue & Charles 1996)
Superluminal jets	seen in J1655-40 and GRS 1915+105 (may not be LMXB)
Late-type secondaries	all K–M except J1655-40, 4U1543-47
Short orbital periods	all in range 4h – 6.5d
Ellipsoidal modulation of secondary	Visible in quiescence; estimate for i
$\sim$75% are BHCs	out of $\sim$20 objects
Abundance anomalies in secondary	Li excess; enhanced alpha-elements in J1655-40
Residual accretion disc	classical disc emission line profiles
Pop I gal distribution	likely massive stellar progenitors

have either exponential or linear decays depending on the peak luminosity and orbital period. This is because the outbursts are analogous to those of dwarf novae in which a thermal-viscous instability triggers a much more luminous state for the accretion disc and matter is accreted onto the white dwarf. The DN outburst is terminated by a cooling front that propagates through the disc (see e.g. Cannizzo 2000 and references therein). However, in SXT outbursts the accretion luminosity is much higher and this is capable of keeping the disc hot by irradiation and hence preventing the onset of the cooling front. In these circumstances the high-viscosity state is maintained and accretion continues until the entire disc is emptied. This then leads to the long recurrence times between SXT outbursts (usually years to decades), although it has been noted by King et al. (1997) that the critical luminosity to ionise the disc depends on whether the central source is point-like (i.e. neutron star) or extended (black hole, with no direct radiation visible).

Shahbaz & Kuulkers (1998) have also shown that, for short orbital periods of $\leq$1 day (i.e. non-evolved systems) the optical outburst amplitude is inversely proportional to the orbital period. It is assumed that, in quiescence, the optical light is dominated by the secondary. For systems with longer orbital periods, the brightness of the Roche-lobe filling secondary increases with period faster than that of the hot disc, thereby producing a smaller outburst amplitude.

3   Measuring Masses

The great importance of SXTs in the field of LMXBs arises from the ability to study the secondary star directly during their extended periods of quiescence, a feature that is (usually) impossible for the $\sim$100 bright, "steady" LMXBs (such as e.g. Sco X-1). With low-mass secondaries and hence high mass ratios, the measured mass functions $(f(M) = (M_X \sin i)^3/(M_X + M_C)^2)$ provide important constraints on M_X. Furthermore, quiescent studies of SXTs allow the orbital inclination to be estimated via the amplitude of the ellipsoidal modulation that is seen in the (presumed Roche-lobe filling) secondary star. And if the rotation speed of the (also presumed co-rotating) secondary can be measured via the rotational broadening of its absorption spectrum, then it is possible in principle to derive a full solution of the SXT orbital parameters. This has been summarised in earlier reviews, such as e.g. Charles (1998b) where appropriate references can be found, and updated results are given in Table 3.

The LMXB section of Table 3 is divided into: BHC SXTs with late-type companions; the two intermediate spectral-type BHCs; and one neutron star SXT. The latter acts as an independent verification of the mass measurement technique. The others imply an average compact object mass $\sim$8–10 M$_\odot$, very substantially in excess of the assumed maximum neutron-star mass ($<$3.2 M$_\odot$; but see Miller et al 1998). Curiously there are no BHCs in the 3–5 M$_\odot$ range, which might be expected if they are formed by accretion-induced collapse of a neutron star. But this may be a selection effect resulting from such systems not exhibiting transient behaviour (which is almost the only circumstance under which an accurate mass estimate can be made). Indeed it is worth noting that, since turning on in 1994, GRO J1655-40 has been active for a significant fraction of the time (see Fig. 1) and the mass estimate is the lowest in the group. This effect (due to irradiation of the disc) is addressed by van Paradijs (1996) and King et al. (1996). Furthermore, the only significant BHC LMXB which is in Table 1, but not Table 3 is GX339-4. Whilst normally considered to be a "steady" source (and hence has no mass estimate), it shows long intervals of inactivity that suggests it too may be a low mass BH.

3.1   Direct Measurements of BH Motion?

This analysis would be greatly helped if orbital velocity information associated with the compact object were available. In spite of the large mass ratios

Table 3. Black-hole X-ray binary dynamical mass estimates

Source	$f(M)$ $(M_\odot)$	q $(=M_X/M_2)$	i	M_X $(M_\odot)$	M_2 $(M_\odot)$
LMXBs					
V404 Cyg	6.08±0.06	17±1	55±4	12±2	0.6
G2000+25	5.01±0.12	24±10	56±15	10±4	0.5
N Oph 77	4.86±0.13	>19	60±10	6±2	0.3
N Mus 91	3.01±0.15	8±2	54^{+20}_{-15}	6^{+5}_{-2}	0.8
A0620-00	2.91±0.08	15±1	37±5	10±5	0.6
J0422+32	1.19±0.02	9.0±2.5	≤30	≥7.2	0.3
J1655-40	3.24±0.14	3.6±0.9	67±3	6.9±1	2.1
4U1543-47	0.22±0.02	-	20–40	5.0±2.5	2.5
Cen X-4	0.21±0.08	5±1	43±11	1.3±0.6	0.4
HMXBs					
Cyg X-1	0.25±0.01	0.4–0.8	27–67	4.8–14.7	11.7–19.2
LMC X-1	0.12±0.04	~0.5	40–65	2.5–6	8–20
LMC X-3	2.3±0.3	2.5–6	50–75	4–7	1–3

and the broad and complex Hα emission profile, a radial-velocity modulation in this profile has been detected in a number of quiescent SXTs such as A0620-00 and Nova Mus 1991 (Haswell & Shafter 1990; Orosz et al. 1994), and also G2000+25 and Nova Oph 1977 (Casares et al. 1995, Harlaftis et al. 1996; Harlaftis et al. 1997). This modulation is in anti-phase with that of the secondary and has an amplitude close to that expected, but regrettably it exhibits a phase shift of ~35 degrees and must therefore be distorted by non-Keplerian components. For this reason it cannot be used to provide dynamical mass information. A close examination of the Hα profiles reveals that they are double peaked (and consistent with accretion disc models), but are often *asymmetric* with a broad blue emission wing. This contamination could be due to a disc wind, the mass transfer stream or a hot spot on the disc.

3.2　Outburst Spectroscopy of GRO J1655-40

It would be even more useful if dynamical information could be obtained during the outburst phase, as the majority of the quiescent SXTs are extremely faint. The remarkable transient GRO J1655-40 (see next section) exhibited extended (non-exponential) outbursts in 1994 and 1996, during which spectroscopy was performed by Soria et al (1998). The red spectra exhibited strong, complex Hα emission, but its substantial asymmetric profile precluded its use as a tracer of the BH's dynamical velocity. However, the brightness of the source allowed blue spectroscopy of the HeIIλ4686 emission line. The broad wings of the double-peaked profile were measured as these are assumed to be representative of the inner disc regions, and these gave a radial velocity curve in anti-phase with the companion star to within 9$\pm$20°. However, the key difficulty with using this for dynamical information is that the HeII mean velocity is -182 km s^{-1}, whereas the secondary absorption lines give -142 km s^{-1} (and with very small errors). This systematic blue shift in the disc (emission) lines supports the disc-wind explanation.

4   The Superluminal Transients

GRO J1655-40 is also known as Nova Sco 1994 and, together with the discovery of GRS1915+105 also in 1994, brought an entirely new type of behaviour to this field. Other transients had been seen to be strong radio sources during outburst, but VLA and VLBI observations of these new objects revealed ejection events that were "superluminal", the first time that such phenomena had been observed within the Galaxy (see Fender and Mirabel, these proceedings). Furthermore, the X-ray light curves of both these objects were unlike any of the "classical" SXTs in that they displayed continuing and extremely variable X-ray activity (GRS1915+105 is still active at the present time). Only GRO J1655-40 (the optically brightest in quiescence of all the SXTs) is amenable to a dynamical study of the secondary, since GRS1915+105 suffers from extremely high interstellar extinction ($A_V \sim$26). However, its variable, K$\sim$14 IR counterpart has been observed both photometrically and spectroscopically (Eikenberry et al. 1998; Bandyopadhyay et al. 1998) to exhibit both short- and long-term quasi-periodicities ($\sim$30 minutes and $\sim$30 days). The similarity of the IR spectrum with that of massive Be X-ray binaries led Mirabel et al. (1997) to propose that GRS1915+105 is not an LMXB but is instead the long-sought BH counterpart of the Be X-ray sources (all of which exhibit X-ray pulsations and are hence neutron stars). More recently, VLT IR spectra of the source (Martì et al. 2000) have been compared with Cyg X-3, a short period (4.8 hr) X-ray binary in which the companion is interpreted as a Wolf-Rayet star (van Kerkwijk 1993), but the nature of the compact object is as yet unclear.

The optical counterpart of GRO J1655-40 has, in quiescence, one of the earliest spectral types (mid-F), together with a high γ-velocity which suggests that it might have been formed from a neutron star that had suffered accretion-induced collapse (Brandt et al 1995). After its initial 1994 outburst, a subsequent optical rebrightening was found to have begun $\sim$6 days before the X-rays, indicating an "outside-in" outburst of the accretion disc (Orosz et al 1997). This substantial delay is explained by the inner disc needing to be re-filled before accretion onto the compact object can take place, it having been evaporated by the hard X-rays from the ADAF flow during quiescence (see Hameury et al. 1997).

The optical brightness of GRO J1655-40 allows high quality photometric light curves to be obtained and combined with high resolution phase-resolved spectroscopy, from which the orbital system parameters can be derived (Orosz & Bailyn 1997; but note the error analysis of van der Hooft et al. 1997, the results of which are incorporated in Table 3). With such an early spectral type, GRO J1655-40 has a low mass ratio ($q\sim3$) which means that the ellipsoidal modulation is sensitive to both q and i (the latter being constrained by the observed grazing eclipse, see Fig. 8 of Orosz & Bailyn 1997). Curiously, the normal star appears to be crossing the Hertzsprung gap and about to ascend the giant branch, hence driving the much higher mass transfer rate (Kolb et al. 1997), but with temporary drops in M that return it to the transient domain.

4.1   Effects of Irradiation

(a) Echo Mapping. In such sufficiently bright systems during outburst as GRO J1655-40, the significant X-ray variability allows another technique to be employed to probe the structure of the binary and that is "echo mapping". This is a well-developed procedure for mapping AGN structure (see e.g. Horne 1999), but its application to LMXBs is limited by the demand for high time resolution at optical/UV wavelengths. This is necessitated by the ~seconds light-travel time across short-period LMXBs which imposes a demand on optical instrumentation currently met by few observatories around the world. Instead Hynes et al. (1998) obtained simultaneous X-ray (RXTE) observations with high speed HST optical/UV FOS spectroscopy to reveal a ~10–20s delay between optical and X-ray at the times of X-ray flaring activity (see fig 2). With J1655-40's relatively long orbital period (2.6d) and well-established orbital ephemeris, this delay is entirely due to reprocessing within the accretion disc and not heating of the secondary's inner face. If such behaviour were observed throughout several orbital cycles then it would be possible in principle to fully map the system geometry, but this is a very difficult observational programme.

(b) Distortion of the Radial-velocity Curve. However, with short-term flares having a clearly visible effect on the accretion disc, it is entirely possible that the overall (very high) L_X has a significant (but non-varying) impact on the atmosphere of the secondary star. The original dynamical study of GRO J1655-40 was performed by Orosz & Bailyn (1997) and should, ideally, have been undertaken during X-ray quiescence, but in order to obtain full orbital phase coverage they employed some spectra taken during outburst. This appeared to be reasonable since the J1655-40's secondary is sufficiently luminous to be easily spectroscopically visible even during outburst. However, could the X-ray irradiation of the secondary have affected the absorption line strengths in a way that systematically distorted the radial-velocity curve?

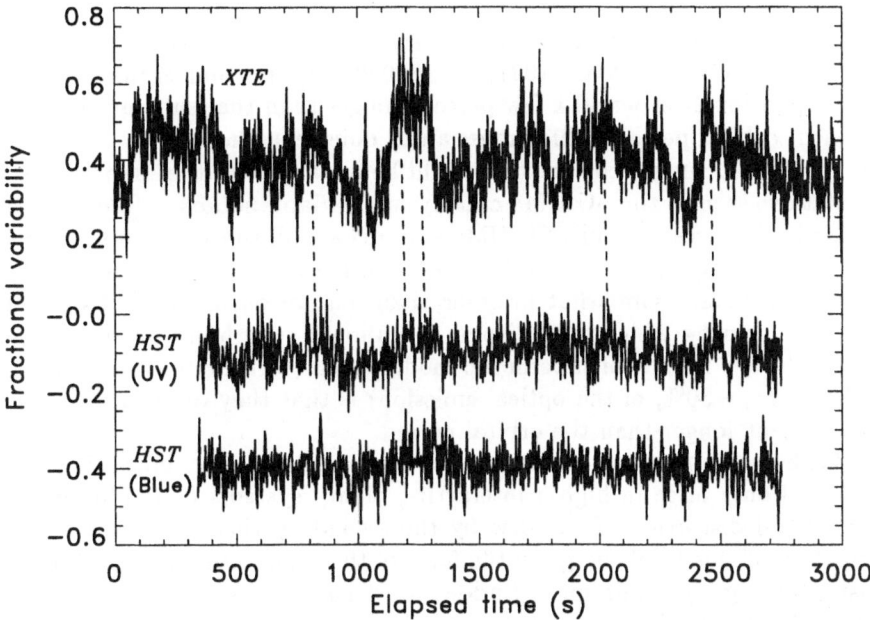

Fig. 2. X-ray (RXTE) and optical/UV (HST) lightcurves of GRO J1655-40 obtained in 1996 by Hynes et al. (1998) which shows the ~10–20s delay of the optical response to the X-ray flares, and are connected by dashed lines.

This was investigated by Phillips et al. (1999) who tested for the presence of such an effect by fitting an elliptical orbit to the Orosz & Bailyn data. Since this yielded a highly significant $e=0.12\pm0.02$ they undertook a more sophisticated analysis by fitting an irradiated model of the secondary. Such effects must be important since the ratio of incident X-ray flux to local stellar flux at its surface is ~7! Their free parameters were q, i and disc opening angle (for shadowing of the secondary's equatorial regions) from which they obtained $q=2.8$ and a compact object mass in the range 4.1–$6.6 M_\odot$. This was tested and (substantially) confirmed when Shahbaz et al. (2000) were able to obtain a completely quiescent radial velocity curve (the results of which are included in Table 1) which constrained q to the range 2.29–2.97 and a compact object mass of 5.5–$7.9 M_\odot$. This still leaves open the possibility of Brandt et al. (1995) that GRO J1655-40, with its high γ-velocity might be an example of a relatively low mass BH formed by accretion-induced collapse of a neutron star. This is also addressed by Israelian et al. (these proceedings).

5   SXTs and SU UMa Superoutbursts

A curious feature in the optical light curves of SXTs during outburst was first noted by Charles et al. (1991) in GS2000+25, and that is the presence of a modulation at a period a few percent longer than the (subsequently determined) orbital period (which must await quiescent photometry and spectroscopy). They suggested that this might be analogous to the "superhump" modulations seen in the SU UMa cataclysmic variable sub-class (see Warner 1995 and references therein). SU UMa systems are all short period CVs (below the period gap) which exhibit both normal and superoutbursts. The superoutbursts are somewhat brighter than normal outbursts and last for ~ 2 weeks, during which they display a hallmark superhump in their light curves. The key feature of superhump modulations (which represent substantial fractions, $\sim 30\%$, of the optical emission) is that they occur at a period a few percent longer than the orbital period.

This behaviour is now accepted to be due to an eccentric, precessing disc which will only arise in high q interacting binary systems (≥ 3; Whitehurst 1988). Tidal distortion of the disc by the secondary then gives rise to the dissipation of this tidal energy in the form of the superhump. Since all SXTs satisfy this high q requirement (indeed most have $q > 5$; see Table 3) then they might also display superhump-type phenomena and this is what Charles et al (1991) suggested for the modulation that they observed during the outburst of GS2000+25. Similar features were subsequently seen in Nova Mus 1991 (Bailyn 1992) and GRO J0422+32 (Kato et al. 1995). The potential importance of this phenomenon, if its interpretation could be confirmed, was that it could provide an independent estimate of q, as Mineshige et al. (1992) had shown how the superhump – orbital period difference was related to q.

However, SXT periods are all much longer (at least a factor 3 or more) than those of SU UMa systems, and hence require much longer observing runs with which to accurately determine the period of any modulations present. Hence, O'Donoghue & Charles (1996) compiled and reanalysed the outburst data from the above 3 sources. They showed that the modulation observed in outburst could *not* be occurring on the orbital period, but had to be due to a process on a slightly longer period. The excess periods of these modulations were also consistent with the mass ratios already obtained for these sources from optical spectroscopy. Interestingly, non-orbital effects during quiescence had already been seen in A0620-00 (see Haswell's 1996 presentation of McClintock and Remillard's data, and Leibowitz et al. 1998) that led Haswell to propose that it was due to an eccentric precessing disc, but at a high inclination so that a partial eclipse occurred only at certain precession phases. The difficulty with this explanation is that it requires an inclination that is inconsistent with the ellipsoidal light curve analysis (Table 3).

Nevertheless, there is an outstanding problem with the presence of superhumps during SXT outbursts, and that is understanding the emission process. In SU UMa systems, the dominant light sources are the high state disc and the tidally dissipated energy within the disc. However, in SXTs these are completely overwhelmed by the X-ray irradiation from the luminous compact object or inner disc (by a factor ~1000), and so a SU UMa-type superhump light source should never be visible. It has therefore been recently suggested by Haswell et al. (2000) that SXT superhumps occur because of variations in the disc projected area (and hence the area available for X-ray reprocessing). One obvious difference between SU UMa and SXT superhumps is then predicted by Haswell et al., and that is the inclination dependence, as the SU UMas show no change of superhump fraction with binary inclination.

6   Future Work

The continuous access to X-ray all-sky monitors is providing a steady stream of new transient sources with which it should be possible to address:

- the distribution of BH masses, with particular interest at the low (~3–5 $M_\odot$) and high (>20 $M_\odot$) ends
- the evolution of the accretion disc during outburst/decay, re-flares and superhumps, and the emission mechanisms using multi-wavelength (γ-ray to radio) campaigns
- the nature of the secondary star in reddened systems using high resolution IR spectroscopy
- activity and long-term variations in the secondary with robotic photometric monitoring

Acknowledgments. I am particularly grateful to Rob Hynes for his assistance in the preparation of the figures for this manuscript.

References

1. Bailyn C.D. (1992) Ap.J. **391**, 298
2. Bałucińska-Church M., Church M.J. et al. (2000) MNRAS **311**, 861
3. Bandyopadhyay R., Martini P. et al. (1998) MNRAS **295**, 623
4. Brandt W.N., Podsiadlowski P.& Sigurdsson S. (1995) MNRAS **277**, L35
5. Cannizzo J.K. (2000) New Astron.Rev. **44**, 41
6. Casares J., Charles P.A. & Marsh T.R. (1995) MNRAS **277**, L45
7. Charles P.A., Kidger M.R. et al. (1991) MNRAS **249**, 567
8. Charles P. (1998a) in *Proc 13th North American Workshop on CVs and Related Objects* ASP Conf Series 137, p220 eds Howell, Kuulkers & Woodward, ASP, San Francisco
9. Charles P. (1998b) in *Theory of Black Hole Accretion Discs*, p1 eds Abramowicz, Björnsson & Pringle, CUP, Cambridge
10. Chen W., Shrader C. & Livio M. (1997) ApJ **491**, 312
11. Eikenberry S.S., Matthews K. et al. (1998) Ap.J. **494**, L61
12. Giacconi R., Gorenstein P. et al. (1967) Ap.J. **148**, L119

13. Gursky H., Giacconi R. et al. (1963) Phys.Rev.Lett. **11**, 530
14. Hameury J.-M., Lasota J.-P., McClintock J.E. & Narayan R. (1997) Ap.J. **489**, 234
15. Harlaftis E.T., Horne K. & Filippenko A.V. (1996) PASP **108**, 762
16. Harlaftis E.T., Steeghs D., Horne K. & Filippenk, A.V. (1997) AJ **114**, 1170
17. Haswell C.A. (1996) in IAU Symp 165, *Compact Stars in Binaries* eds van Paradijs, van den Heuvel & Kuulkers. Kluwer, Dordrecht
18. Haswell C.A., King A.R. et al. (2000) MNRAS (submitted)
19. Haswell C.A. & Shafter A.W. (1990) Ap.J. **359**, L47
20. Herrero A., Kudritzki R.P. et al. (1995) A&A **297**, 556
21. Horne K. (1999) in *Quasars and Cosmology* ASP Conf Series 162, p189 eds Ferland & Baldwin. ASP, San Francisco
22. Hynes R.I. et al. (1998) MNRAS **299**, L37
23. Kato R., Mineshige S. & Hirata R. (1995) PASJ **47**, 31
24. King A.R., Kolb U. & Burderi L. (1996) Ap.J. **464**, L127
25. King A.R., Kolb U. & Szuszkiewicz E. (1997) Ap.J. **488**, 89
26. Kolb U., King A.R., Ritter H. & Frank J. (1997) Ap.J. **485**, L33
27. Kuulkers E. (1998) New Astron. Rev. **42**, 613
28. Lasala J., Charles P.A. et al. (1998) MNRAS **301**, 285
29. Leibowitz E.M., Hemar S. & Orio M. (1998) MNRAS **300**, 463
30. Martì J., Mirabel I.F. et al. (2000) A&A **356**, 943
31. Miller J.C., Shahbaz T. & Nolan L.A. (1998) MNRAS **294**, L25
32. Mineshige S., Hirose M. & Osaki Y. (1992) PASJ **44**, L15
33. Mirabel I.F., Bandyopadhyay R. et al. (1997) Ap.J. **477**, L45
34. O'Donoghue D. & Charles P.A. (1996) MNRAS **282**, 191
35. Orosz J.A. et al. (1994) Ap.J. **436**, 848
36. Orosz J.A. & Bailyn C.D. (1997) Ap.J **477**, 876 (and Ap.J. **482**, 1086)
37. Orosz J.A., Remillard R.A., Bailyn C.D. & McClintock J.E. (1997) Ap.J. **478**, L83
38. Phillips S.N., Shahbaz T. & Podsiadlowski Ph. (1999) MNRAS **304**, 839
39. Shahbaz T., Charles P.A. & King A.R. (1998) MNRAS **301**, 382
40. Shahbaz T. & Kuulkers E. (1998) MNRAS **295**, L1
41. Shahbaz T., Groot P. et al. (2000) MNRAS **314**, 747
42. Soria R., Wickramasinghe D.T. et al. (1998) Ap.J. **495**, L95
43. Van der Hooft F. et al. (1997) MNRAS, **286**, L43
44. Van Kerkwijk M.H. (1993) A&A **276**, L9
45. Van Paradijs J. (1996) Ap.J. **464**, L139
46. Warner B. (1995) in *Cataclysmic Variable Stars*, 187: CUP, Cambridge
47. Whitehurst R. (1988) MNRAS **233**, 529

Neutron Star Mass Determinations

M.H. van Kerkwijk

Utrecht University, P.O. Box 80000, 3508 TA Utrecht, The Netherlands

Abstract. I review attempts made to determine the properties of neutron stars. The focus is on the maximum mass that a neutron star can have, or, conversely, the minimum mass required for the formation of a black hole. There appears to be only one neutron star for which there is strong evidence that its mass is above the canonical $1.4\,M_\odot$, viz., Vela X-1, for which a mass close to $1.9\,M_\odot$ is found. Prospects for progress appear brightest for studies of systems in which the neutron star should have accreted substantial amounts of matter.

1 The Minimum Mass Required to Form a Black Hole

For the study of black holes, the relevance of neutron stars is mostly that below a certain maximum mass, degeneracy pressure due to nucleons is sufficient to prevent an object from becoming a black hole. Unfortunately, the equation of state of matter at densities above nuclear-matter density is rather uncertain and theoretical estimates of the minimum mass required to form a black hole range from just over $1.4\,M_\odot$ to about $2.5\,M_\odot$ [3,10].

Constraints on the equation of state can be obtained from a variety of observed properties of neutron stars. Among the more direct measurements that have been used or proposed are: (i) the maximum mass inferred from dynamical measurements in binaries; (ii) the minimum spin period among millisecond pulsars; (iii) identification of kHz quasi-periodic oscillations with the orbital frequency in the last stable orbit; (iv) identification of quasi-periodic oscillations with the Lens-Thirring precession period; (v) influence of gravitational light bending on X-ray light curves in radio pulsars; (vi) model-atmosphere fitting of X-ray lightcurves during X-ray bursts (resulting from thermonuclear run-aways on the neutron-star surface); (vii) gravitational redshift from γ-ray spallation lines in accreting systems; (viii) gravitational redshift and surface gravity from model-atmosphere analysis of spectra of isolated neutron stars. Less direct measurements include: (ix) matching observed neutron-star temperatures and inferred ages to cooling curves; (x) neutrino light curves in supernova explosions; (xi) pulsar glitches; (xii) moment of inertia from accretion torques in combination with knowledge of the magnetic field strength from cyclotron lines; (xiii) comparing X-ray fluxes between states in which a neutron star is accreting and in which matter is stopped at the magnetosphere and "propelled" away. For references and somewhat more detail, see [20]. The strongest constraints on the equation of state are still set by dynamical mass measurements, so I will restrict myself to those below.

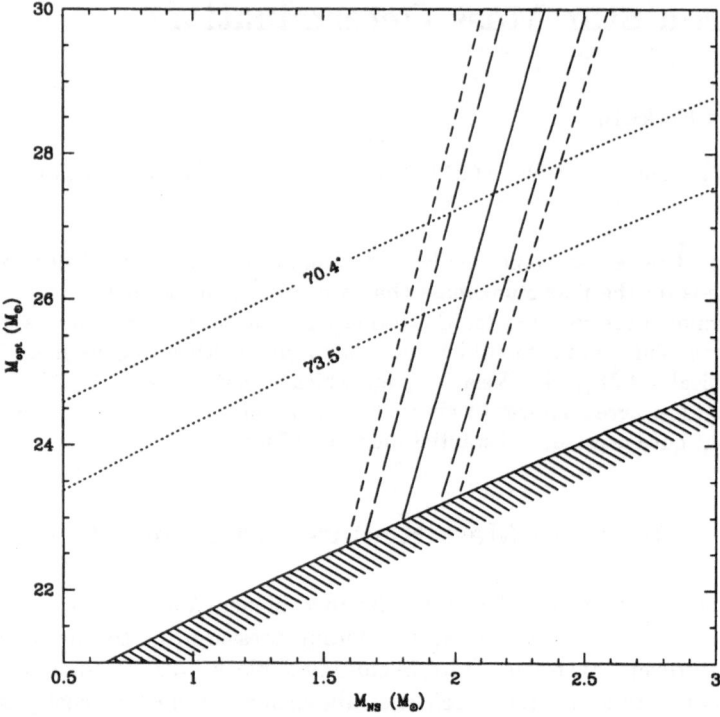

Fig. 1. Constraints on the mass of Vela X-1 and its supergiant companion HD 77581 [1,2]. The constraint on the mass ratio from the X-ray pulse delay and optical radial-velocity curves is indicated by the solid line. The long and short-dashed lines next to it indicate the 95% and 99% confidence limits, respectively. The lines stop at the region excluded by the pulse-timing mass function (to go below it would require $\sin i > 1$). The 95% and 99% confidence lower limits on the inclination derived from the duration of the X-ray eclipse are indicated by the two dotted lines.

2   Dynamical Measurements

Most mass determinations have come from radio timing studies of pulsars; see [16] for an excellent review. The most accurate ones are for pulsars that are in eccentric, short-period orbits with other neutron stars, in which several non-Keplerian effects on the orbit can be observed: the advance of periastron, the combined effect of variations in the second-order Doppler shift and gravitational redshift, the shape and amplitude of the Shapiro delay curve shown by the pulse arrival times as the pulsar passes behind its companion, and the decay of the orbit due to the emission of gravitational waves. The most famous of the double neutron-star binaries is the Hulse-Taylor pulsar, PSR B1913+16, for which recent measurements give $M_{\mathrm{PSR}} = 1.4411 \pm 0.0007\,M_{\odot}$ and $M_{\mathrm{comp}} = 1.3874 \pm 0.0007\,M_{\odot}$ [14,13]. Almost as accurate masses have been inferred for PSR B1534+12, for which the pulsar and its companion are found to have very similar mass: for both, $M = 1.339 \pm 0.003\,M_{\odot}$ [11].

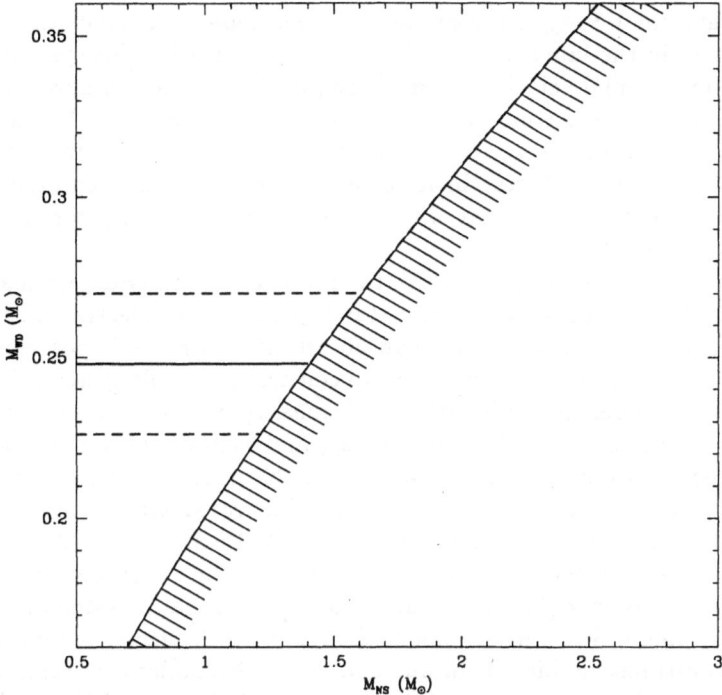

Fig. 2. Constraints on the mass of PSR B1855+09 and its white-dwarf companion [7,16]. The horizontal line is at the best fit white-dwarf mass derived from the amplitude of the Shapiro delay curve, and the short-dashed lines reflect the 95% confidence uncertainties on that measurement. The lines stop at the limit derived from the pulse timing mass function, at $i = 90°$; to be to the right or below it would require $\sin i > 1$. The barely visible short-dashed curve just left of it reflects the 95% confidence lower limit on the inclination set by the shape of the observed Shapiro delay curve (the upper limit would not be visible in this graph).

Neutron-star masses can also be determined for some binaries containing an accreting X-ray pulsar, from the amplitudes of the X-ray pulse delay and optical radial-velocity curves in combination with constraints on the inclination (the latter usually from the duration of the X-ray eclipse, if present). This method has been applied to about half a dozen systems [6,8,18], but the masses are generally not very precise.

So far, for all but one of the neutron stars, the masses are consistent with being in a surprisingly narrow range, which can be approximated with a Gaussian distribution with a standard deviation of only $0.04\,M_\odot$ [16]. The mean of the distribution is $1.35\,M_\odot$, close to the "canonical" value of $1.4\,M_\odot$.

The one exception is the X-ray pulsar Vela X-1, which is in a 9-day orbit with the B0.5 Ib supergiant HD 77581. For this system, a rather higher mass of around $1.8\,M_\odot$ has consistently[1] been found ever since the first detailed study in the late seventies [21,19]. A problem with this system is that the measured radial velocities show strong deviations from a pure Keplerian radial-velocity curve, which are correlated within one night, but not from one night to another. A possible cause could be that the varying tidal force exerted by the neutron star in its eccentric orbit excites high-order pulsation modes in the optical star which interfere constructively for short time intervals.

We were granted time at ESO to improve the mass determination of this possibly very massive neutron star from 200 new spectra, taken in as many nights. These cover more than 30 orbits, and make it possible to average out the velocity excursions and to constrain possible systematic effects with orbital phase. In combination with measurements from our old photographic plates, earlier CCD spectroscopy, and high-resolution IUE spectra, we derived a 95% confidence constraint on the mass of the neutron star of $M_{\rm NS} = 1.87^{+0.23}_{-0.17}$ [1]. Our constraints are illustrated graphically in Fig. 1. One sees that even at 99% confidence, $M_{\rm NS} > 1.6\,M_\odot$. It should be noted, however, that from the data it appears that while the excursions in radial velocity are mostly random, there is also a component that is systematic, locked to orbital phase. Since we do not understand these effects, it may be that our mass estimate is biased. In our trials with excluding the worst-affected phase ranges, however, we consistently found that the fitted mass became even higher [19,1].

3 Trying for Bias

The narrow range in masses inferred for most neutron stars might be seen as evidence for a relatively low maximum neutron-star mass. It could also be, however, that it reflects the formation mechanism. Indeed, from models, it appears that supernova explosions result in neutron stars with masses preferentially in two narrow peaks, one around $1.3\,M_\odot$ and one around $1.65\,M_\odot$ [15]. It is tempting to associate the latter with Vela X-1, but for honesty it should be noted that from the same calculations it is expected that only lighter neutron stars can be formed by stars which lose their envelope during their evolution, as would likely have happened for the progenitor of Vela X-1.

If the narrow range indeed reflects the formation process, it seems worthwhile to focus especially on systems in which the neutron star is expected to have accreted a lot of matter since it was formed. Such systems are the low-mass X-ray binaries and their descendents, the pulsars with low-mass white dwarf companions. In the latter systems, the white dwarfs typically have masses of $0.3\,M_\odot$. However, in order for mass transfer to have happened,

---

[1] One study based on IUE spectra of HD 77581 appeared to find a lower neutron-star mass [12]. However, this was found to be due to a bug in the cross-correlation software used [1].

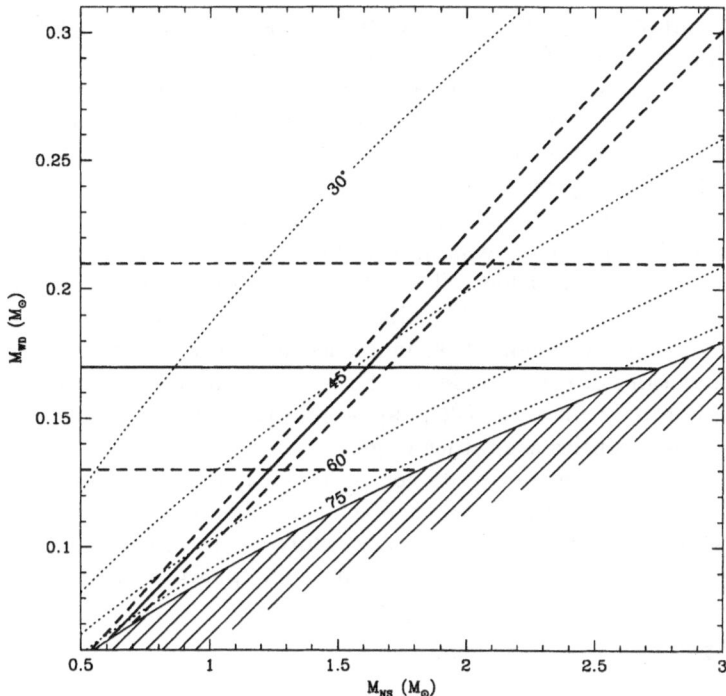

Fig. 3. Constraints on the mass of PSR J1012+5307 and its white-dwarf companion [17,4]. The horizontal solid line reflects the white-dwarf mass inferred from the surface gravity measured from the optical spectrum of the white dwarf. The slanted solid line is the constraint set by the mass ratio inferred from the optical radial-velocity and radio pulse-delay curves. For both curves, the 95% confidence regions are indicated by the associated short-dashed lines. All lines stop at the limit derived from the pulse timing mass function, at $i = 90°$; to be to the right or below it would require $\sin i > 1$. Dotted lines indicate the relations expected for some other values of the inclination.

these stars need to have evolved to at least the end of the main sequence life. For this to happen in a Hubble time, their masses need to have been at least $0.8\,M_\odot$ initially. Since the mass transfer in these systems is thought to be stable, the neutron star should thus have accreted more than $0.6\,M_\odot$, i.e., have become substantially more massive than it was initially.

From radio timing measurements, it is generally more difficult to measure masses for these systems, because the orbits are circular and relatively wide. However, for one system, PSR B1855+09, the orbit is very close to edge on, and it has been possible to derive constraints from the Shapiro delay [7,16]. These constraints are indicated in Fig. 2.

Another way of determining the masses in these systems uses optical spectroscopy of the white-dwarf companion. From a model-atmosphere fit to the spectrum, one can determine the effective temperature and surface gravity. From the latter, the white-dwarf mass follows, assuming a theoretical mass-radius relation. If one can also measure the radial-velocity orbit, and determine the mass ratio (in combination with the pulse-delay orbit), then one has a constraint on the neutron-star mass. So far, the only pulsar for which this has been possible is PSR J1012+5307 [17,4], whose companion is particularly bright ($V \simeq 20$). The present constraints are shown in Fig. 3.

One sees that for both systems, the mass determinations are at present not precise enough to provide meaningful additional constraints on the maximum mass a neutron star can have. The situation unfortunately is no better for determinations in low-mass X-ray binaries. A problem for those is that the X-ray sources generally do not pulse, and hence all information has to be gleaned from observations of the companions. So far, a meaningful determination has been possible only for Cyg X-2 [5,9]. A relatively high mass is found, but again the uncertainties are rather large.

Despite the above, prospects for progress appear brightest for further studies of systems in which the neutron star should have accreted a substantial amount of matter. Some other white-dwarf companions are bright enough, and more may be found in the on-going pulsar searches.

References

1. Barziv O., Kaper L., Van Kerkwijk M. H., Telting J., Van Paradijs J., 2000, in preparation
2. Bildsten L., Chakrabarty D., Chiu J., et al., 1997, ApJS 113, 367
3. Datta, B., 1988, Fund. Cosmic Phys. 12, 151
4. Callanan P. J., Garnavich P. M., Koester, D., 1998, MNRAS 298, 211
5. Casares J., Charles P. A., Kuulkers E., 1997, ApJ 493, L39
6. Joss P. C., Rappaport S. A., 1984, ARA&A 22, 537
7. Kaspi V. M., Taylor J. H., Ryba M., 1994, ApJ 428, 713
8. Nagase F., 1989, PASJ 41, 1
9. Orosz J. A., Kuulkers E., 1999, MNRAS 305, 132
10. Srinivasan G., 2000, these proceedings
11. Stairs I. H., Arzoumanian Z., Camilo F., Lyne A. G., Nice D. J., Taylor J. H., Thorsett S. E., Wolszczan A., 1998, ApJ 505, 352
12. Stickland D., Lloyd C., Radziun-Woodham A., 1997, MNRAS 286, L21
13. Taylor J. H., 1992, Phil. Trans. R. Soc. London A 341, 117
14. Taylor J. H., Weisberg J. M., 1989, ApJ 345, 434
15. Timmes F. X., Woosley S. E., Weaver Th. A., 1996, ApJ 457, 834
16. Thorsett S. E., Chakrabarty D., 1998, ApJ 512, 288
17. Van Kerkwijk M. H., Bergeron P., Kulkarni S. R., 1996, ApJ 467, L89
18. Van Kerkwijk, M. H., van Paradijs, J., Zuiderwijk, E. J., 1995, A&A 303, 497
19. Van Kerkwijk M. H., Van Paradijs J., Zuiderwijk E. J., Hammerschlag-Hensberge G., Kaper L., Sterken C., 1995, A&A 303, 483
20. Van Paradijs J., 1998, in Buccheri R., Van Paradijs J., Alpar M. A. (eds), The Many Faces of Neutron Stars. Kluwer Academic Publishers, Dordrecht, 279 (astro-ph/9802177)
21. Van Paradijs J., Zuiderwijk E. J., Takens R., Hammerschlag-Hensberge G., Van den Heuvel E.P.J., De Loore C., 1977, A&AS 30, 195

On the Limiting Mass of Neutron Stars

G. Srinivasan

Raman Research Institute, C.V. Raman Ave., Bangalore-560 080, India

Abstract. The maximum mass of neutron stars is one of the most important predictions of the general relativistic theory of stellar structure. This paper highlights the difficulties in making a definitive prediction, and also reviews the progress made so far.

The first attempt to model neutron stars was by Oppenheimer and Volkoff [1]. Earlier Chandrasekhar has shown that there is a limiting mass to the configuration of white dwarfs in which the gravitational pressure is balanced by the pressure of degenerate electrons. Soon after that Gamow argued that in sufficiently massive stars neutron cores will form when the thermonuclear sources of energy had been exhausted. Oppenheimer and Volkoff wanted to know whether Gamow's conjecture would be right for arbitrarily massive stars ("to investigate whether there is some upper limit to the possible size of such a neutron core"). In a sense the answer to this question was implicit in Chandrasekhar's theory of cold degenerate stars. The *Chandrasekhar's limiting mass* for neutron stars is 5.73 $M_\odot$. To recall, this is the upper limit to the mass of a Newtonian star supported against gravity by the degeneracy pressure of neutrons.

Oppenheimer and Volkoff realised that for a neutron star the modifications of Newtonian gravity due to General Relativity would have to be taken into account. To calculate the radius of a star of a given mass one has to integrate the equation of hydrostatic equilibrium with an assumed equation of state, viz., pressure as a function of density. In Newtonian gravity this equation is

$$\frac{dP}{dr} = \frac{-GM(r)\rho(r)}{r^2}.$$

(1)

The corresponding equation in General Relativity is

$$\frac{dP}{dr} = \frac{-G\left[M(r) + 4\pi r^3 P(r)/c^2\right]\left[\rho(r) + P(r)/c^2\right]}{r^2\left[1 - \frac{2GM(r)}{rc^2}\right]}.$$

(2)

Here ρ is the mass-energy density, and the other symbols have the usual meaning. Equation (2) incorporates modifications arising due to the fact that in Einstein's theory all forms of energy contribute to gravity, and reduces to equation (1) in the limit $c \to \infty$. With these modifications incorporated, Oppenheimer and Volkoff went on to construct models of neutron stars in exactly the same way Chandrasekhar had constructed models of white dwarfs, with the only difference being that the particles which contributed to the pressure in the present case were the neutrons.

While there was no minimum mass in their theory, they concluded that there are no equilibrium configurations for neutron stars with mass greater than 0.7 $M_\odot$. This value for the limiting mass is much smaller than the *Chandrasekhar limiting mass* of 5.73 $M_\odot$ for two reasons. Firstly, in the relativistic theory the relevant mass is the *gravitational mass* and not the rest mass - a distinction one had not made in the Newtonian theory. Secondly, in relativistic gravity the maximum mass is reached at a finite value of the central density (5×10^{15} g cm^{-3}), unlike in Chandrasekhar's theory where the central density is infinite at the limiting mass.

How good is this estimate of the maximum mass? Not very good. And the reason is the following. Although Oppenheimer and Volkoff had treated gravitation properly, they had assumed, following Chandrasekhar, that the neutrons could be regarded as an ideal fermi gas. Whereas this assumption is exactly valid for the electrons, it is not valid for the neutrons. Since a proper treatment of the interaction between the nucleons continues to be at the heart of the problem, the following clarification is in order. A degenerate electron gas has the peculiar property that as the density increases it becomes more *ideal*. This may be seen as follows: the kinetic energy scales inversely as the *square* of the interparticle spacing, whereas the potential energy scales inversely as the interparticle spacing. It is because of this that interaction effects become less important at high densities, i.e., the gas becomes more ideal. But this is not true for a gas of neutrons and protons interacting via *nuclear forces*. Indeed, interaction effects dominate at high densities. This is also true for a gas of atoms interacting via, say, the van der Waals force.

In view of this, to model neutron stars better than Oppenheimer and Volkoff one has to use a more realistic equation of state (EOS). During the past three decades there have been numerous efforts to achieve this objective. Interestingly, some of the pioneering efforts preceded the discovery of the neutron star by a decade! In this brief review it is not possible to attempt even a summary of the vast literature. Instead, it would be more worthwhile if I were to present the "canonical" picture of the interior of a neutron star, indicate points of disagreement, and leave you with an idea of what is the best one can say at present about the limiting mass of neutron stars.

Since neutron stars are "cold" objects ($k_B T <<$ excitation energies $\sim$ MeV) one may regard it as the *ground state of matter*. Therefore, the outer layers will consist of Fe^{56} arranged as a ferromagnetic lattice (the ground state at zero pressure). As one goes deeper into the solid crust one will encounter very neutron-rich nuclei. This is because at the enormous pressures reached there Fe^{56} is no longer the most stable nucleus - the electrons at the top of the fermi sea will combine with the protons in the nuclei to produce neutrons. In fact, at the bottom of the crust the nuclei will be so neutron rich that neutrons will ooze out of the nuclei. Consequently, the lattice will be immersed not only in a gas of electrons - like in terrestrial metals -but also in a *fluid* of neutrons.

At a depth where the density reaches the "nuclear density" $\sim 2.5 \times 10^{14}$ g cm^{-3} the adjacent nuclei will merge into one another and one will essentially have a fluid of neutrons, with an admixture of protons and electrons (whose percentage will be determined by the condition of beta equilibrium).

There is broad agreement up to this point in the description. Given this there is a zoo of equations of state. What one wants is the pressure as a function of density. Since the pressure is related to energy density, one has to estimate the various contributions to the energy at a given baryon density: the energy of the nuclei, the coulomb energy of the lattice, the binding energy of the crustal neutron fluid (if one is above the neutron drip density of 4.3×10^{11} g cm^{-3}), etcetera. Different investigators have over the years brought to bear different degrees of sophistication in estimating the above mentioned contributions. Fortunately, during the last decade or so there has been an impressive convergence of the results. For an authoritative account of this and other aspects relevant to our discussion we refer the reader to, for example, Baym [2].

After this lengthy digression let us return to the main topic of our discussion. To recall, we noted that to get a good estimate of the limiting mass of neutron stars one has to model a neutron star properly and derive a realistic EOS. Since there now exist a variety of modern EOS let us see what predictions they make for the maximum mass. The derived gravitational mass as a function of the central density is shown in Figure 1 for a number of EOS. As may be seen, the majority of them predict a maximum mass around 2 M$_\odot$ - these are *stiff equations of state* which include three-body forces while calculating the energy per nucleon in the fluid core. The reason why the maximum mass predicted by these is significantly greater than the value of 0.7 M$_\odot$ obtained by Oppenheimer and Volkoff is clear - at the super-nuclear densities reached in the fluid interior the repulsive nature of the nuclear force plays an important role in supporting the star.

The figure also shows an EOS whose prediction is at variance. In fact, it predicts a maximum mass $\sim 1.4 \ M_\odot$ which is disconcertingly close to the measured masses of radio pulsars. Let us try to understand why this is so.

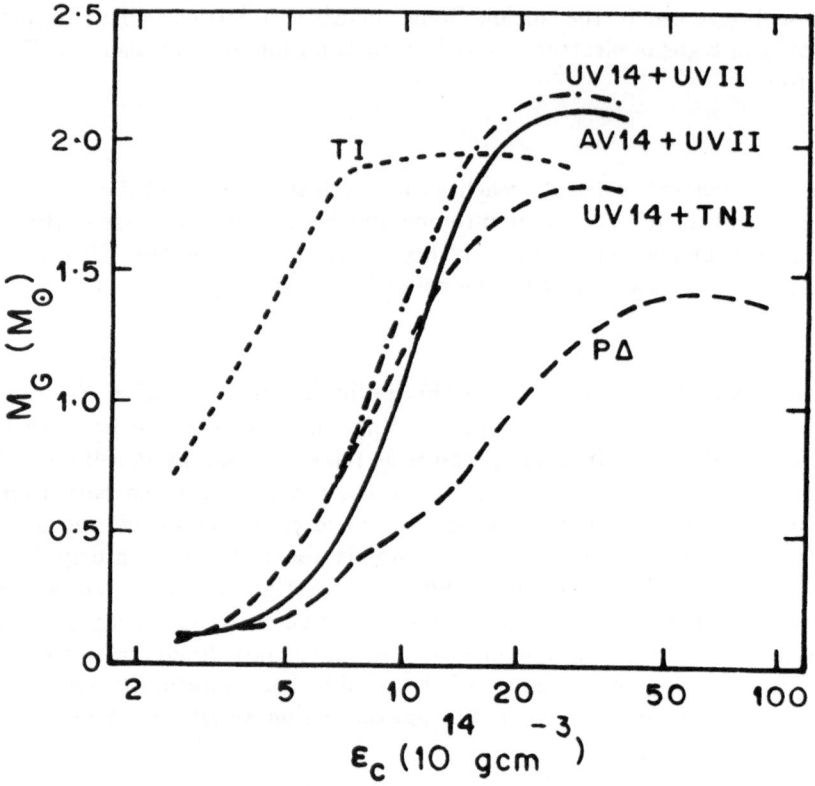

Fig. 1. The gravitational mass vs. the central density for several EOS (adapted from Baym, [2]). The four curves which predict a maximum mass $\sim 2 \ M_\odot$ are for rather stiff EOS. The curve labeled P_Δ is a soft EOS, and is discussed in the text.

In the discussion so far we have considered only three types of constituent particles, viz., neutrons, protons and electrons, all obeying Fermi-Dirac statistics. It is reasonable to ask whether particles such as π and K mesons can also exist in a condensed star. If they naturally occur then there could be several interesting consequences. For example, π mesons could be spontaneously produced as follows. They come in three charge species - neutral, positively or negatively charged, and their rest mass is ~ 140 MeV. As already mentioned, one expects beta equilibrium in the fluid core. This requires that the chemical potential of the neutrons is equal to the sum of the chemical potentials of the protons and electrons, viz., $(\mu_n = \mu_p + \mu_e)$. At a depth where the density reaches a critical value such that the chemical potential of the electrons exceeds the rest mass of the π meson, the neutron can decay to a proton and a π^-. To put it differently, a new channel opens up for the decay of the neutron, and π mesons will be spontaneously produced. Naively, this is likely to happen at densities slightly greater than 2.5×10^{14} g cm^{-3}. More realistically, if one allows for the interaction of the π meson with the medium one will expect this instability to set in around twice the nuclear density. K mesons can also be spontaneously produced, and a rough estimate yields a value for the critical density $\sim$ three times the nuclear density.

One might wonder what all this has to do with the maximum mass of neutron stars! The spontaneous occurrence of π and K mesons may be profoundly important because they obey Bose-Einstein statistics. A fundamental property of a gas of Bose particles is that below a certain critical temperature they will condense into the zero-momentum state. This phenomenon is known as *Bose-Einstein condensation.* Since particles in the zero-momentum state do not contribute to the pressure, and since the majority of mesons will be in this state, the occurrence of such a condensation will greatly *soften* the equation of state. The soft EOS which predicts a maximum mass ~ 1.4 M$_\odot$ (see Figure 1) admits such a possibility.

Whether such a Bose-Einstein condensation occurs in neutron stars in nature is still controversial. Given the uncertainties in the theoretical predictions it is perhaps wise to appeal to the available observational evidence. Two points may be made in this context. An important consequence of pion and kaon condensation is that they (particularly the former) will have very high neutrino luminosities, which in turn will result in ultra-rapid cooling of the star [2]. The detection of soft X-rays (presumably black body emission) by ROSAT from a number of relatively old neutron stars appears to rule this out. The second piece of evidence pertains to the recently measured mass of Vela X-1. From a comprehensive set of observations Barziv et al.[3] derive a mass limit for the neutron star ≥ 1.7 M$_\odot$. If confirmed by further observations, such a value would be difficult to accommodate in a soft EOS scenario. In view of these pieces of circumstantial evidence one is forced to - at least tentatively - favour a stiff EOS.

Given the above mentioned theoretical uncertainties regarding the very nature of the central regions of a neutron star, one might approach the problem differently: can one appeal to some fundamental principles and try to set at least a conservative upper limit to the mass of neutron star? The answer is "Yes". One might impose the following two *minimal* requirements:

1. The EOS should satisfy the *microscopic stability* condition viz., $\frac{dP}{d\rho} \geq 0$. Matter would collapse if this is violated.
2. The EOS satisfies the *causality* condition $\frac{dP}{d\rho} \leq c^2$. This is the requirement that the velocity of sound be less than the velocity of light.

Soon after the discovery of neutron stars a number of authors derived bounds on the mass subject to these very general constraints, and we mention two such attempts. Nauenberg and Chapline [4] approached the problem from the point of view of the equilibrium and stability of neutron stars in general relativity, while Rhoades and Ruffini [5] adopted the conventional approach of integrating the Oppenheimer-Volkoff equation. Both pair of authors assumed the following: if one knows the pressure P_o at a given mass-energy density ρ_m, then the pressure P at any higher value of density ρ must satisfy the inequality

$$P \leq P_0 + (\rho - \rho_m)c^2 . \tag{3}$$

ρ_m *may be taken to be the largest value of density for which the pressure can be confidently predicted.* Nauenberg and Chapline took $\rho_m = 5 \times 10^{14}$ g cm^{-3} and $P_o = 7 \times 10^{33}$ dyn cm^{-2} and derived a value for the limiting mass equal to 3.6 M$_\odot$. Rhoades and Ruffini "matched" the above EOS at a density of 4.6×10^{14} g cm^3 to the well known EOS due to Harrison et al. [6] and derived a limiting mass of 3.2 M$_\odot$. Since these estimates are often quoted as the "theoretical limit" it is worth pointing out that the bound on the mass is sensitive to the value assumed for the *matching density* ρ_m.

The following parameterized formula for the maximum mass due to Hartle and Sabbadini [7] brings this out clearly:

$$M < 4.8 M_\odot \left(\frac{2 \times 10^{14} \text{ g cm}^{-3}}{\rho_m} \right)^{\frac{1}{2}} . \tag{4}$$

There is a further qualification. So far we have not considered the rotation of the star. It is intuitively obvious that since rotation has a stabilizing effect the upper mass limit for a rotating star with the stiffest EOS consistent with causality will be greater. Friedman and Ipser [8] give the following limit:

$$M^{rot} < 6.1 M_\odot \left(\frac{2 \times 10^{14} \text{g cm}^{-3}}{\rho_m} \right)^{\frac{1}{2}} . \tag{5}$$

Thus rotation increases the maximum mass by only $\sim 20\%$. This is because the neutron star is not centrally condensed (like a white dwarf is) and therefore does not admit differential rotation.

It is clear from the above discussion that further progress cannot be made unless one can answer the following questions with confidence: What are the constituent particles in the core of a neutron star, and what is the interaction potential? Is the resultant equation of state *soft* or *ultrastiff*? Several ongoing heavy ion collision experiments will shed light on this in the coming years. In the mean time one should pay special attention to theoretical predictions that help to constrain the EOS. Fortunately, general relativity makes one such prediction.

A most remarkable result of general relativity is that *all rotating, self-gravitating perfect fluids are unstable to non-axisymmetric perturbations which radiate away their angular momentum* [9, 10]. The limiting angular velocity arising from this instability depends on the EOS. If neutron stars in binaries are spun up by accretion then it is conceivable that the shortest period to which they can be spun up could be limited by the above mentioned instability. It is an intriguing fact that the two most rapidly spinning pulsars PSR 1937 + 214 and PSR 1957 + 20 have frequencies within 3% of one another. In the canonical picture the spin frequencies of these *recycled pulsars* will be determined by their magnetic fields. It could, of course, be a coincidence that these two pulsars have nearly the same frequencies. Alternatively, the limiting frequency determined by the non-axisymmetric instability may be close to the observed frequencies of these two pulsars (4033 and 3910 s^{-1}, respectively). If this is the case, and if their masses are at least 1.4 $M_\odot$, then the EOS must be extremely *stiff*. As more millisecond pulsars are discovered one should see an excess of them at the shortest observed period if the above mechanism is relevant and operative. The reason for this qualification is that viscosity can damp out the non-axisymmetric instability driven by gravitational radiation. But given the considerable uncertainty concerning the viscosity of neutron star matter one may tentatively accept the above inference that the EOS of neutron stars is very stiff.

To conclude, it appears at present that it is unlikely that the maximum mass of neutron stars is less than $\sim$ 2 $M_\odot$. It is probably not very much more than this either.

References

1. Oppenheimer, J.R. & Volkoff, G.M. (1939) Phys. Rev. **55**, 374.
2. Baym, G. (1991) in *Neutron Stars: Theory and Observations*, 21. J. Ventura and D. Pines (eds.), Kluwer Academic Publishers.

3. Barziv, O., Kaper, L., M.H. van Kerkwijk, J.H. Telting, J.A. van Paradijs (2000), submitted to A&A
4. Nauenberg, M. & Chapline, G. (1973) Astrophys. J. **179**, 277.
5. Rhoades, C. & Ruffini, R. (1973) in *Black Holes*, 1972 *Les Houches Lectures*, Gordon & Breach.
6. Harrison, B.K. & Wheeler, J.A. (1958). See Harrison, B.K. et al. 1965. *Gravitation Theory and Gravitational Collapse*, University of Chicago Press, Chicago.
7. Hartle, J.B. & Sabbadini, A.G. (1977) Astrophys. J. **213**, 831.
8. Friedman, J.L. & Ipser, J.R. (1987) Astrophys. J. **314**, 594.
9. Friedman, J.L. (1978) Comm. Math. Phys. **62**, 247.
10. Friedman, J.L. & Ipser, J.R. (1992) Phil. Trans. R. Soc. Lond. A. **340**, 391.

Keck Observations
of Black Hole X-Ray Transients

Emilios T. Harlaftis[1] and Alexei V. Filippenko[2]

[1] Institute of Astronomy and Astrophysics, National Observatory of Athens,
P.O. Box 20048, Athens - 118 20, Greece
[2] Department of Astronomy, University of California
Berkeley, CA 94720-3411, USA

Abstract. The advent of the large effective apertures of the Keck telescopes has resulted in the determination with unprecedented accuracy of the mass functions and mass ratios of faint ($R \approx 21$ mag) X-ray transients (GS 2000+25, GRO J0422+32, Nova Oph 1977, Nova Vel 1993).

1 Introduction

Most (~70%) X-ray novae harbor black-hole candidates in low-mass X-ray binaries and offer a testing ground for examining the physics around such exotic objects. Unlike classical novae, X-ray novae are accretion-driven events with a typical rise of 8–10 mag in a few days and a subsequent decline over several months. The most significant advances have come from observations in quiescence when the companion star can be detected and thus the compact object "weighed" (see, e.g., Charles 2000). The 1990s were dominated by efforts to measure the masses, thus producing the first observed and theoretical mass distribution of black holes (Bailyn et al. 1998; Miller et al. 1998; Fryer 1998). The masses can be fully determined from the mass function, mass ratio, and binary inclination.

2 Keck Observations of Black Hole X-Ray Binaries

The Keck telescopes have provided the best measurements of mass functions and mass ratios of X-ray transients. Utilizing the Doppler effect produced by the shifting photospheric lines due to the orbital motion of the companion star (donor) around the black hole, Filippenko et al. (1999, and references therein) produced the four most accurate mass functions, reducing the uncertainties by almost an order of magnitude compared to previous results (GS 2000+25, GRO J0422+32, Nova Oph 1977, Nova Vel 1993). Figure 1 shows the great improvement that the large Keck aperture offers in relation to 4-m class telescopes in extracting radial-velocity curves of the motion of the donor star around the black hole.

Indeed, further work on the Keck data by Harlaftis et al. (1999, and references therein) has determined mass ratios for the first time for binaries as

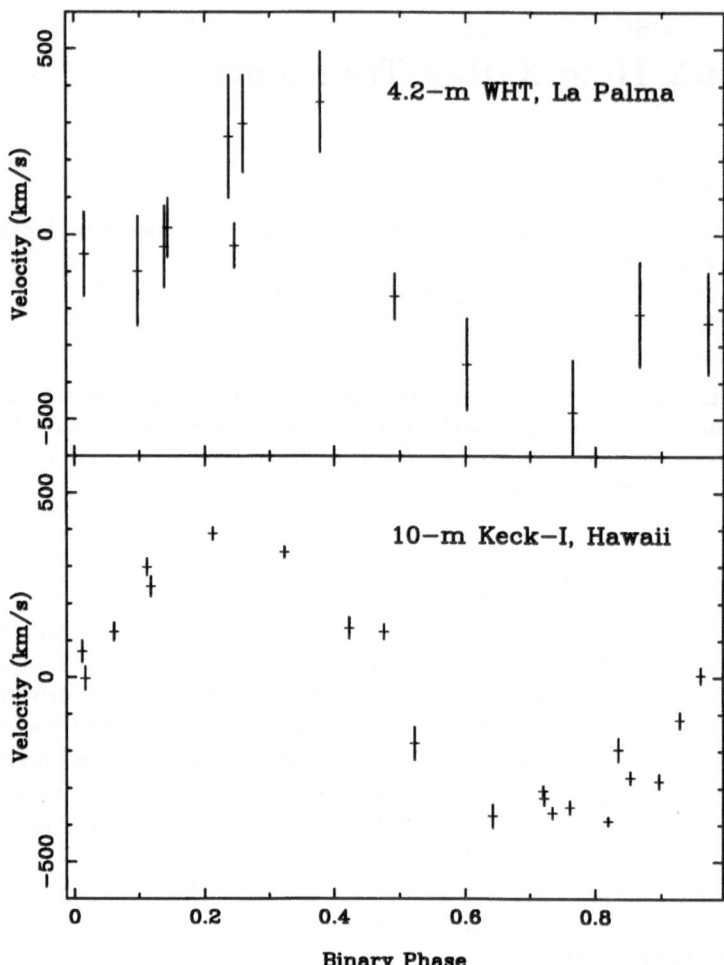

Fig. 1. *Bottom:* The radial-velocity curve of the companion star to the black hole GRO J0422+32 as extracted from spectra near Hα (Keck-I/LRIS) in just one night (Harlaftis et al. 1999). *Top:* The radial-velocity curve of the same star as extracted from 4.2-m WHT/ISIS near-infrared spectra (8450–8750Å) in 3 nights (Casares et al. 1995). The reduction in the individual uncertainty measurements is a factor of four using Keck. The sinusoidal fit gives $K_c = 338 \pm 39$ km s^{-1} with the WHT data and $K_c = 372 \pm 10$ km s^{-1} with the Keck data for the radial velocity semi-amplitude of the companion star. This means a better accuracy in the estimate on the lower limit of the mass of the black hole from $f_x = (PK_c^3)/(2\pi G) = 0.85 \pm 0.30$ $M_\odot$ to 1.21 ± 0.06 $M_\odot$ in the low-inclination system GRO J0422+32.

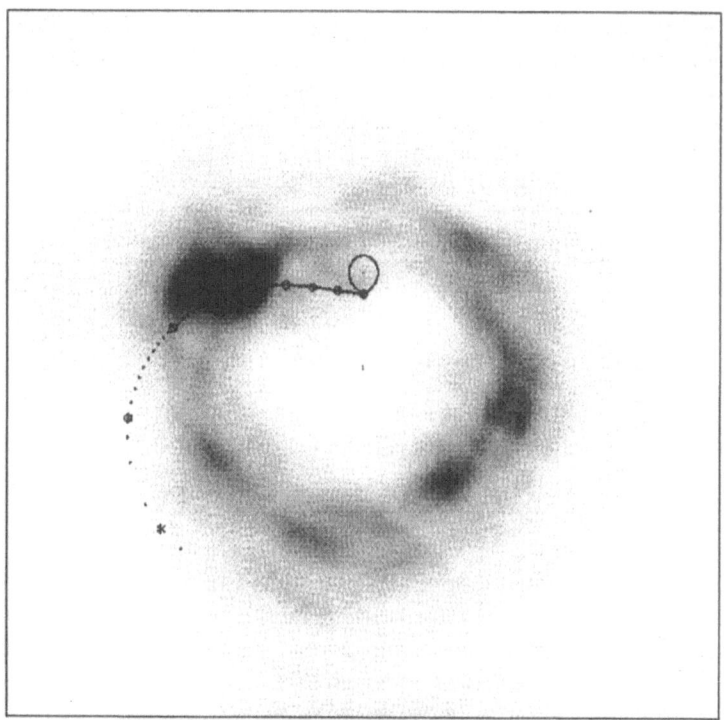

Fig. 2. The Hα Doppler image of the accretion disk surrounding the black hole GS 2000+25, as obtained from 13 Keck-I spectra (1/2 hour each over the 8.3 hour binary period). By projecting the image in a particular direction, one obtains the Hα emission-line profile as a function of velocity; for example, projecting toward the top results in the profile at orbital phase 0.0, which has a blueshifted peak. The path in velocity coordinates of gas streaming from the dwarf K5 companion star is illustrated, and the bright spot at the upper left results from collision with the accretion disk. The image was reconstructed from the phase-resolved spectra by applying Doppler tomography, a maximum entropy technique (Harlaftis et al. 1996).

faint as $R \approx 21$ mag (GRO J0422+32, GS 2000+25, and Nova Oph 1977). The mass ratio $q = M_2/M_1$ is determined by measuring the rotational broadening of the absorption lines of the companion, $v \sin i$, through the relation

$$\frac{v \sin i}{K_c} = 0.46 \left[(1 + q)^2\, q \right]^{1/3}.$$

The rotational broadening is determined using a χ^2 minimalization procedure where the solution consists of a set of three variables, namely spectral type, rotational broadening, and veiling factor. In particular, spectra of various spectral-type main-sequence stars are processed so as to simulate the observed spectra. The results of the technique are given in Table 1 (f is the fraction of light contributed by the companion star at red wavelengths). Determination of the inclination (inferred from the ellipsoidal modulation of the companion star) can then fully describe the masses of the binary components.

Table 1. Keck-deduced parameters

	Oph 1977	GRO J0422+32	GS 2000+25	Vel 1993
K_c (km s^{-1})	441±6	372±10	520±5	475±6
f_x	4.65±0.21	1.13±0.09	5.01±0.15	3.17±0.12
Spectral type	K5V±2	M2V$^{+2}_{-1}$	K5V$^{+1}_{-2}$	K8V±2
f %	30±3	61±4	94±5	
$v \sin i$ (km s^{-1})	50$^{+17}_{-23}$	90$^{+22}_{-27}$	86±8	
q	0.014$^{+0.019}_{-0.012}$	0.116$^{+0.079}_{-0.071}$	0.042±0.012	

The accretion disk in its quiescent state has mainly been undetected so far by X-ray satellites but can be studied in the optical and near-infrared. An imaging technique, Doppler tomography, shows the accretion disks in GS 2000+25 and Nova Oph 1977 to be present. Mass transfer from the companion star continues vigorously to the outer disk as evidenced by the "bright spot," the impact of the gas stream onto the outer accretion disk (Fig. 2; Harlaftis et al. 1996, 1997). The lithium line at 6707 Å was only detected in GS 2000+25 (see Martin et al. 1994 for lithium in X-ray binaries).

References

1. Bailyn, C. D., Jain, R. K., Coppi, P., Orosz, J. A. (1998) ApJ, 499, 367
2. Casares, J., et al. (1995) MNRAS, 276, 35
3. Charles, P. A. (2000), these proceedings
4. Filippenko, A. V., et al. (1999) PASP, 111, 969
5. Fryer, C. L. (1999) ApJ, 522, 413
6. Harlaftis, E. T., Collier, S. J., Horne, K., Filippenko, A. V. (1999) A&A, 341, 491
7. Harlaftis, E. T., Horne, K., Filippenko, A. V. (1996) PASP, 108, 762
8. Harlaftis, E. T., Steeghs, D., Horne, K., Filippenko, A. V. (1997) AJ, 114, 1170
9. Martin, E., et al. (1994) ApJ, 435, 791
10. Miller, J. C., Shahbaz, T., Nolan, L. A. (1998) MNRAS, 294, L25

Measuring the Mass of the Black Hole in GS2000+25 Using IR Ellipsoidal Variations

Dawn M. Leeber, Thomas E. Harrison, and Bernard J. McNamara

New Mexico State University, Las Cruces, NM 88003, USA

Abstract. Soft X-ray Transients (SXTs) are binary systems that are believed to consist of a black hole and a normal late type dwarf star which fills its Roche Lobe. We have used GRIM II on the ARC 3.5 meter telescope at Apache Point Observatory to obtain infrared photometry of GS2000+25 (QZ Vul). By modeling the SXT ellipsoidal variations with WD98, we can determine the orbital period and inclination of the system. The inclination for a best fit circular orbit is $75°$, and when combined with the observed mass function, corresponds to a primary mass of $6.55 M_\odot$. More data is needed to better define the minima, fill in the small gaps in the light curve, and explore the possibility of an eccentric orbit.

1 What are Soft X-Ray Transients?

Soft X-ray Transients (SXTs) are *transient* binary systems that consist of a black-hole primary and a distorted (i.e. Roche lobe filling), late type, cool dwarf secondary. They display large and sudden X-ray and optical outbursts ($L_x = 10^{38}\, erg/s$; $\Delta V \sim 7\,\mathrm{mags}$), separated by long intervals of quiescence. Their outbursts result from a sudden, dramatic increase in the accretion rate onto the compact object. SXTs show *soft* spectra compared to other X-ray sources. The secondary stars are visible and dominate the systemic infrared luminosity in quiescence, allowing us to observe and model the ellipsoidal variations of the secondary star as it orbits the primary.

2 IR Ellipsoidal Variations: Observations and Modeling

The most difficult parameter to determine when estimating the mass of the primary is the orbital inclination, i. Since the known SXTs are not eclipsing, i can only be determined through the modeling of ellipsoidal variations. Ellipsoidal variations result from the rotational and tidal distortions of the Roche lobe filling secondary star and its non-uniform surface brightness distribution (limb darkening, gravity brightening). In the infrared, the secondary stars of SXTs are brighter, and the variations are much less contaminated by any residual accretion disk or hot spot than in the optical regime.

To construct the GS2000+25 J-band light curve shown in Fig. 1, differential photometry was performed on data taken with the GRIM II infrared imager on the ARC 3.5 meter telescope at Apache Point Observatory. The

58   Dawn Leeber et al.

newest version of the University of Calgary's Wilson-Devinney light curve modeling code, WD98, was used to model the data (Dr. Josef Kallrath, private communication). The model in Fig. 1 includes a semi-detached system with a K5V secondary, circular orbit, reflection effect, logarithmic limb darkening coefficients, and a J-band Kurucz atmosphere.

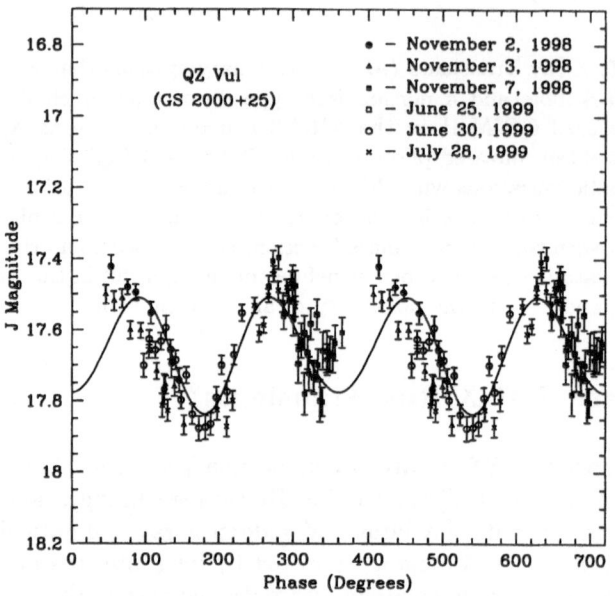

Fig. 1. GS2000+25 J-band light curve and model. Various points represent data obtained between 11/98 and 7/99, and the solid line is the $i = 75°$ circular orbit model from WD98 described in the text. Error bars are 1σ. The data show a maximum to minimum amplitude of ~ 0.4 mags.

3   Discussion

The current WD98 model indicates an inclination angle of 75° and a resulting mass of 6.55 $M_{\odot}$. No outburst x-ray eclipses were observed, so 75° represents the upper limit of allowable inclinations. More data is needed to fill in the gaps in the light curve and better define the minima. Each night of data has been carefully normalized, but until we have a data set that covers both a minima and maxima, the true maximum to minimum amplitude can not be conclusively determined. Note that the circular orbit model does not fit the light curve perfectly. The possibility of an eccentric orbit will be explored once the minima are more clearly defined. If this system were to have an *elliptical orbit, it* would be an unexpected, yet important discovery.

YALO Observations of 4U 1543-47

Jerome A. Orosz[1], Charles D. Bailyn[2], Raj K. Jain[2], and Jenny Greene[2]

[1] Astronomical Institute, Utrecht University, PO BOX 80 000. 3508 TA Utrecht, The Netherlands
[2] Department of Astronomy, Yale University, PO POX 208101, New Haven, CT 06520-8101, USA

Abstract. The black hole binary 4U 1543-47 has been observed extensively (using B, V, and I) during the 1998 and 1999 seasons by the YALO (Yale - AURA - Lisbon - Ohio State) 1 meter telescope at Cerro Tololo. The folded light curves show clean amplitudes of 0.05 mag in B, 0.03 mag in V, and < 0.02 mag in I. We discuss models of the light curves and present updated and refined system parameters.

1 Introduction

4U 1543-47 is a bright X-ray transient that was suspected to be a strong black hole candidate based on its X-ray spectrum, which is composed of an ultrasoft component and a hard power-law tail. The extensive optical study[1] provided dynamical evidence that the compact object is a black hole. Orosz et al. [1] determined a spectroscopic orbital period of $P = 1.123 \pm 0.008$ days and a K-velocity of $K_2 = 124 \pm 4$ km s^{-1} for the A-star secondary. The mass function of the compact object is $f(M) = 0.22 \pm 0.02\,M_\odot$.

The mass of the compact object depends on the orbital inclination i and the mass ratio $Q \equiv M_1/M_2$: $M_1 = f(M)(1 + 1/Q)^2/\sin^3 i$. Alternatively, one can compute the mass of the compact object when given i and the mass of the secondary star M_2: $f(M) = M_1 \sin^3 i/(M_1 + M_2)^2$.

A monitoring program of 4U 1543-47 was begun on the YALO (Yale-AURA-Lisbon-Ohio State) 1 meter telescope at CTIO during the start of the 1998 season with the goal of obtaining more precise light curves in the B, V, and I filters. We report here some preliminary results of this campaign.

2 Results

Period: A Lomb-Scargle periodogram of the shows that the frequency of the tallest peak is 1.7911 cycles/day. This corresponds to a period of 0.5583 days, which is half of the spectroscopic period of $P_{\text{spect}} = 1.123 \pm 0.008$ days[1]. The "false alarm" probability for this peak is 6.53×10^{-6}. The B and I periodograms have peak frequencies consistent with the peak frequency in V. For the following we adopt a period of $P = 1.1163 \pm 0.0005$ days, were the error was estimated from the spread of the periods derived from the three filters. This refined period is consistent with the spectroscopic period.

Ellipsoidal Models: The folded and binned light curves are shown in Figure 1. The folded light curves are clearly double-waved. The amplitude of the light curve is the largest in B and the smallest in I, similar to what is seen in GRO J1655-40 [2]. In contrast to GRO J1655-40, the light curve amplitudes are rather small for 4U 1543-47: roughly 0.05 mag in B and 0.03 mag in I.

For mass ratios larger than about 5, the 99.99% confidence range on the inclination implied from the model fits is roughly $18° \leq i \leq 26°$. For mass ratios closer to 1, the range is shifted by about 2° higher. We used a Monte Carlo procedure [3] to compute the confidence limits on the component masses. The 90% confidence region for the black hole mass is $4.01 \leq M_1 \leq 6.73\,M_\odot$. If we assume the A-star is on the main sequence ($2.3 \leq M_2 \leq 2.6\,M_\odot$), then $5.45 \leq M_1 \leq 7.73$.

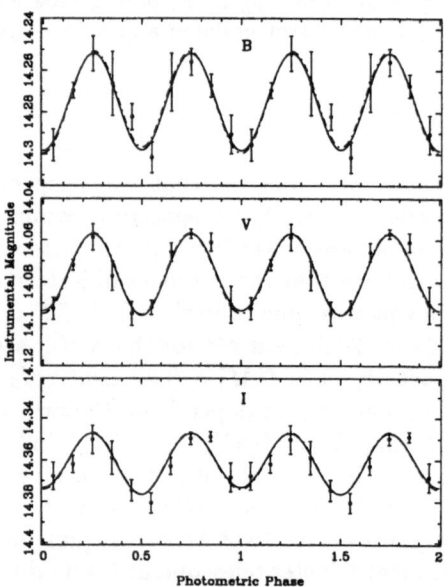

Fig. 1. The folded light curves and the ellipsoidal models. he solid lines are for a model with $i = 21.5°$ and $Q = 10$ ($\chi^2 = 27.99$). In this case $M_1 = 5.41\,M_\odot$ and $M_2 = 0.54\,M_\odot$. The dash-dotted lines are for a model with $i = 24°$ and $Q = 2.7$ ($\chi^2 = 30.15$). In this case $M_1 = 6.14\,M_\odot$ and $M_2 = 2.27\,M_\odot$.

References

1. Orosz, J. A., Jain, R. K., Bailyn, C. D., McClintock, J. E., & Remillard, R. A. 1998, ApJ, 499, 375
2. Orosz, J. A., & Bailyn, C. D. 1997, ApJ, 477, 876
3. Orosz, J. A., & Wade, R. A. 1999, MNRAS, 310, 773

Optical Spectroscopic Observations of Black-Hole Candidates

Michelle Buxton[1] and Stephane Vennes[2]

[1] Research School of Astronomy & Astrophysics, Australian National University, Private Locked Bag, Weston Creek, ACT, Australia, 2602

[2] Astrophysical Theory Center, Australian National University, John Dedman Building, Canberra, ACT, 2000

1   Observations and Data Reduction

All observations were made using the Double-Beam Spectrograph on the 2.3m telescope at Siding Spring. See Table 1 for a log of observations. Both the red and blue sides were utilised together with the D1 dichroic mirror. All spectra were reduced using the standard reduction techniques in IRAF.

2   Results

GX 339-4 : The asymmetric line profile of Hα was clearly evolving during each night of observations with the main peak shifting between the blue and red side. A double-peaked profile could be discerned in some spectra but was not resolved sufficiently to measure radial velocities. Therefore, we used a V/R analysis [1] to quantify the change in line-profile. Conducting a period search on the ratios using a least-squares method, we found a primary period at 42.86h and a secondary period at 14.93h, very close to the published orbital period of 14.86h [2].

LMC X-1 and LMC X-3 : We have found the radial velocities of the nebula lines [OIII] $\lambda 5006$ and Hβ to be 252.6 ± 1.0 kms^{-1} and 251.5 ± 1.8 kms^{-1}, respectively, consistent with Hutchings Crampton & Cowley [3]. The published period for LMC X-1 is 4.23d [4]. Our period search found a period of 3.78d for HeI $\lambda 4471$ and 4.19d for HeII $\lambda 4544$. There is a very large scatter which may be due to X-ray heating of the secondary star surface. Modelling this emission should lead to more accurate radial velocities and, hence, improve the amplitude and systematic velocity of the system. The resulting periods from our period search for LMC X-3 were 1.73d for Hγ, 1.72d for Hβ and 1.69d for Hα which agree well with that of Cowley et al. [5] of 1.70d.

XTE J1550-564 : The spectra are dominated by a strong, double-peaked Hα emission line and the Paschen series in emission [6]. The continuum is reddened and steadily decreases in flux by a factor of four by May 1999. The interstellar NaD line in absorption is also present. The ratios of Hα/HeI$\lambda 6678$

fluxes decreases as the soft X-rays decrease. V-I increases as the X-rays go down, as does V-R, but B-V remains constant. This implies that the system is becoming redder during this period.

GS 1354-64 : The overall spectrum is heavily reddened. Prominent lines are Hα, Hβ and Hγ together with HeII λ5411, 4686 and HeI λ4471. The Hα profile is clearly double-peaked with a separation of ~ 600 kms^{-1} [7]. HeI λ6678 and λ7065 are also present and may also be double-peaked, but this is less certain due to the lower signal-to-noise.

Table 1. Log of Observations

Object Name	HJD (-2450000)	Grating (l/mm)	Exposure (sec)
GX 339-4	960-964	300/316, 1200	1800-3600
	1045-1048	1200	2000
	1248-1249	300/316	900-1800
	1262-1265	300/316	900
	1286-1287	1200	1800
	1310	300/316	1800
LMC X-1 & LMC X-3	1183-1189	1200	600-1800
XTE J1550-564	1249	300/316	900
	1262-1265	300/316	900
	1310	300/316	1800
GS 1354-64	836-837	300/316	1800-2700

References

1. Thorstensen J.R., 1986, AJ, 91, 940.
2. Callanan P.J., Charles P.A., Honey W.B. & Thorstensen J.R., 1992, MNRAS, 259, 395.
3. Hutchings J.B., Crampton D. & Cowley A.P., 1983, ApJ, 275, L43.
4. Hutchings J.B. et al., 1987, AJ, 94, 340.
5. Cowley A.P. et al., 1983, ApJ, 272, 118.
6. Buxton M., Vennes S., Ferrario L. & Wickramasinghe D., 1998, IAUC 6815.
7. Buxton M., Vennes S., Wickramasinghe D. & Ferrario L., 1999, IAUC 7197.

The Center of the Galaxy:
Evidence for a Massive Black Hole

Andreas Eckart, Thomas Ott, and Reinhard Genzel

Max-Planck-Institut für extraterrestrische Physik,
Giessenbachstraße, D-85740 Garching, Germany

Abstract. Recent spectroscopic and imaging data provide new evidence for the presence of a massive black hole at the center of the Milky Way. The most recent results are based on new near-infrared observations of the central stellar cluster of our Galaxy conducted with the infrared spectrometer ISAAC at the ESO VLT UT1 and the MPE speckle camera SHARP at the ESO NTT. These data demonstrate clearly that there is no strong CO band head absorption originating in the northern part (S1/S2 area) of the central stellar cluster at the position of Sgr A*. This makes it likely that these K~14.5 stars are O9 - B0.5 stars with masses of 15 to 20 $M_\odot$. We also report the detection of Brγ line emission at the position of the central stellar cluster which could be associated with the 'mini-spiral' rather than with the Sgr A* cluster itself. The combined data on radial velocities and proper motions show that the overall stellar motions are very close to isotropy. They indicate a central mass concentration with a density $\geq 10^{12.6}$ $M_\odot$ pc^{-3} that dominates the potential between 0.01 and 1 pc. The derived mass ranges between 2.6 and 3.3 × 10^6 $M_\odot$ for a distance of 8.0 kpc. Such a mass concentration cannot be stable and must be present in the form of a black hole.

1 Introduction

One of the best cases for the presence of a massive black hole is the center of our Galaxy. There both the gas and stellar dynamics indicate the presence of a large unresolved central mass (Eckart and Genzel 1996, 1997, Genzel et al. 1997, Ghez et al. 1998, Genzel et al. 1999). Investigations of the motions of gas and stars have provided evidence for the existence of massive black holes in the nuclei of many galaxies (Richstone et al. 1998, Magorrian et al. 1998, Kormendy and Richstone 1995). The measured mass concentration cannot be stable and therefore is most likely present in the form of a massive black hole (Maoz 1998). Since the Galactic Center is very close - only 8 kpc - we can obtain line-of-sight velocities (through spectroscopy) and/or proper motions of individual stars that are within only a few light days of the radio/near-infrared position of Sgr A* (Menten et al. 1997). The locations corresponding to the maximum velocity dispersion and to the maximum stellar surface density agree with the position of the compact radio source Sgr A* to within ±0.1" (Ghez et al. 1998). Combined with stellar surface density counts these data provide convincing qualitative evidence for the presence of a central point mass ranging between 2.6 and 3.3×10^6 $M_\odot$ (Sellgren et al. 1990, Krabbe et al. 1995, Haller et al. 1996, Genzel et al. 1996, 1997, Eckart and Genzel 1996, 1997, Ghez et al. 1998, Genzel 1999).

Recent data provide first spectroscopic information on the central ~20 light days. Genzel et al. (1997) reported R=$\lambda/\Delta\lambda$~35 speckle spectroscopy measurements on individual objects in the central ~1" diameter stellar cluster at the position of Sgr A*(IR). In combination with other data this spectroscopic information can be used to derive a lower limit to the mass associated with the compact radio source. Here we present new ISAAC R$\geq$3000 K-band spectroscopy of the Sgr A* stellar cluster in the 2.058 μm He I, 2.165μm Brγ emission lines, and the 2.29 μm CO bandhead absorption lines. The combination of these spectroscopic data taken in excellent seeing (0.3" to 0.5") with our new speckle image reconstructions based on SHARP NTT data strengthen the case for a compact mass and add to our understanding of the stellar population near the center of the Galaxy.

The spectroscopic observations were carried out in the first half nights of 30 June and 1 July, 1999, using the infrared (0.9-5μm) spectrometer ISAAC (Moorwood et al. 1998) at one Nasmyth focus of the ESO VLT UT1 (*Antu*). Diffraction limited imaging data was obtained using the MPE speckle camera SHARP at the ESO NTT between 18 to 21 June. Also during this run we had excellent seeing conditions. Details of the observations are given in Eckart, Ott, Genzel (1999; see also the spectroscopic results obtained with NIRSPEC on Keck II by Figer et al. 2000).

2 Spectroscopic Results

Spectroscopic observations of the Galactic Center require excellent seeing to separate the small flux contribution of the typically K=14.5-15.0 sources from the very bright neighboring IRS 16 complex which contains stars as bright as K=10. The excellent seeing conditions at the VLT allowed us to obtain this information with no or only little correction for the seeing wing contribution of these bright neighbouring objects.

2.1   CO Absorption Lines

The new June 1999 SHARP speckle image reconstruction is compared with a section of the two dimensional ISAAC spectroscopic exposure on the CO(2-0) bandhead absorption line in Fig. 1. This comparison demonstrates the excellent seeing we had at the VLT and allows us to identify the individual sources that contributed to the flux density in the 0.6" slit. One can distinguish between the northern and southern part of the Sgr A* stellar cluster as well as a star to the south with obvious bandhead absorption. In Fig. 2 we show spectra of the northern and southern part of the Sgr A* cluster as well as the spectrum of a star just 1.12" south of the central position. The spectra clearly show the complete absence of strong CO bandhead absorption for the northern S-sources close to the position of Sgr A* and the detection of a late-type star just ~0.6" south of S10 and S11 (Genzel et al. 1997). Very weak bandhead absorption on the northern Sgr A* cluster is in agreement with an expected contribution from the underlying stellar cluster. The bandhead absorption on the southern part of the central stellar cluster is probably due to a significant flux density contribution from the late type star just 0.6" to the south and 0.5" to the east of S10 and S11. In addition there are a few weaker sources with separations from S10 and S11 of less than 0.4" that could give rise to contaminating flux. These measurements are in full agreement with our initial results that we obtained via R~35 speckle spectroscopy on the individual objects S1, S2, S8, and S11. It also indicates that most of the other S-sources that now fell in our slit cannot be stars with strong CO bandhead absorption.

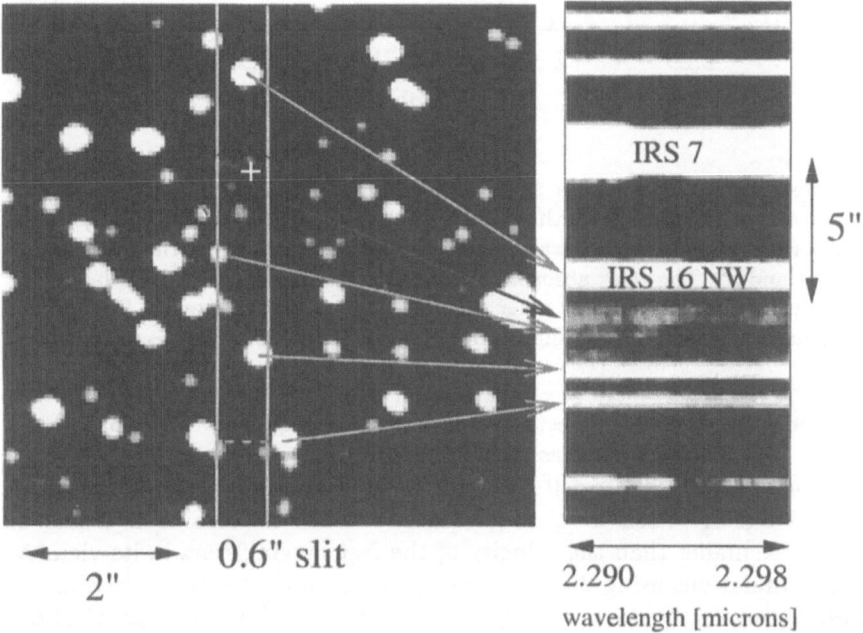

Fig. 1. Comparison between the results of the June 1999 SHARP epoch at the NTT (left) and the central portion of a two dimensional ISAAC spectroscopic exposure on the CO(2-0) bandhead absorption line (right). The infrared seeing during the 5 minute ISAAC exposure was of the order of 0.3". One can clearly distinguish between the northern and southern part of the small stellar cluster surrounding the position of Sgr A* indicated by a cross in the SHARP image.

From this data one can conclude that the m_K~14.5 sources in the central Sgr A* cluster are most likely moderately luminous (L~5,000 to 10,000 $L_\odot$)

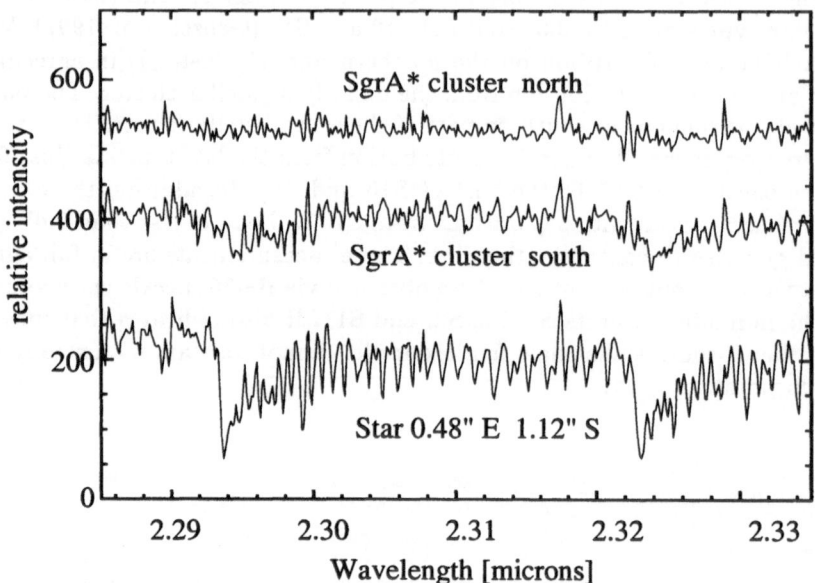

Fig. 2. Spectra of the northern and southern part of the Sgr A* cluster as well as the late type star 1.12" south of the center. No CO(2-0) and CO(3-1) band-head absorption is measured towards the northern part containing the fast moving sources S1 and S2. The weak absorption features towards the south could be due to contaminating flux from sources other than S9, S10 and S11 (see text). The spectra at the southern positions were shifted down by 150 and 300 units, respectively.

early-type stars. If these objects are on the main sequence they would have to be O9 - B0.5 stars with masses of 15 to 20 $M_\odot$.

Backer (1996, 1999) and Reid et al. (1999) have shown that the proper motion of Sgr A* itself is $\leq$ 16-20 km s^{-1} which is close to 2 orders of magnitudes smaller than the velocity of the fast moving stars in its vicinity. N-body simulations using 20 $M_\odot$ as an upper limit of the mass distribution of these high velocity stars result in a lower limit of 10^3 $M_\odot$ for Sgr A* (Reid et al. 1999, see also Genzel et al. 1999, 1997). If this mass is enclosed within the radio size of Sgr A* ($\leq$ 1 AU) this already implies a central mass density larger than 10^{18} $M_\odot$ pc^{-3}.

2.2   Recombination Lines

Our high spatial and spectral resolution data clearly show the presence of Brγ and He I emission which is apparently spatially coincident with the location of the Sgr A* central stellar cluster. From our Brγ data (see Fig. 3) we find a line width of <120 km s^{-1} and a velocity gradient of about 35 km s^{-1} between the southern part (S10/S12-region) and the northern part (S1/S2-region) of the cluster. In both slit settings this line emission appears to be connected to the more extended line emission over the remaining central cluster. This fact combined with the small line width at any position in that region indicates that the emitting gas is not necessarily associated with the Sgr A* stellar cluster. If the emission would be associated with the cluster we would expect a larger line width due to the higher gravitational potential indicated by the rapid motions of the stars. However, we cannot exclude at the present stage of the data reduction any broad and weak emission components that would indicate a higher velocity dispersion.

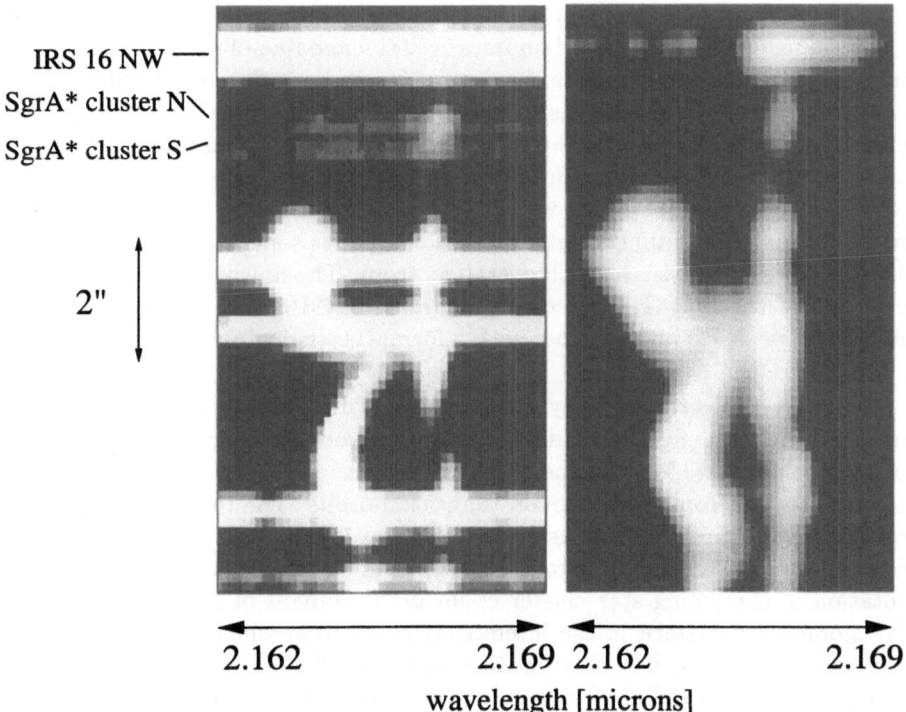

Fig. 3. The central portion of a 5 minute ISAAC exposure centered on the Brγ emission line. On the left we show the line plus continuum map - on the right the continuum subtracted wavelength position map. Some sources are indicated on the left. There is a clear detection of Brγ emission towards the central stellar cluster at the position of Sgr A*. At the top of the right panel the line emission of IRS 16 NW is shown. The lower wavelength part of the spectrum includes some residuals from the continuum subtraction.

3   New Results on the Stellar Dynamics

The new diffraction limited speckle imaging data at the NTT was obtained 10 days before our VLT run. These data mainly serve as a new proper motion epoch (Genzel et al. 1997, Eckart and Genzel 1996, 1997, Ghez 1998) but also allow us to further investigate the structure and variability of sources (Ott, Eckart, Genzel 1999). Due to the short time difference between the speckle imaging and the spectroscopic measurements we know the brightness and exact positions of all the prominent Sgr A* cluster members that contribute to the observed flux in the slit we used during our spectroscopic measurements. In Fig. 4 we show our new proper motion determination of the fastest moving source S1 close to the position of the compact radio source Sgr A*.

Genzel et al. (1999) report a new analysis of the stellar dynamics in the Galactic center, which is based on improved sky and line-of-sight velocities for more than one hundred stars in the central few arcseconds from the black-hole candidate SgrA*. This analysis shows that the overall stellar motions are very close to being isotropic. For those 32 stars for which we determined all three velocity components the absolute, line of sight and sky velocities are in good agreement. This is consistent with a spherical star cluster. Likewise the sky-projected radial and tangential velocities of all 104 proper motion stars in our sample are also consistent with overall isotropy. The anisotropy-independent estimate of the Sun-Galactic center distance based on the stellar velocities ranges between 7.8 and 8.2 kpc, with a formal statistical uncertainty of ± 0.9 kpc.

The sky-projected velocity components of the young, early type stars, however, indicate significant deviations from isotropy which is strongly dependent on radius. Most of the bright HeI emission-line stars at separations from $1''$ to $10''$ from SgrA* are on tangential orbits. The HeI stars and most of the brighter members of the IRS16 complex largely follow a clockwise (as seen on the sky) and counter-rotating, coherent rotation pattern. This overall rotation of the young star cluster could be a remnant of the original angular momentum pattern in the interstellar cloud from which these stars were formed. The fainter, fast moving stars within $\approx 1''$ from SgrA* appear to be largely moving on radial or very elliptical orbits. We have so far not detected deviations from linear motion (i.e. acceleration) for any of them. Most of the SgrA* cluster members are on clockwise orbits as well. VLT spectroscopy and NTT speckle-spectroscopy both suggest that they may be early type stars. The motion of the SgrA* cluster stars is consistent with a scenario in which they are those members of the early-type-star cluster that happened to have small angular momentum and thus sank to the immediate vicinity of SgrA*.

The Leonard-Merritt projected mass estimators and Jeans equation modeling (Genzel et al. 1999) confirm previous conclusions (from isotropic models) that a compact central mass concentration (of density $\geq 10^{12.6}$ $M_\odot pc^{-3}$) dominates the potential between 0.01 and 1 pc. Depending on the modeling

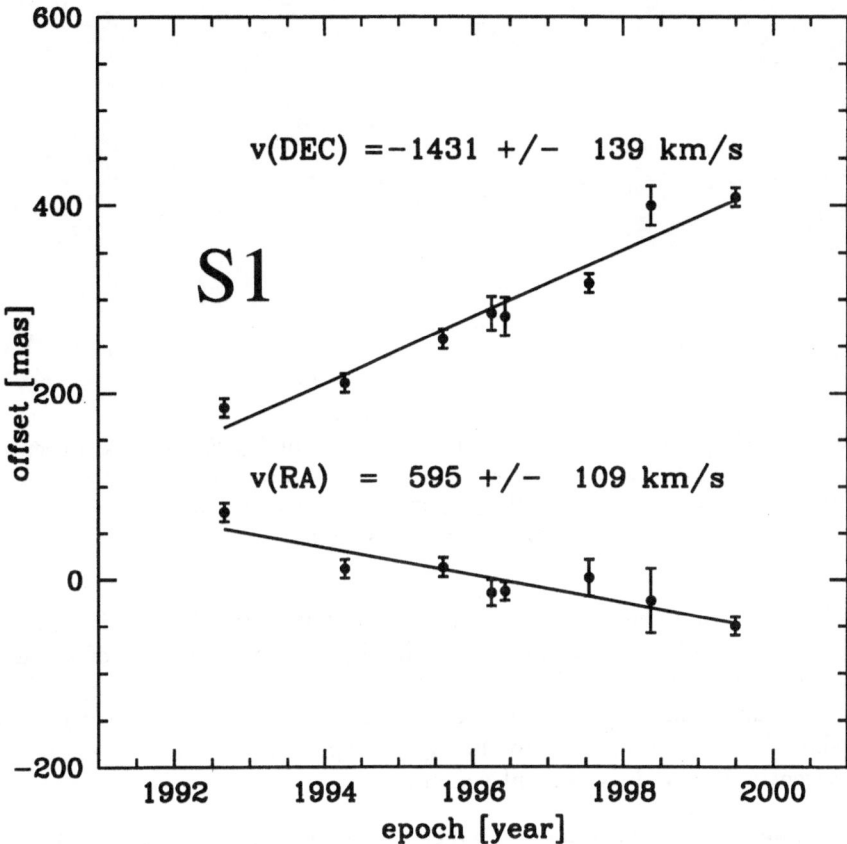

Fig. 4. Proper motion of the fast moving object S1 close to the position of the compact radio source Sgr A*. The plot includes all data between 1992 and 1999. The declination graph has been moved upwards by 300 mas.

method used the derived mass ranges between 2.6 and 3.3 $\times$ 10^6 $M_\odot$ for a distance of 8.0 kpc.

4   Conclusions

Our new spectra prove the lack of strong CO bandhead absorption on the fast moving stars in the direct vicinity of the compact radio source Sgr A*. This indicates that SgrA* must be linked to a large compact mass. The most recent data confirm and strengthen recent work on the central mass distribution (cf. Eckart and Genzel 1996, 1997, Genzel et al. 1997, Genzel and Eckart 1997, Ghez et al. 1998, Maoz 1998). It appears that the most probable configuration

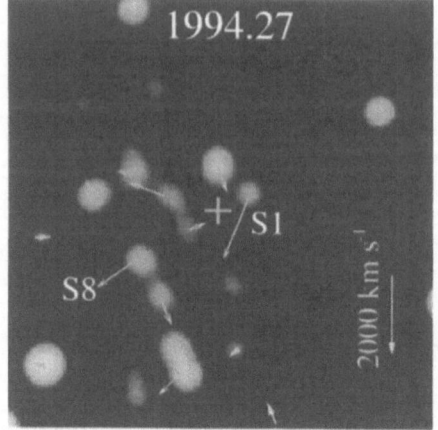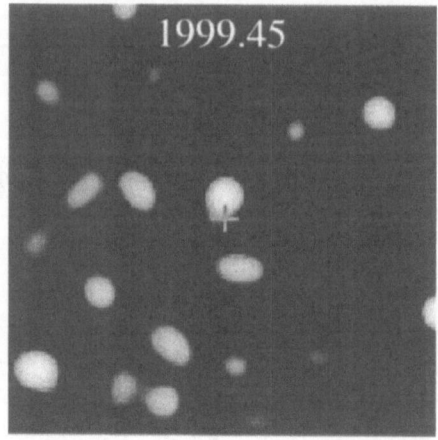

Fig. 5. Two images of the central stellar cluster surrounding Sgr A* taken over 5 years apart. The cross indicates the position of Sgr A*. The arrows in the 1994 image indicate the direction and value of the proper motion velocity. Their end-points are at the current positions of the stars in the 1999 image.

of the central mass concentration is a massive, but currently inactive black hole. At the measured mass and mass density any dark cluster of stellar remnants (neutron stars, stellar black holes), low luminosity stars (e.g. white dwarfs) or sub-stellar objects would have a lifetime less than about 10^7 years. This short lifetime in addition to the high compactness is inconsistent with any currently observed dynamical system.

Acknowledgements: We thank A. Ghez, M. Morris, and S.R. Stolovy for interesting discussions and comments. We are also grateful to the VLT UT1- and the NTT-team and especially to U. Weilenmann and H. Gemperlein for their interest and technical support of SHARP at the NTT.

References

1. Backer, D.C. 1996, in Unsolved Problems of the Milky Way, eds.
2. Backer, D.C. & Sramek, R.A., 1999, Ap.J. 524, 805 L.Blitz and P.Teuben, Proc. of IAU 169 (Kluwer:Dordrecht), 193
3. Eckart, A. and Genzel, R. 1996, NatE 383, 415
4. Eckart, A. and Genzel, R. 1997, MNRAS 284, 576
5. Eckart, A., Ott, T., Genzel, R. 1999, A&A in press
6. Figer, D.F., Becklin, E.E., McLean, I.S., Gilbert, A.M., Graham, J.R., Larkin, J.E., Levenson, N.A., Teplitz, H.I., Wilcox, M.K., Morris, M., 2000, submitted to Ap.J.
7. Genzel, R., Thatte, N., Krabbe, A., Kroker, H. and Tacconi-Garman, L.E. 1996, Ap.J.472, 153
8. Genzel, R., Eckart, A., Ott, T. and Eisenhauer, F. 1997, MNRAS 291, 219

9. Genzel, R., Pichon, C., Eckart, A., Gerhard, O. and Ott, T., 1999, submitted to MNRAS.

10. Ghez, A., Klein, B., Morris, M. and Becklin, E., 1998, Ap.J. 509, 678

11. Haller, J.W., Rieke, M.J., Rieke, G.H., Tamblyn, P., Close,L. and Melia,F. 1996, Ap.J. 456, 194

12. Kormendy, J. and Richstone, D. 1995, Ann.Rev.Astr.Ap.1995, 581

13. Krabbe, A. Genzel, R., Eckart, A., Najarro, F., Lutz, D. et al. 1995, Ap.J.Lett. 447, L95

14. Lucy,L.B., 1974, A.J. 79, 745

15. Magorrian, J. et al. 1998, A.J. 115, 2285

16. Maoz, E. 1998, Ap.J. 494, L131

17. Menten, K.M., Eckart, A., Reid, M.J. and Genzel, R. 1997, Ap.J. 475, L111

18. Moorwood, A., et al. 1998, The Messenger 94, 7

19. Ott, T., Eckart, A. and Genzel, R. 1999, Ap.J. in press

20. Reid, M.J., Readhead, A.C.S., Vermeulen, R.C., Treuhaft, R.N. 1999, Ap.J. 524, 816.

21. Richstone, D. et al. 1998, Nat 395, 14

22. Sellgren, K., McGinn, M.T., Becklin, E.E. and Hall, D.N.B. 1990, Ap.J. 359, 112

Towards Complete Stellar Orbits
Around the Galaxy's Central Black Hole:
The First Acceleration Measurements

A. M. Ghez, T. Kremenek, A. Tanner, M. Morris, and E. Becklin

University of California, Los Angeles CA 90095-1562, USA

Abstract. Four years ago, we initiated a proper motion study of the Galaxy's central stellar cluster using diffraction-limited K[2.2μm]-band images obtained with the W.M. Keck I 10-m telescope. With relative positional accuracies of ∼3 milliarcsec, we have been able to measure stars moving with velocities up to 1400 km s^{-1} and, for the first time ever, accelerations of 1-3 mas yr^{-2} (0.1-0.3 cm s^{-2}) at projected distances from the apparent radio counterpart of the black hole, Sgr A*, of 0.004 pc (0.1 arcsec). These measurements have not only provided us with direct dynamical evidence for a 2.6×10^6 M$_\odot$ central black hole, but have also permitted us to begin to constrain the orbits of these stars, which appear to be bound to the central mass. Continued study of these orbits (which possibly have periods as short as 10 years) will allow us to infer the radial distribution of dark matter within a few hundredths of a parsec around the black hole.

1   Introduction

In 1995, we began a 2μm diffraction-limited proper-motion program with the Keck 10-m telescope of the inner $5'' \times 5''$ of stars in the Galaxy's central stellar cluster in order to assess whether or not our Galaxy contains a supermassive central black hole. The first stage of this experiment was to obtain two-dimensional velocity measurements, from which a statistical estimate of the mass distribution could be determined. Analysis of the data obtained between 1995-1997 yielded two-dimensional velocity measurements for 90 stars, with velocities as high as 1400 km s^{-1}, and clearly demonstrated the existence of a $2.6 \pm 0.2 \times 10^6$ M$_\odot$ black hole at the center of our Galaxy (Ghez et al. 1998). Since 1997 an additional 6 maps have been obtained, extending the time baseline to 4 years. With the new data, the velocity uncertainties are reduced by a factor of 3 and a similar mass estimate is derived. This is consistent with earlier work (Eckart et al. 1997; Genzel et al. 1997), although in our experiment the source confusion is reduced by a factor of 9, the number of stars with proper motion measurement in the central 25 arcsec^2 is doubled, and the accuracy of the velocity measurements in the central 1 arcsec^2 is improved by a factor of 15 over a comparable time period. In addition to simply increasing the time baseline for velocity measurements, we have advanced this experiment in two significant ways which are reported here: (1) The first Keck adaptive optics images of the galactic center have been obtained, allowing us to obtain a more complete census of stars in this region, and (2) the first measurements of stellar accelerations in this field have now been achieved.

2   From Speckle Imaging to Adaptive Optics

First light with the Keck II Adaptive Optics (AO) system, which contains a 349-actuator deformable mirror and a 256x256 NICMOS array science camera, was obtained on 1999 February 4. We observed the Galactic Center on 1999 May 26 as part of the early science program, in order to test the system's ability to image a crowded field while correcting on a relatively faint (R = 13.7 mag) guide star located 30 arcsec off-axis (Wizinowich et al. 2000). Figure 1 shows a comparison of the new AO image with a typical speckle image during the same month. Although the cores of both the speckle and AO point spread functions are approximately diffraction-limited, the halos are greatly reduced in the AO image, leading to a factor of three higher Strehl ratio. With AO imaging, we can now identify stars much closer to the brighter stars, which increases the sample by a factor of two when the limiting magnitude of stars included is not changed (K $\leq$ 16 mag). It should also be noted that the AO image represents only 75 sec of integration on source, whereas the speckle image has a total integration time of 1600 sec. With longer integration times, AO should allow us to (1) probe yet a larger sample of fainter stars, (2) place stringent limits on Sgr A*, and (3) explore the possibility of a gravitational lensing experiment (Alexander & Sternberg 1999). For the time being, the AO map has permitted us to increase the number of stars in the proper motion study.

3   From Velocities to Accelerations

With relative positional accuracies of ~3 milliarcsec, we can now fit the motions of stars with a second-order polynomial as opposed to a simple linear fit, which was done in our earlier work. Figure 2 shows the two stars with the clearest measurements of an acceleration term. These two stars are also distinguished as the two fastest moving stars and the two closest to the nominal position of Sgr A*. The magnitudes of the two-dimensional acceleration vectors are 1.2 & 2.7 mas yr^{-2}, respectively, for S0-1 & S0-2, or equivalently 0.15 & 0.33 cm s^{-2}, consistent with the order of magnitude expected for objects in orbit around a central mass concentration of $2.6 \times 10^6 \, M_\odot$, allowing for projection effects. For comparison, one anticipates, in the case of a face-on circular orbit of radius 0."1 (the average projected radius for the two stars), an acceleration of 2.4 cm s^{-2}. Furthermore, the direction of the projected acceleration vectors, which should intersect at the location of the dynamical center, is consistent with a large mass concentration at the position of Sgr A*.

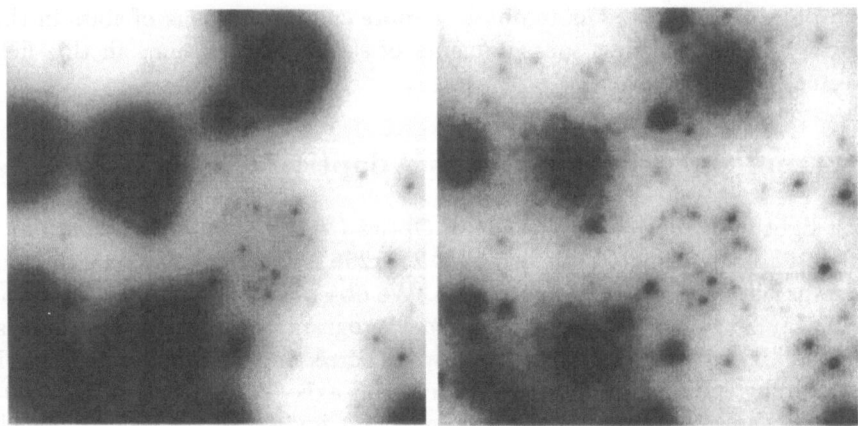

Fig. 1. A $\sim 3'' \times 3''$ region showing the Sgr A* cluster (the faint stars located just to the right of the center of the field of view). Both images were taken in 1999 May at 2.2μm (K-band), however the image on the left was produced by shift-and-adding Keck I speckle data and the image on the right was obtained with the new Keck II adaptive optics system. The adaptive optics image represents a large improvement (see text for details).

It is now possible to begin to explore individual orbits. Currently we have demonstrated that these two stars' motions can indeed be fit by bound orbits around a central mass of $2.6 \times 10^6\,M_\odot$ under the assumption that the true focus is coincident with the location of Sgr A* (see Figure 3). High quality orbital fits are found for the two stars, with periods ranging from 15 - 550 and 35 - 1200 years for S0-2 and S0-1, respectively. Although the fits are not yet unique, they have dramatically constrained the possible orbits; less than 1% of the explored parameter space provides a good fit to the data.

4   The Future

With the new AO system, we hope to soon add measurements of the radial velocities, giving us the three-dimensional velocity vectors. With systematic proper motion and radial motion measurements, we will be able to derive the true orbital parameters for the individual stars. The ultimate goal is to be able to release the two critical parameters that are currently being held fixed: the total mass and the true focus. These measurements are particularly important as they would allow us to address the following questions:
[1] Is the $2.6 \times 10^6\,M_\odot$ inferred for the central dark mass concentrated in a single black hole, or is a measurable fraction of it present in the form of a compact cloud of stellar remnants, dark matter particles (Salati & Silk 1989, Tsiklari & Viollier 1998; Gondolo & Silk 1999), or other forms of dark matter? The current statistical estimates on the distribution of dark matter at the galactic center only confine it to radii less than ~ 0.01 pc (0.25 arcsec).

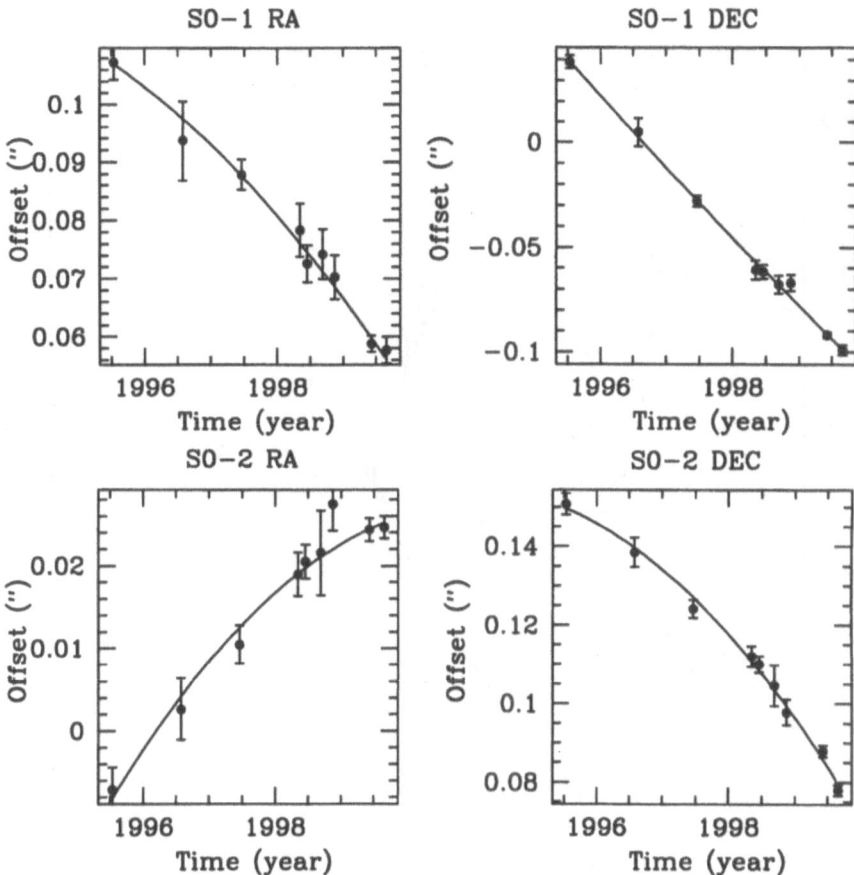

Fig. 2. East-West and North-South positional offsets from the nominal location of Sgr A* are plotted as a function of time for the two stars, S0-1 & S0-2, that have the most significant acceleration measurements. It should be noted that the offset range shown is scaled to the points in each plot and therefore varies from $\sim 0.''04$ to $\sim 0.''15$. The solid line in each plot shows the model used to assess the acceleration term: a second order polynomial with a time zero-point fixed to 1995. Both the magnitude and direction of these accelerations are consistent with a $2.6 \times 10^6 M_\odot$ black hole located at the position of Sgr A*.

However, we now can begin to use the orbits of stars such as S0-1 and S0-2, which are located only about $0.''1$ (0.004 pc) away from Sgr A*, to either push the limits on the volume of the central dark matter concentration down by an order of magnitude, or to resolve any mass distribution that might be present. Every individual star with a measurable orbit becomes a separate and independent probe of the interior mass, so by comparing stellar orbits displaying a range of semi-major axes, but modest eccentricities, we will be

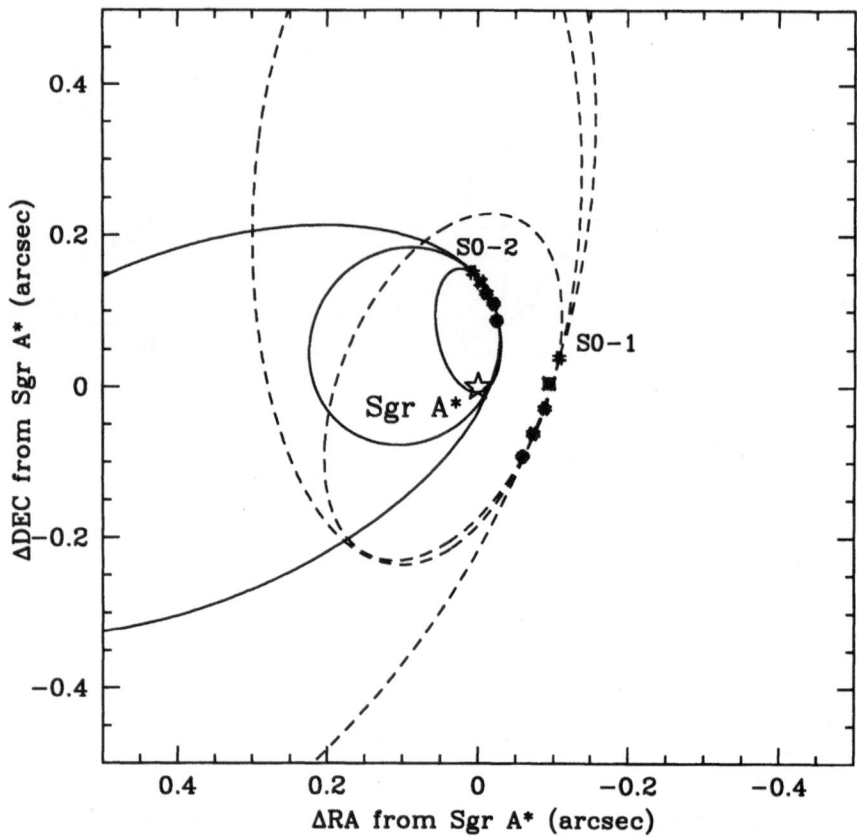

Fig. 3. A $1'' \times 1''$ region, centered on the nominal position of Sgr A*, showing the motion of S0-1 & S0-2 and possible orbital solutions. Only the measurements obtained in ~June of each of the 5 years are shown and both stars appear to be moving clockwise about Sgr A*. In the orbital analysis, two constraints are applied, a central mass of $2.6 \times 10^6 M_\odot$ and a true focus located at the position of Sgr A*. Displayed are orbital solutions with periods of 17, 80, 505 years for S0-2 and 63, 200, 966 years for S0-1.

able to infer the presence of an entourage of dark matter around the black hole having a total mass as small as 10% of the mass of the black hole. [1]

---

[1] We note that orbital precession caused by an extended mass distribution can in principle have a contribution from general relativistic effects for stars that pass within ~$10R_{sh}$, should they survive the tidal forces. However, unraveling the general relativistic effects would take a long time and none of the stars currently being tracked are predicted to get so close to Sgr A*.

[2] Is Sgr A* the central black hole? The orbits will pinpoint the dynamical center with a factor of 20 better accuracy than is possible with velocity information alone. With velocities only, the dynamical center can be located to within $\pm$ 100 mas (1σ), which shows that the dynamically-determined location of the dark-matter concentration is statistically consistent with the position of the unusual radio source Sgr A*. However, this is inadequate to definitively associate Sgr A* with the black hole, given the complexity of the region.

[3] Are there deviations from Newtonian gravitation at size scales on which it has not heretofore been accurately tested? While few will doubt that Newtonian gravitation is accurate at scales of $10^{16} - 10^{17}$ cm, or thousands of AU, the nearly elliptical orbits that we are now measuring will provide the first real test on such size scales. The periods of normal binaries with such separations are far too long ($\sim 10^5$ yr) for such orbits to be accurately surveyed, but the time-scale contraction wrought by a few million solar mass black hole makes such orbits measurable on a feasible time scale. We note that this question is closely coupled with question 1; any deviation from purely Keplerian orbits will be extremely interesting in one of the two stated respects, and multiple orbit determinations covering a range of size scales will sort out the dominant physics.

[4] Are the orbits isotropically distributed, or are the luminous stars inhabiting the central cusp characterized by a dominant radial or tangential velocity component? If the orbits are distributed anisotropically, is that a result of long-term dynamical processes or is it a dynamical remnant of the star formation process? With the orbits of a sizeable sample of stars, it is possible to constrain their dynamical evolution.

[5] What is the distance to the Galactic Center? These measurements will provide an independent measurement with an accuracy of a few percent (Salim & Gould 1999).

References

1. Alexander, T. & Sternberg, A. 1999, ApJ, 520, 137
2. Eckart, A. & Genzel, R. 1997, MNRAS, 284, 576
3. Genzel, R., Eckart, A., Ott, T. & Eisenhauer, F. 1997, MNRAS, 291, 219
4. Ghez, A. M., Klein, B. L., Morris, M., & Becklin, E. E. 1998, ApJ, 509, 678
5. Gondolo, P. & Silk, J. 1999, PhRvL 83, 1719
6. Salati, P. & Silk, J. 1989, ApJ, 338, 24
7. Salim, S., & Gould, A. 1999, ApJ, 523, 633
8. Tsiklauri, D. & Viollier, R.D. 1998, ApJ, 500, 591
9. Wizinowich, P., Acton, D. S., Shelton, C., Stomski, P., Gathright, J., Ho, K., Lupton, W., Tsubota, K., Lai, O., Max, C., Brase, J., An, J., Avicola, K., Olivier, S., Gavel, D., Macintosh, B., Ghez, A., & Larkin, J. 2000, PASP, in press

Evidence for Massive Black Holes in Nearby Galactic Nuclei

Tim de Zeeuw

Leiden Observatory, Postbus 9513, 2300 RA Leiden, The Netherlands

Abstract. Masses of black holes in nearby galactic nuclei can be measured in a variety of ways, using stellar and gaseous kinematics. Reliable black-hole masses are known for several dozen objects, so that demographic questions can start to be addressed with some confidence. Prospects for the near future are discussed briefly.

1 Introduction

Active galaxies and quasars are powered by physical processes in an accretion disk surrounding a massive black hole [37,54]. The observed number of active galaxies increases towards high redshift z, so that many 'normal' galaxies must have been active in the past [59]. This implies that inactive massive central black holes must lurk in the nuclei of nearby normal galaxies. In the past decade, much work has been done to measure the masses of these black holes, to establish the relation between black-hole mass and the global/nuclear properties of the host galaxy, and to understand the role these objects play in driving internal dynamical evolution [47,64].

A black hole of mass M_{BH} in a galactic nucleus dominates the gravitational potential inside the so-called radius of influence which is usually defined as $r_{BH} = GM_{BH}/\sigma^2$, where G is the gravitational constant, and σ is the characteristic velocity dispersion in the host galaxy. In physical units

$$r_{BH} \sim 0.4 \left(\frac{M_{BH}}{10^6 M_\odot}\right) \left(\frac{100\,\text{km/s}}{\sigma}\right)^2 \text{pc}. \tag{1}$$

For a galaxy at distance D, r_{BH} corresponds to an angular size

$$\theta_{BH} \sim 0''\!.1 \left(\frac{M_{BH}}{10^6 M_\odot}\right) \left(\frac{100\,\text{km/s}}{\sigma}\right)^2 \left(\frac{1\,\text{Mpc}}{D}\right). \tag{2}$$

Inside the sphere of influence, the black hole generates a central cusp in the density distribution of the stars. The resulting density and luminosity $\propto r^{-\gamma}$ with $3/2 \le \gamma \le 9/4$ [53]. The typical velocities scale $\propto r^{-1/2}$ with radius.

While many early-type galaxies have cusps in their central luminosity profiles [9,18,33], these by themselves are not proof of the presence of a black hole, as other processes can generate such cusps [34]. For this reason most studies concentrate on obtaining spectroscopy at the smallest angular scales, to find evidence for the expected Keplerian rise in the velocities of stars

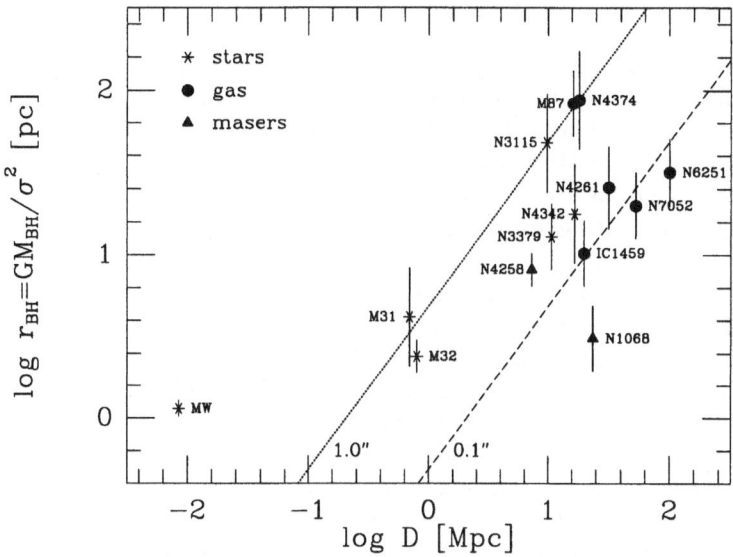

Fig. 1. The radius of influence $r_{BH} = GM_{BH}/\sigma^2$ of central black holes with well-determined masses versus the distance of the host galaxy. The masses were derived from stellar kinematics (stars), gas kinematics (dots) and VLBI measurements of masers (triangles). The dotted line corresponds to an angular size $\theta_{BH} = 1\rlap{.}''0$, and the dashed line corresponds to $0\rlap{.}''1$. The error bars represent the quoted uncertainty in M_{BH} [2,8,12,15,16,20,21,24,28,30,38,42,43,49,63].

and gas. This requires high spatial resolution. For example, the $3 \times 10^9 M_\odot$ black hole in the nucleus of the galaxy M87 [31] in the Virgo cluster has $\theta_{BH} \approx 1\rlap{.}''5$. The general approach is to determine the *luminous* mass in stars from the observed surface brightness distribution, and to compare this with the *dynamical* mass derived from kinematic measurements of stars or gas [32,55]. If one can show that the mass density inside a certain radius is larger than anything that can be produced by normal dynamical processes, then the object is considered to be a black hole. In a few cases it is possible to find direct evidence for the presence of a relativistic object (§3).

Figure 1 shows r_{BH} as defined in Eq. (1) for the best published black-hole mass determinations versus the distance to the host galaxy. Lines of constant angular resolution θ_{res} run diagonal. The early determinations clustered near $\theta_{res} \approx 1\rlap{.}''0$ [56], but HST has pushed this to $\theta_{res} \approx 0\rlap{.}''1$. Detailed modeling has shown that, depending on the internal dynamical structure of the host galaxy, the effects of the black hole often are visible only inside projected radii that are significantly smaller than r_{BH} (e.g., eq. (4.2) in [52]). This suggests that measured masses corresponding to $\theta_{BH} \sim \theta_{res}$ should be treated with caution, as they are likely to be overestimates (see §4). At present, only VLBI *measurements can probe the regime* $\theta_{res} < 0\rlap{.}''1$.

2   Stellar Dynamical Modeling

The nearest galactic nucleus is the center of our own Galaxy. Despite the large foreground extinction, it is possible to resolve individual stars in the Galactic center in the infrared, and to measure not only their radial velocities, but also their proper motions [15,28], and accelerations [29]! Dynamical modeling of this remarkable data provides unequivocal proof that our Galaxy contains a central black hole of nearly three million solar masses.

The dynamics of the nuclei of nearby early-type galaxies can be probed with stellar absorption-line spectroscopy of the integrated light. This generally requires long exposure times. The orbital structure in these systems is rich [47,64], so a true inward increase of the mass-to-light ratio M/L must be distinguished from radial variation of the velocity anisotropy. This can be done by measuring the shape of the line-of-sight velocity distribution [27,41]. The orbital structure is related to the intrinsic shape of the galaxy. This can be constrained by measurements along multiple position angles [5,44], or, even better, by integral-field spectroscopy (§5).

Determination of the black-hole mass and the orbital structure requires construction of dynamical models. There has been a steady increase in the sophistication of model construction in the past decade. Early isotropic spherical models were replaced by anisotropic spheres, and then by axisymmetric models with a special orbital structure (phase-space distribution function $f = f(E, L_z)$ where E is the orbital energy and L_z is the angular momentum component parallel to the symmetry axis, e.g., [52]). More recently axisymmetric models with the full range of possible anisotropies, and multiple components have been used. The first such study was done for M32, and included ground-based data along four position angles and eight FOS pointings [44]. The model was constructed by a version of Schwarzschild's [57] numerical orbit-superposition method, which fits the surface brightness distribution as well as all kinematic observables [13].

Another example of this approach is provided by the E7/S0 galaxy NGC 4342 in Virgo. This is a low-luminosity object, seen nearly edge-on, with a prominent nuclear stellar disk. Cretton & van den Bosch [12] used ground-based major-axis kinematics from the WHT and five FOS pointings [7], and compared these with general axisymmetric models containing a spheroid, a stellar disk, a nuclear disk and a black hole. Free parameters in the modeling were M_{BH} and the stellar mass-to-light ratio Υ. Contours of constant χ^2 in the (Υ, M_{BH})-plane show that the best fit is obtained for $M_{BH} \sim 3 \times 10^8 M_\odot$, but with a significant uncertainty (Figure 2).

To date, black-hole masses have been derived in this way for about fifteen objects. The full analysis has been published only for M32 [44], N4342 [12], and NGC3379 [24], all based on ground-based and FOS data. Black-hole masses based on three-integral axisymmetric modeling of STIS absorption-line spectroscopy have been reported for a dozen objects by Gebhardt's group [25], but the data and the models have not yet been published.

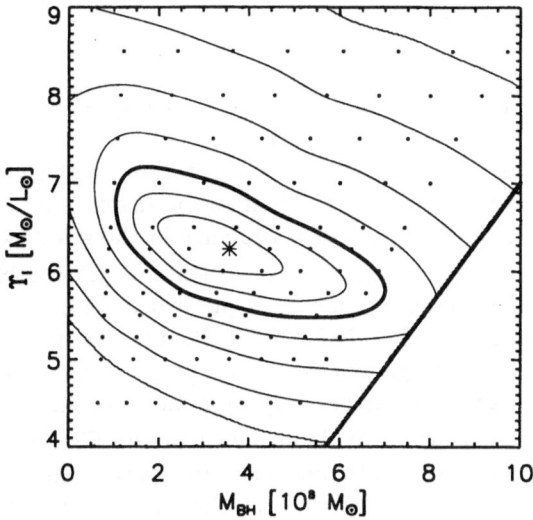

Fig. 2. The black hole in NGC 4342. Contours indicate the goodness-of-fit χ^2 to the observed photometry and kinematics as a function of black hole mass M_{BH} and Υ_I, the I-band stellar mass-to-light ratio. Dots indicate dynamical models that were constructed. The asterisk denotes the best fit. The first three contours surrounding it are the 68.3%, 95.4% and 99.7% confidence levels, while the subsequent contours correspond to a factor of two increase in χ^2 [12].

While in some cases axisymmetric models are consistent with the data, in others they clearly are inadequate (§5). The nucleus of M31 has been known to be asymmetric since 1974 [36]. Observations with the integral-field spectrograph TIGER revealed that the black hole resides in the secondary peak in the brightness distribution [3], and led to models with an asymmetric distribution of stars [62]. Measurements with OASIS on the CFHT [2] have shown that HST spectroscopy with FOC [60], FOS and STIS along the apparent symmetry axis of the eccentric structure has in fact missed the kinematic major axis. A comprehensive non-axisymmetric model that fits this data will teach us much about the structure and formation of this nearby nucleus.

Some nearby galactic nuclei are shrouded in dust, and their internal kinematics is best probed at longer wavelengths. This is now possible through the availability of near-IR spectrographs, which employ the CO bandhead at 2.3μ to derive the stellar kinematics. Anders [1] used the MPE-built 3D integral-field unit to show that the derived nuclear kinematics in the largely dust-free nearby galaxy NGC 3115 agrees with the kinematics measured at shorter wavelengths [16], demonstrating that this approach works. 3D observations of the luminous merger remnant NGC 1316 (Fornax A) confirmed the central σ of $\approx$230 km/s obtained at short wavelengths from the ground and with FOS [58], suggesting $M_{BH} \lesssim 10^8 M_\odot$ [14]. Instruments such as SINFONI will probe nearby dusty nuclei with a resolution similar to HST [46].

3 Gas Kinematics

Optical emission lines. The nuclei of active early-type galaxies, and those of
most spirals, contain extended optical emission-line gas. Its kinematics can
be used to constrain the central mass distribution. In this case the exposure
times can be relatively short, but care must be taken to model possible non-
circular motions and the effects of turbulence in the gas disks.

High spatial resolution emission-line kinematics (FOC, FOS and STIS), together with careful modeling, has been published for six cases: M87, NGC
4261, NGC 4374, NGC 6251, NGC 7052, IC 1459 [8,20,21,38,42,63]. These
very luminous early-type galaxies cover a modest range in total luminosity,
but the derived black-hole masses vary by a factor of ten.

An example of this approach is provided by the E3 galaxy IC 1459. This
giant elliptical hosts a compact nuclear radio source, has a counter-rotating
stellar core ($\sim 10''$), a shallow cusp in the luminosity profile, and a blue
central point source [9,23]. FOS kinematics of the emission-line gas associated
with the nuclear dust-lane reveals a disk in rapid rotation (Figure 3), with
significant velocity dispersion. Detailed models for the gas motions which include the effects of turbulence show that $M_{\rm BH} \approx 2.5 \times 10^8 M_\odot$ [63], which
is a factor of ten smaller than the mass suggested by axisymmetric $f(E, L_z)$
models of the ground-based stellar kinematics along the major axis. A general
stellar dynamical model that incorporates major-axis STIS spectroscopy and
ground-based spectroscopy along four position angles is under construction.

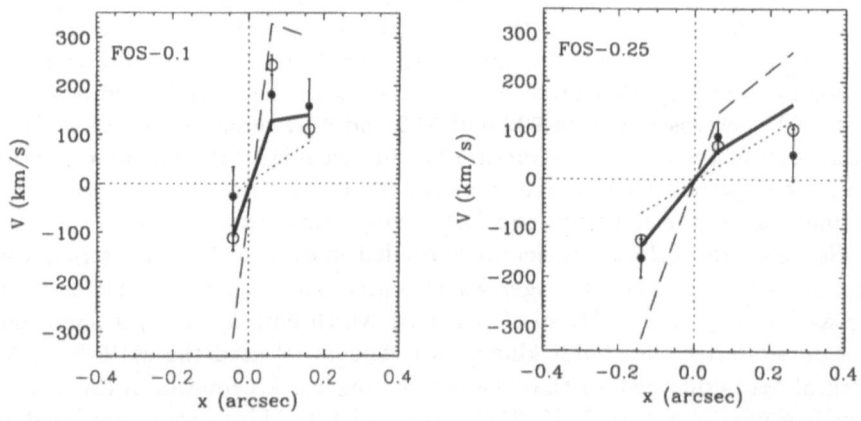

Fig. 3. Emission-line gas kinematics of the nucleus of IC1459, derived from three
FOS pointings with the $0\overset{''}{.}1$ (left) and three with the $0\overset{''}{.}25$ apertures (right). Rotation velocities V were derived from Hα+[NII] (open circles) and Hβ measurements
(dots). The heavy solid line is the prediction of a model with $M_{\rm BH} = 1.0 \times 10^8 M_\odot$.
Dotted and dashed curves are for $M_{\rm BH} = 0$ and $7 \times 10^8 M_\odot$, respectively [63].

Kinematics of emission-line gas will provide many black-hole masses in the near future. Ongoing HST programs include Hα emission-line spectroscopy with STIS for 21 radio-loud ellipticals (PI Baum) and 54 Sb/Sc spirals (PI Axon). These studies generally use three parallel slits to measure possible deviations from simple circular motion.

Masers. Nuclear maser emission can be measured with VLBI techniques in a few spiral galaxies. This achieves the highest spatial resolution to date, but is possible only when the circumnuclear disk is nearly edge-on. A search of ≈700 nuclei revealed maser emission in 22 cases, and disk-like kinematics in six of these [50]. The best black-hole mass determinations are for NGC 1068, NGC 4258, and NGC 4945 [30,49].

X rays. Recently, it has become possible to measure the profile of the Fe Kα line at 6.4 keV in the nuclei of nearby Seyfert galaxies [51]. The width of this line approaches ≈10^5 km/s, which is direct evidence that the emitting gas must be near the Schwarzschild radius of a relativistic object. The mass of the black hole, and possibly its spin, can be derived from the detailed shape of the line profile [19,45]. The latest generation of X-ray telescopes, in particular *Chandra* and XMM, will extend this work to more objects.

Reverberation mapping. The observed time-variation of broad Hβ emission lines from Seyfert nuclei can be used to derive the radius of the broad-line region by means of the light travel-time argument [6]. In combination with simple kinematic models for the motion of the broad-line clouds, this provides an estimate of the mass of the central object responsible for these motions. Early mass determinations appeared to be systematically lower than those obtained by other means [32], but this discrepancy has now disappeared [26].

4 Demographics

For the past five years, our understanding of black-hole demography was summarized in a diagram of black-hole mass M_{BH} versus absolute bulge luminosity L_{bulge}, where the 'bulge' was taken to be the entire galaxy for an elliptical or lenticular, or the actual bulge in case of a spiral [35,55]. Early work based on simple axisymmetric Jeans models and ground-based kinematics [39] was interpreted as evidence for a tight correlation of M_{BH} with L_{bulge}, and hence with the total mass of the bulge. Much effort was spent in trying to reproduce this correlation in models of galaxy formation, and in relating it to the energy production in quasars [55].

More appropriate dynamical models combined with HST kinematics have generally resulted in a downward revision of the earlier masses [25,32,40]. Figure 4a is the resulting demography diagram for the best determinations. It shows a rough correlation, but with a large range in M_{BH} at fixed L_B. The masses are consistent with various black-hole formation scenarios [40,61], and they agree with the quasar light prediction for reasonable efficiencies [55,59].

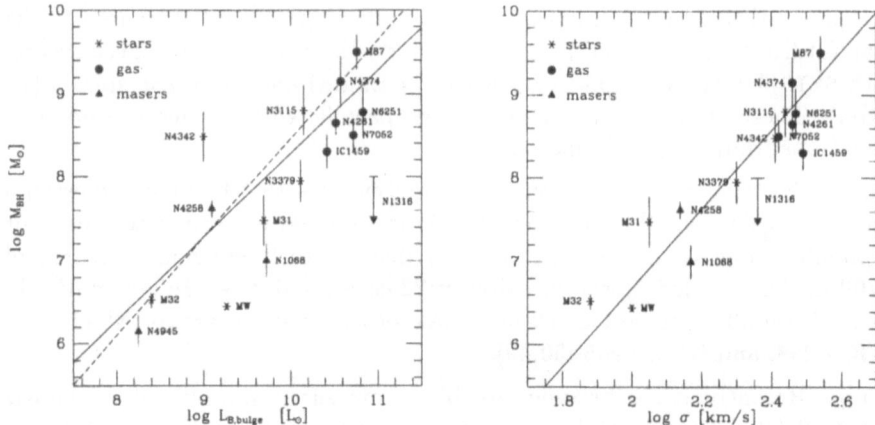

Fig. 4. Demography of black holes. Left: M_{BH} versus total absolute luminosity $L_{B,bulge}$ of the host bulge or spheroid. Coding of points is as in Fig. 1. The dashed line is the correlation based on $f(E, L_z)$ models [39]. The solid line is the prediction for adiabatic growth models [40]. Recently published measurements generally give lower masses, indicating mild radial anisotropy. They display a significant scatter at fixed $L_{B,bulge}$. There is an observational bias against detecting small black-hole masses in luminous galaxies. Right: M_{BH} versus velocity dispersion σ of the host bulge or spheroid. The solid line is the relation proposed in [48].

However, there is an observational bias against detecting small black holes in large galaxies, i.e., for cases where the radius of influence is simply smaller than the spatial resolution currently achievable.

Absolute galaxy luminosity correlates with velocity dispersion [17], so M_{BH} is expected to correlate with σ, measured outside the radius of influence of the black hole. As illustrated in Figure 4b, the scatter in this correlation is considerably less than in the M_{BH} versus L_{bulge} correlation [22,25]. A good fit is provided by $M_{BH} = 1.30(\pm 0.36) \times 10^8 M_\odot (\sigma/200 \text{ km s}^{-1})^{4.72(\pm 0.36)}$, where σ is the velocity dispersion at $R_e/8$, and R_e is the effective radius [48]. This suggests yet another link between global and nuclear galaxy properties, but it is not understood what causes this link, or whether there are other parameters involved. Some of the galaxies in the diagram are not axisymmetric, so the black hole mass derived from axisymmetric modeling may need revision.

Substitution of the M_{BH} versus σ correlation in Eq. (2) provides the following estimate of the formal radius of influence of the black hole:

$$\theta_{BH} \sim 0\overset{''}{.}4 \left(\frac{\sigma}{100 \text{ km/s}} \right)^{2.7} \left(\frac{1 \text{ Mpc}}{D} \right). \tag{3}$$

As noted in §1, the actual radius of influence may be smaller. Assuming that Eq. (3) is valid for all galaxies (which is not proven) allows one to use a ground-based velocity dispersion measurement to estimate the minimum resolution needed to measure a reliable black-hole mass.

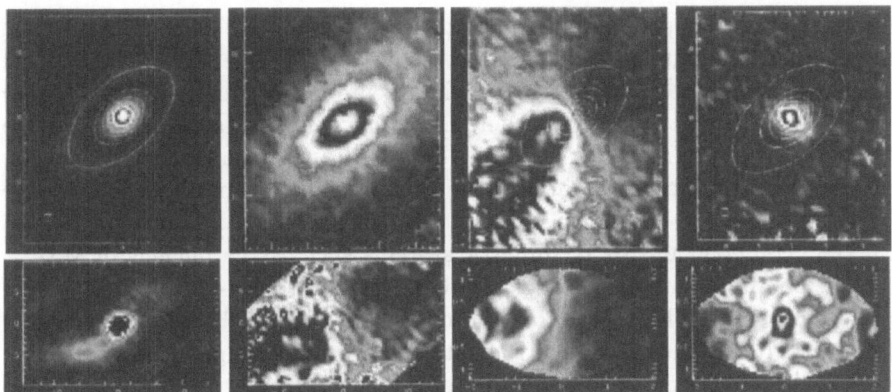

Fig. 5. Integral-field spectroscopy of the E6 galaxy NGC 3377. Top, from left to right: reconstructed intensity, stellar Mg b index, mean velocity, and velocity dispersion. Bottom, from left to right: gas intensity (OIII λ5007) and velocity, and stellar mean velocity and velocity dispersion of the nuclear region. The first six panels are based on a two-hour exposure with SAURON. The field shown is $30'' \times 39''$. The last two panels are based on a 3.5 hour exposure with OASIS, with $0''.16$ sampling and a $4'' \times 4''$ field. The stars show a striking rotating disk pattern with the spin axis misaligned $\approx 10°$ from the photometric minor axis, indicating the galaxy is triaxial. The gas also reveals non-axisymmetric structures and motions [10,11].

5 Next Steps

There is strong evidence for the existence of massive black holes in most galactic nuclei, based on different measurement approaches. Our understanding of black-hole demographics is still limited, because the number of reliable mass measurements is relatively modest. Significant improvement is expected through spectroscopic surveys of galactic nuclei with emission-line gas, and through further work on stellar absorption-line kinematics.

To make further progress, high-resolution observations of the nuclei obtained with STIS on board HST, and with adaptive optics from the ground (e.g., OASIS on the CFHT, SINFONI on the VLT), need to be complemented with the wide-field kinematics of the host galaxy, in order to measure its intrinsic shape and orbital structure [5,64]. For this reason, the dynamics groups of the universities at Lyon, Leiden, and Durham have built the special-purpose integral-field spectrograph SAURON for the WHT, with high throughput, and a field of view of $33'' \times 41''$ sampled at $0''.94 \times 0''.94$ [4,65], and are using it to observe a representative sample of 80 nearby early-type galaxies.

Figure 5 shows integral-field kinematics for the E6 galaxy NGC 3377 [10,11]. The SAURON maps of the stellar and gaseous motions reveal that this galaxy is not axisymmetric. The minor axis rotation persists in the high-spatial resolution measurements with OASIS. NGC 3377 has a steep-cusped central luminosity profile, and the expectation based on general dynamical arguments is that this galaxy should be nearly axisymmetric, at least in the inner regions [47,64]. It will be interesting to probe this galaxy at even higher spatial resolution, to establish whether the nucleus is axisymmetric and the derived black-hole mass correct, or whether the non-axisymmetry persists, and our understanding of the nuclear dynamics needs modification. Either way, the codes for numerical model construction need to be generalized to triaxial geometry, and to asymmetric systems such as the nucleus of M31.

The ongoing systematic programs will reveal the black-hole demographics as a function of Hubble type, radio properties, intrinsic shapes, and internal dynamics. It will also establish the rate of occurrence of gaseous and stellar nuclear disks, and of nuclear star clusters, and the importance of black-hole driven secular dynamical evolution for shaping galaxies.

It is a pleasure to acknowledge comments and contributions by Nicolas Cretton, Roeland van der Marel, Gijs Verdoes Kleijn, and the SAURON team.

References

1. Anders S., 1999, PhD Thesis, Ludwig–Maximilians Univ., München
2. Bacon R., 2000, in *NGST Science and Technology*, eds E.P. Smith & K. Long, ASP Conf. Ser., **207**, p. 333
3. Bacon R., Emsellem E., Monnet G., Nieto J.L., 1994, AA, **281**, 691
4. Bacon R., et al., 2000, MNRAS, submitted
5. Bak J., Statler T.S., 2000, preprint (astro-ph/0003468)
6. Blandford R.D., McKee C.F., 1982, ApJ, **255**, 419
7. van den Bosch F.C., Jaffe W., van der Marel R.P., 1998, MNRAS, **293**, 343
8. Bower R., et al., 1998, ApJ, **492**, L111
9. Carollo C.M., Franx M., Illingworth G.D., Forbes D., 1997, ApJ, **481**, 710
10. Copin Y., 2000, PhD Thesis, ENS, Lyon
11. Copin Y., Cretton N., et al., 2000, in prep.
12. Cretton N., van den Bosch F.C., 1999, ApJ, **514**, 704
13. Cretton N., de Zeeuw P.T., van der Marel R.P., Rix H.W., 1999, ApJS, **124**, 383
14. Davies R.L., in *Imaging the Universe in Three Dimensions*, eds W. van Breugel & J. Bland–Hawthorn, ASP Conf. Ser., **195**, p. 134
15. Eckart A., Genzel R., 1997, MNRAS, **284**, 576
16. Emsellem E., Dejonghe H., Bacon R., 1999, MNRAS, **303**, 495
17. Faber S.M., Jackson R., 1976, ApJ, **204**, 668
18. Faber S.M., et al., 1997, AJ, **114**, 1771
19. Fabian A.C., Rees M.J., Stella L., White N.E., 1989, MNRAS, **238**, 729
20. Ferrarese L., Ford H.C., Jaffe W., 1996, ApJ, **470**, 444
21. Ferrarese L., Ford H.C., 1999, ApJ, **515**, 583
22. Ferrarese L., Merritt D.R., 2000, astro-ph/0006053
23. Franx M., Illingworth G.D., 1988, ApJ, **327**, L55
24. Gebhardt K., et al., 2000a, AJ, **119**, 1157
25. Gebhardt K., et al., 2000b, astro-ph/0006289
26. Gebhardt K., et al., 2000c, astro-ph/0007123

27. Gerhard O.E., 1993, MNRAS, **265**, 213
28. Ghez A.M., Klein B.L., Morris M., Becklin E.E., 1998, ApJ, **509**, 678
29. Ghez A.M., 2000, AGM, **16**, 42
30. Greenhill L.J., Gwinn C.R., Antonucci R., Barvainis R., 1996, ApJ, **481**, L23
31. Harms R.J., et al., 1994, ApJ, **435**, L35
32. Ho L., 1998, in *Observational Evidence for Black Holes in the Universe*, ed. S.K. Chakrabarti (Dordrecht: Kluwer), p. 157
33. Jaffe W., Ford H.C., O'Connell R.W., van den Bosch F.C., Ferrarese L., 1994, AJ, **108**, 1567
34. Kormendy J., 1993, in IAU Symp. 153 *Galactic Bulges*, eds H. Dejonghe & H.J. Habing (Dordrecht: Kluwer), p. 209
35. Kormendy J., Richstone D.O., 1995, ARA&A, **33**, 581
36. Light E.S., Danielson R.E., Schwarzschild M., 1974, ApJ, **194**, 257
37. Lynden-Bell D., (1969), Nature, **233**, 690
38. Macchetto F.D., Marconi A., Axon D.J., Capetti A., Sparks W., Crane P., 1997, ApJ, **489**, 579
39. Magorrian J., et al., 1998, AJ, **115**, 2285
40. van der Marel R.P., 1999, AJ, **117**, 744
41. van der Marel R.P., Franx M., 1993, ApJ, **407**, 525
42. van der Marel R.P., van den Bosch F.C., 1998, AJ, **116**, 2220
43. van der Marel R.P., de Zeeuw P.T., Rix H.W., Quinlan G.D., 1997a, Nature, **385**, 610
44. van der Marel R.P., Cretton N., de Zeeuw P.T., Rix H.W., 1998, ApJ, **493**, 613
45. Martocchia A., Karas V., Matt G., 2000, MNRAS, **312**, 817
46. Mengel S., Eisenhauer F., Tezca M., Thatte N., Röhrle C., Bickert K., Schreiber J., 2000, SPIE, **4005**, 301
47. Merritt D.R., 1999, PASP, **111**, 129
48. Merritt D.R., Ferrarese L., 2000, astro-ph/0008310
49. Miyoshi M., et al., 1995, Nature, **373**, 173
50. Moran J., 2000, talk at Oort Workshop April 2000, Leiden University
51. Nandra K., George I.M., Mushotzky R.F., Turner T.J., Yaqoob J., 1997, ApJ, **477**, 602; 1999, ApJ, **523**, L17
52. Qian E.E., de Zeeuw P.T., van der Marel R.P., Hunter C., 1995, MNRAS, **274**, 602
53. Quinlan G.D., Hernquist L., Sigurdsson S., 1995, ApJ, **440**, 554
54. Rees M.J., ARA&A, **22**, 471
55. Richstone D.O., et al., Nature, **395A**, 14
56. Rix, H.W.: 1993, in *IAU Symposium 153, Galactic Bulges*, eds H. Dejonghe & H.J. Habing (Dordrecht: Kluwer), p. 423
57. Schwarzschild M., 1979, ApJ, **232**, 236
58. Shaya E., 1999, priv. comm.
59. Soltan A., 1982, MNRAS, **200**, 115
60. Statler T.S., King I.R., Crane P., Jedrzejewski R.I., 1999, AJ, **117**, 894
61. Stiavelli M., 1998, ApJ, **495**, L91
62. Tremaine S.D., 1995, AJ, **110**, 628
63. Verdoes Kleijn G.A., van der Marel R.P., Carollo C.M., de Zeeuw P.T., 2000, AJ, **120**, in press
64. de Zeeuw P.T., 1996, in *Gravitational Dynamics*, eds O. Lahav, E. Terlevich & R.J. Terlevich (Cambridge University Press), p. 1
65. de Zeeuw P.T., et al., 2000, ING Newsletter, 2, 11

Effects of Anisotropy
on the Central Dark Mass in NGC 3115.
New Results from Integral Field Spectroscopy

Stephan W. Anders, Niranjan Thatte, and Reinhard Genzel

Max Planck Institut für extraterrestrische Physik
D-85748 Garching, Germany

Abstract. We report new results on the stellar kinematics and the mass distribution of the galaxy NGC 3115 based on NIR integral field spectroscopic data. Investigations using long slit spectroscopic data have yielded strong evidence for the presence of a massive dark object of ca. 10^9 solar masses. NGC 3115 therefore appears to be a prominent candidate for hosting a black hole in its center.
We demonstrate that with integral field spectroscopy the rotation and velocity dispersion can be much better constrained by sampling in both spatial dimensions. This yields revised and more secure results.

1   Observations and Data

We carried out our observations of NGC 3115 in the K–Band using MPE's integral field spectrometer **3D** together with the tip–tilt adapter **ROGUE** at the ESO–2.2m telescope in La Silla/Chile. The seeing conditions were excellent. Our NIR–spectra provide rotation and dispersion fields using the CO $v = 2 - 0$ bandhead feature at 2.3 μm as well as LOSVDs fully consistent with previously measured results. Furthermore the Gauss–Hermite–coefficients h3 could also be reproduced (van der Marel et al. 1994 [2]).

2   Modelling

Based on the 3D–spectra we modelled the stellar kinematics of NGC 3115 and derived mass profiles based on the model (Anders 1999 [1]). In figure 1 the thin lines represent the observed spectra for the central 0".3 (left picture), for the point at a distance of 0".85 NE along the photometric major axis (middle picture) and for the point in a distance of 0".42 on the minor axis in NW direction. Clearly the K–band absorption CO–bandheads can be seen. The thick lines show the modelled results with various template stars (Schreiber 1998 [3]). The best fits to the spectral features were obtained using late K– and early M–stars. In a first simulation we chose the stellar velocity to obey a Keplerian equation. The second model uses a spatial anisotropic dispersion with its tangential and azimuthal components equal. It results in a constant

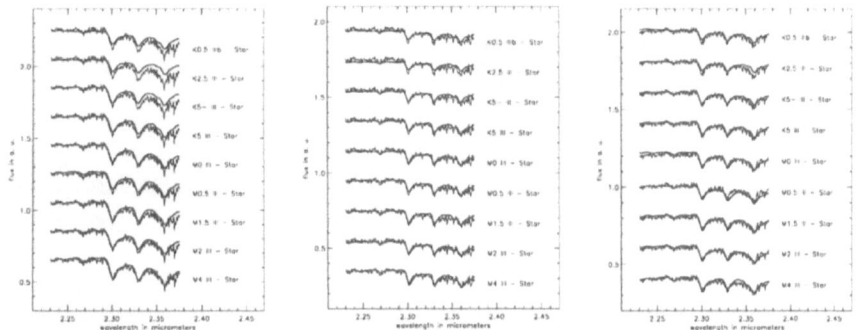

Fig. 1. Thin lines: 3D–K–band–spectra of NGC 3115 of the central region (left image), of the point 0".85 NE on the photometric major axis (middle image) and of the point 0".42 NW on the minor axis (right image). The thick lines represent the results of the model calculations for various template stars.

value for the anisotropy parameter β and the various components are defined by the equation

$$\sigma_{rad} = \frac{\sigma_{0rad}}{r^\alpha} \qquad \sigma_{tan,az} = \frac{\sigma_{0tan,az}}{r^\alpha}$$

The anisotropic model best reproduces the observed two dimensional form of the rotation and dispersion fields as can be seen in figure 2 and are also consistent with the measured Gauss–Hermite–coefficients. In addition the model leads to consistent values of the anisotropy parameter and to plausible distribution functions. It also provides a good representation of the velocity dispersions obtained with the HST (Kormendy et al. 1996 [4]). Using the Boltzmann–Equation we could obtain mass profiles and M/L ratios and from the models (figure 3). Whereas the usage of isotropic dispersions lead to central masses of $10^9 M_\odot$ consistent with previous results (Kormendy et al. 1996 [4]) the anisotropic models reduce this to ca. $2 \times 10^7 M_\odot$.

References

1. Anders, S. W. (1999), Beugungsbegrenzte Nahinfrarot–Feldspektroskopie und Stellare Kinematik in der Galaxie NGC 3115, thesis, Ludwig–Maximilians Universität München
2. R. P. van der Marel et al.(1994), Velocity profiles of galaxies with claimed black holes - I. Observations of M31, M32, NGC 3115 and NGC 4594, MNRAS **268**, 521 - 543
3. Schreiber, N. M. (1998), Near–infrared imaging spectroscopy and mid–infrared spectroscopy of M82: revealing the nature of star formation activity in the archetypal starburst galaxy, thesis, Ludwig–Maximilians Universität München
4. J. Kormendy et al.(1996), HUBBLE SPACE TELESCOPE SPECTROSCOPIC EVIDENCE FOR A $2 \times 10^9 M_{Sun}$ BLACK HOLE IN NGC 3115, ApJ Letters **459**, L57 - 60

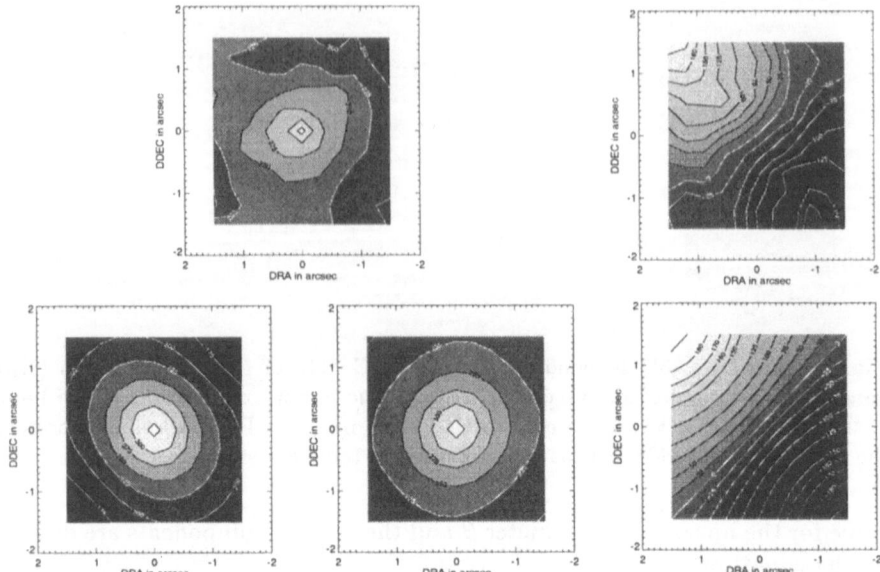

Fig. 2. Measured dispersion (upper left pic.) and rotation (upper right pic.) fields together with the modelled results. The simulation with isotropic Keplerian dispersion can be studied in the lower left picture. The lower middle map show the corresponding result for the model with the spatial anisotropic dispersions. The calculated rotation field is represented in the lower right figure.

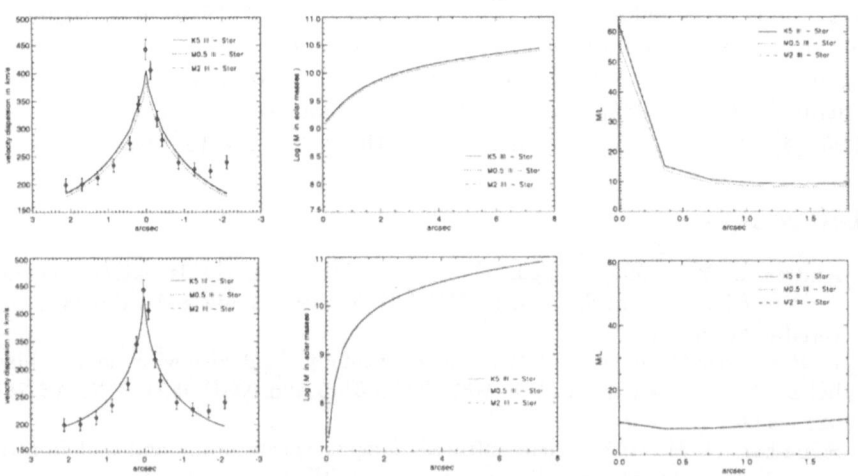

Fig. 3. Reproduction of HST–data for the model with the isotropic Keplerian velocity dispersion (upper left) and with the anisotropic one (lower right). The corresponding mass profiles as a function of distance are shown in the middle column and the mass–to–light ratios in the right column.

Black-Hole Results from STIS

Rob P. Olling[1], D. Merritt[1], C.L. Joseph[1], and M. Valluri[2]

[1] Rutgers University Department of Physics & Astronomy
   136 Frelinghuysen Rd. Piscataway, NJ 08854-8019, USA
[2] University of Chicago Department of Astronomy and Astrophysics
   5640 S. Ellis Ave. Chicago, IL 60637, USA

Abstract. The Space Telescope Imaging Spectrograph (STIS) has obtained high resolution (with $\Delta V \gtrsim 20$ km s^{-1} and $FWHM \sim$0.1-0.2") spectra of the nuclei of about 15 nearby galaxies in a search for supermassive black holes. This talk concentrates on the data reduction process and the difficulties which have to be overcome to obtain reliable kinematic measurements from STIS observations. Our analysis is based on standard STIS-pipeline software (GSFC's IDL version of the CALSTIS package). We have added significant functionality to address the under-sampling of the spatial part of the point-spread function and the presence of cosmic rays in the images. Typically $\sim$20% of the total counts are cosmic-ray hits, in 5% of the pixels.

We also present preliminary results of the nuclear kinematics for several galaxies for which Rutgers astronomers are lead investigators. In NGC 2841 we obtain a clear signature in the kinematics of a black hole with mass of several tens of million solar masses, the first black hole detection in this galaxy. In M 87 we measure the stellar velocity dispersion at a radius of 0.3", similar to ground-based data, but with far less contamination from adjacent parts of the image.

1   Properties of the Raw Data

To determine kinematic parameters from a galaxy spectrum, instrumental effects must be removed. To avoid defects artifacts and cosmic-rays, an object is observed with several (≥ 2) slightly different pointings. The true galaxy spectrum can then be recovered by comparing the individual exposures, after they have been registered onto an identical pixel grid (in spatial and wavelength coordinates).

The optics of long-slit spectrographs (like STIS') are never perfect. As a result, the spatial location of a source $[L(\lambda)]$ on the CCD changes slowly as a function of wavelength (Fig. 1, top panel). We measured the width of the point-spread function (PSF) as a function of "wavelength" from an exposure of the K giant HR 7615. At places where the star is centered on a pixel center, the resulting cross-dispersion profile (CDP) is sharply peaked (small FWHM). When it is located between two pixels, the resulting profile is significantly broader (Fig. 1, middle panel, full line). Such broadening is expected if the PSF is critically- or under-sampled. Depending on its CDP width, the galaxy spectrum may be similarly affected.

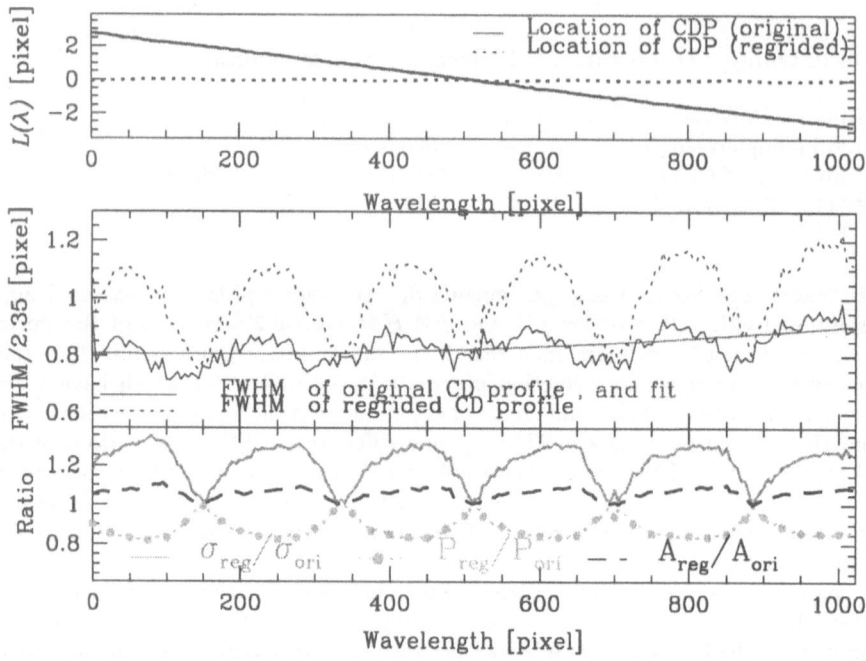

Fig. 1. Dependence of cross-dispersion properties as a function of "wavelength". *Top panel*: Location before (full line) and after (dotted line) regridding. *Middle panel*: the width of the CDPs (same line coding). *Lower panel*: the ratio of CDP properties after-to-before regridding [CDP widths (full line), peaks (dots) and areas (dash)]. Linear interpolation was used to regrid the CDPs.

To register individual exposures onto an identical grid, each exposure has to be regridded. However, regridding amplifies the difference between the widths of on-pixel and between-pixel cross-dispersion profiles if the CDP is poorly sampled (Fig. 1, middle panel, dotted line). Like the width, the area and the peak of the CDP are also affected by regridding (Fig. 1, lower panel). From Figure 1 we can also infer that an incident spectrum of constant flux results in an output spectrum with a ~20% intensity variation with a period of ~175 pixels (~3300 km s^{-1} for STIS' G750M grating centered on the Ca II triplet region). The undulation is badly modeled by a high-order polynomial. The resulting intensity modulations are different for each exposure as they have been dithered (shifted w.r.t. each other). As a result, the undulations will be out of phase so that averaging of exposures can produce unpredictable results, especially when just a few exposures are available. Since each set of exposures will produce different $L(\lambda_i)$'s, the effects are unique to each observing program. We are simulating these effects and take appropriate

action to avoid their pitfalls. Our efforts should enhance the reliability of
the resultant spectra, and the kinematical information we derive from them.

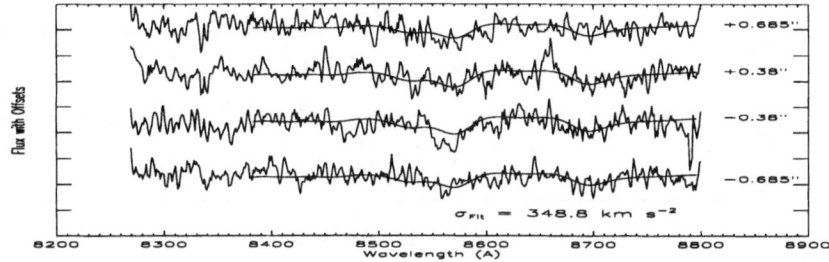

Fig. 2. M 87 spectra based on 10 orbits of integration. The data were binned into
4 spectra that were sampled over 6 rows (0.304") centered at +0.685, +0.38, -0.38,
and -0.685 " from the center. LOSVDs with $\sigma \sim 350$ km s^{-1} are superimposed.

Figure 1 indicates that our ability to derive kinematical parameters from
a spectrum may be limited due to the presence of an undulating continuum.
Also note that any remaining modulation in the template star will propagate
to the derived LOSVDs if the under-sampling is not handled properly, even if
the CDP of the galaxy itself is sufficiently wide. There are several approaches
that can be followed to mediate these effects: 1) one can pre-smooth the CDPs
to avoid the under-sampling undulations, 2) one can observe only galaxies
that have broad CDPs, 3) one can apply higher order interpolation schemes to
register the individual exposures, and 4) one can model the effects described
in the previous section and correct for it. Obviously, the first two schemes
are less satisfactory as they do not exploit the full capabilities of HST/STIS.
We therefore analyze our data as outlined in 3) and 4).

Additional problems arise due to cosmic rays (CRs) in $\sim 5\%$ of the pixels
in each one-half orbit integration. On-board rebinning doubles the fraction of
CR-pixels, and so does each one-dimensional *linear* regridding. Thus, a data-
collecting strategy optimized to minimize read noise (one orbit integrations
and on-board rebinning) and a 2D linear regridding strategy leads to science
images that contain 80% CR-pixels. A better strategy (half-orbit integrations,
rebinning in wavelength, and a 1D linear regridding) leads to only 20% CR
pixels.

High-order regridding is required to avoid the undulations due to under-
sampling, but spreads CRs over even more pixels. Thus, CRs *must* be removed
before regridding. We experiment with two CR rejection schemes: 1) locate
and mask CRs in individual exposures, 2) create CR-masks by comparing
the regridded/aligned individual exposures. To obtain a calibrated spectrum,
we then re-map the masked exposures to λ-position space using one of two
procedures: a) align the exposures in the CDP direction, create the sum, and
map the summed image to λ-position space, and b) map each exposure to

λ-position space and create the sum. The simulations we discussed earlier are also designed to include the effects of the data handling steps outlined above.

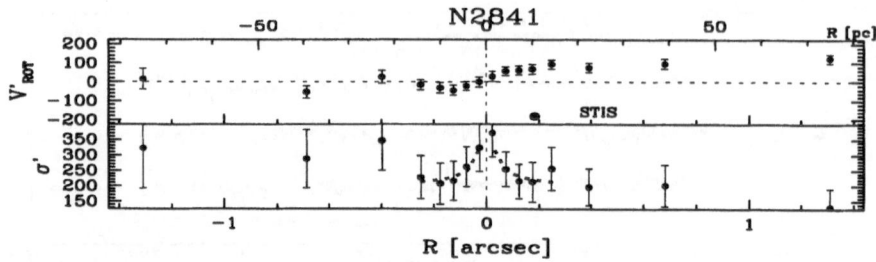

Fig. 3. The rotation curve and the stellar velocity dispersion derived from STIS spectra (FCQ based). V'_{rot} and σ' are "Gauss-Hermite corrected." The dotted line is the predicted velocity dispersions due to a massive central black hole.

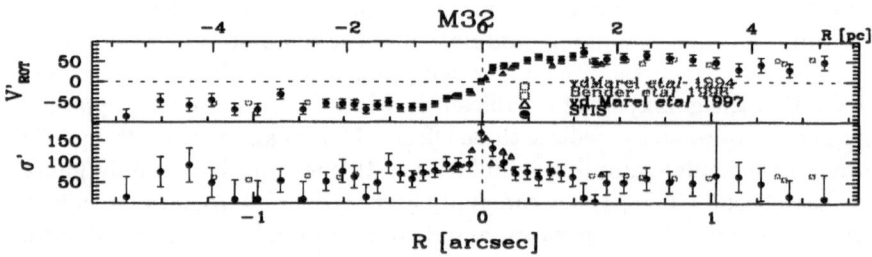

Fig. 4. The rotation curve and the stellar velocity dispersion derived from STIS spectra (FCQ based). "Gauss-Hermite corrections" are also applied.

2   First Results

We report preliminary results for M 87, NGC 2841 and M 32. We have used Merritt's (1997) MPL algorithm to determine the line of sight velocity profiles (LOSVDs) for M 87. M 32 and NGC 2841 have been analyzed using Bender's (1990) FCQ method. In M 87 we determined the stellar velocity dispersion (σ) at a radius twice closer to the central black hole than the best previous work (Fig. 2). For M 32 we can determine the kinematic parameters and LOSVDs with great accuracy: our results are consistent with previous published work but with much improved confidence (Fig. 4). We report a henceforth unknown massive black hole in the center of the spiral galaxy NGC 2841 (Fig. 3). We are currently analyzing these data with sophisticated three-integral stellar dynamical models.

Gas/Stars Coupling in Early-Type Galaxies: Diagnostics for Supermassive Black Holes

Eric Emsellem

Centre de Recherche Astronomique de Lyon, 9 av. Charles André, 69561 Saint-Genis Laval Cedex, France

Abstract. Gaseous and/or stellar kinematics obtained via long-slit spectrography are routinely used to reveal the presence of supermassive black holes (SBHs) in galactic nuclei. Both have their advantages and drawbacks, including the difficulty of building realistic stellar kinematical models or to deal with the contribution of non-gravitational forces in the gas kinematics. In this paper, I present 2D spectrographic and some HST/WFPC2 observations of the gas and stellar components in a small sample of early-type galaxies, demonstrating the need for both high resolution and bidimensional spatial coverage to understand the structure and dynamics of their central regions, and in particular to test the presence of SBHs.

1 Introduction

New discoveries of supermassive black holes at the centre of galaxies are regularly reported in the literature, although the supporting observations are still far from reaching the Schwarzschild radius. Long-slit spectrography has thus been extensively used to obtain the stellar or gas kinematics in the central regions [4]. Stellar orbits are a good tracer of the underlying gravitational potential, but self-consistent dynamical models are difficult to build without some "classical" *adhoc* constraints (e.g. constant mass-to-light ratio, axisymmetry). Gas orbits may definitely look simpler to treat, as dissipation restricts the physical phase space (assuming a stable configuration). But this is not counting with non-gravitational forces, which often play a non-negligible role in the central hundreds of parsecs. The task may also become hopeless when dust extinction significantly alters the light distribution, or if the ionised gas component is too weak for the emission lines to be detected.

This may explain why, up to now, supermassive black holes have almost never been traced using **both** the gaseous and stellar components. The second severe limitation comes from the use of one-dimensional kinematical profiles to determine the mass of the central dark object.

In this paper, I present some preliminary results from an observational program whose aim is to study the coupling between the stars and the gas in the central regions of a few early-type galaxies.

2   Observations

We have obtained spectrographic data cubes for a small sample of 12 early-type galaxies using the TIGER integral field unit at CFHT. Stellar and gas distribution/kinematics were derived from a **blue** (around the Mg triplet) and a **red** (including [NII]/Hα/[SII]) configuration, respectively. The subtraction of the stellar contribution was found critical for the emission line analysis. Similarly, a proper determination of the stellar velocity and dispersion fields required the subtraction of the [NI]λ5200Å doublet from the **blue** spectra.

The complete set of maps (including line strengths) will be published in a forthcoming paper (Emsellem et al. 2000). I only present here a few examples which illustrate the difficulty of interpreting long-slit kinematics in the context of the search for supermassive black holes.

3   The Need for High-Resolution 2D Gas and Stellar Kinematics

3.1   Evidence from the Gas Kinematics

NGC 5838 is a nearly edge-on S0 galaxy ($M_B \sim -19.9$, $i \sim 70\,\mathrm{deg}$) at an assumed distance of 19 Mpc. It contains a significant amount of dust mostly arranged in ring-like structures in the central 20", and a LINER type nucleus. The WFPC2/HST images clearly show the dust features surrounded by a rather boxy bulge. We detected an ionised gas disc in the inner 4" with TIGER: the Hα and [NII] emission lines are difficult to detect in the outer part due to the dominant Hα absorption feature, and the presence of dust.

The measured kinematics are impressing: velocities of ± 360 km s^{-1} reached at about 1".5 from the centre, and a central dispersion of 280 km s^{-1} (Fig. 1). Taken at face value, and considering the simple case of circular orbits, this requires a dynamical mass of about 4×10^9 $M_\odot$ in the inner arcsecond. This in turn implies a very large $M/L_I = 12$ for such a modest galaxy. There are three main exits from this problem: the distance has been underestimated (as $M/L \propto D^{-1}$), the orbits are highly non-circular, or there is a central dark mass of a few $\times 10^9$ $M_\odot$. As we need to find a factor of at least four, we could safely conclude that NGC 5838 contains a massive dark nucleus.

However, the stellar kinematics tell us more. Indeed, a simple isotropic Jeans model, using a constant $M/L_I = 12$ perfectly fits the observed TIGER velocity and dispersion fields, as well as the major-axis kinematics up to 4 kpc [2], with no need for a nuclear dark mass. This strongly suggests that, indeed, the mass-to-light ratio is unusually large, which calls for other causes (e.g. dark halo dominated).

3.2   Where is the Black Hole?

Another case is NGC 4374 (M84) for which convincing arguments have been put forward to support the hypothesis of a supermassive black hole of $\sim$

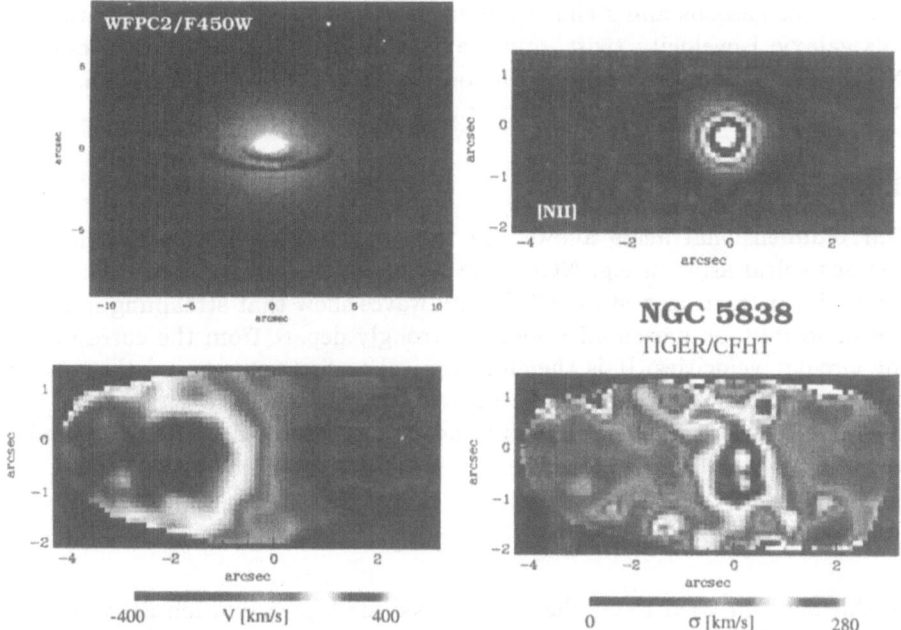

Fig. 1. NGC 5838. Top left: HST/WFPC2 image showing the dust lanes in the central 5". Top right: intensity map of the [NI]λ6584 emission line. Bottom left and right: [NII] velocity and dispersion fields.

1.5×10^9 M$_\odot$ [1]. This includes an impressive velocity curve which peaks at about 400 km s^{-1} at 0".1. But a simple Jeans model with such a massive black hole fails to fit the central kinematics. There is no sign of any peak in the dispersion at the centre, as predicted by the models. The internal dynamics must then be "special" enough to be able to "hide" the black hole. It is also worth noting that the velocities are unusually low, only reaching about 20 km s^{-1} at 2": this implies an extremely high radial anisotropy in the centre, which in turn would be difficult to explain together with the presence of a dominating point mass.

3.3 Axisymmetric Models?

I also wish to point out the importance of two-dimensional kinematics, this time mentioning some SAURON[1] results. In the case of NGC 3377, the presence of a supermassive black hole of 2×10^8 M$_\odot$ has been inferred from axisymmetric models [3] compared with major-axis kinematical profiles obtained via long-slit spectrography. But our SAURON maps clearly show,

---

[1] The SAURON consortium (Lyon/Leiden/Durham) was created to aim at the mapping of the stellar populations and kinematics of a sample of early-type galaxies and spiral bulges. See [5] for details.

both in the gaseous and stellar components, the non-axisymmetric nature of this galaxy: isovelocity twist, minor-axis rotation, spiral arms. The case of NGC 3377, which has always been considered as a "good" candidate, must be revisited in this context, to confirm the estimated black hole mass.

3.4 Spiral Arms

Our bidimensional fields allowed us to detect quite a number of nuclear gaseous spiral arms in e.g. NGC 4278, M104, NGC 2974. Models including a non-linear treatment of $m = 2$ density waves show that streaming motions are important, as measured velocities strongly depart from the corresponding circular velocities. It is therefore critical to first understand the nature of the observed kinematics, before attempting an accurate estimate of the central dark mass. This point should be kept in mind when trying to model gas velocity profiles.

4   Conclusions

In this paper, I tried to emphasize the need for *high resolution 2D* kinematical maps of *both* the gaseous and stellar components in the central regions of galaxies, if we wish to properly determine the mass of a putative supermassive black hole. One rescuing argument could be that, at high intrinsic resolution, the gravitational potential is dominated by the central dark mass and becomes geometrically and dynamically simpler. This may convince some when we take the case of M32, for which WFPC2 and STIS data have been obtained with an intrinsic resolution of 0.4-0.8 pc. But this is certainly not convincing if we consider the case of M31 and its complex nuclear morphology. Nuclei at larger distances do not look simpler as resolution increases.

Supermassive black holes seem to be a common feature of nearby galaxies. Now that we wish to accurately determine their masses, as well as to correlate them with other global properties of the host galaxies, we definitely need to go a step further.

This paper includes work from a number of colleagues I wish to thank here: the SAURON team, Paul Goudfrooij (STScI), and Pierre Ferruit (CRA Lyon).

References

1. Bower, G. A., et al., 1998, ApJL, **492**, 111
2. Dressler, A., Sandage, A., 1983, ApJ, **265**, 664
3. Kormendy, J., et al., 1998, AJ, **115**, 1823
4. Ho, L. C., 1998, in *Observational Evidence for Black Holes in the Universe*, ed. S. K. Chakrabarti, p. 157
5. Peletier, R., et al., 1999, in *Imaging the Universe in three dimensions, Walnut Creek*, USA, ASP Conf. Series, in press

Molecular Gas in Nuclei of the Seyfert Galaxies NGC3227 and NGC1068

Eva Schinnerer[1], Andreas Eckart[2], and Linda Tacconi[2]

[1] California Institute of Technology, Pasadena, CA 91125, USA
[2] Max-Planck-Institut für extraterrestrische Physik,
85740 Garching, Germany

Abstract. New subarcsecond data on NGC 3227 and NGC 1068 obtained with the Plateau de Bure millimeter interferometry indicate large enclosed masses and the possibility of warped molecular gas disks.

We obtained new high resolution mm-interferometric observations of the ^{12}CO (1-0) and ^{12}CO (2-1) line emission of the Seyfert 1 galaxy NGC 3227 and the Seyfert 2 galaxy NGC 1068 at a so far unprecedented angular resolution of 0.6" and 0.7". We modeled the data with planar elliptical orbits representing gas motions in a bar potential and tilted rings representing a warped molecular disk (Schinnerer, Eckart, Tacconi 1999).

For the first time in NGC 1068 we detect millimetric molecular gas emission at radial separations of ≈0.18" (13 pc) from the nucleus (Schinnerer et al. 2000). This indicates an enclosed mass of ∼10^8 M$_\odot$, not correcting for inclination effects. Allowing for a probable contribution from a compact nuclear stellar cluster this value is consistent with a black hole mass of 1.7×10^7 M$_\odot$ estimated from nuclear H$_2$O-maser emission. Modeling of the data with planar elliptical orbits representing gas motions in a bar potential shows that the shift of the line-of-nodes observed in the molecular line velocity field at a radius of about 10" is due to the NIR stellar bar. The analysis of the rotation curve shows that the spiral arms are at the position of an ILR. The low velocity dispersion in the nuclear region and the spiral arms indicates that the gas is distributed in a thin disk with a thickness of about 10 pc and 100 pc FWHM, respectively. The molecular gas emission in the inner 5" is resolved into a ring-like distribution with two bright knots located east and west of the nuclear continuum emission. Our kinematic modeling shows that this ring can be explained either by gas motions due to a nuclear bar potential or a warping of the molecular gas disk. In the absence of observational evidence for a ∼ 1" scale nuclear bar in NIR images we favor the warp model. This model is consistent with physical considerations and observational facts: The 3-dimensional geometry of the warped CO disk provides an explanation for the obscuration of the AGN, extinction of light from the nuclear stellar cluster, and also for the observed NIR/MIR polarization. It also naturally provides a cavity for the ionization cone, since the model predicts that the

warped CO disk would become almost edge-on at a radial distance of about
70 pc.

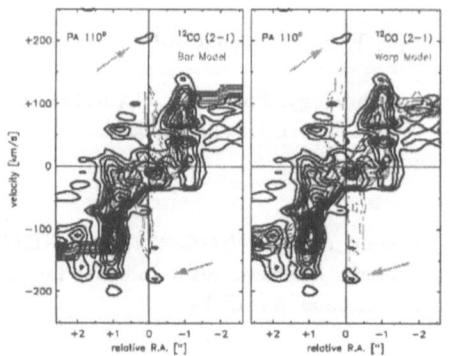

Fig. 1. Comparison of the bar and warp fit for NGC 1068. The arrows indicate the
molecular gas close to the nucleus.

For NGC 3227 we obtained mapped the ^{12}CO (1-0) , ^{12}CO (2-1) , and
HCN (1-0) molecular line emission using the Plateau de Bure interferometer
(Schinnerer, Eckart, Tacconi 2000). In this case we achieved an angular resolution in the ^{12}CO (2-1) line of about 0.6" corresponding to only about 80 pc
at a distance of 17.3 Mpc. The ^{12}CO emission is resolved into an asymmetric
nuclear ring with a diameter of about 3". The HCN line emission is mostly
unresolved at our resolution of $\sim$ 2.4" and contains all of the single dish flux.
In the central arcsecond the gas shows apparent counter rotation. This behavior can be best explained by a warping of the inner molecular gas disk rather
than gas motion in a nuclear bar potential. We detected molecular gas at a
distance from the Seyfert nucleus of only $\sim$13 pc with a velocity of about 75
km/s with respect to the systemic velocity and find that within the central
arcsecond the rotation curve is rising again. These measurements indicate a
lower limit on the enclosed mass of about 2×10^7 M$_\odot$ in the inner 25 pc.

References

1. Schinnerer, E. , Eckart, A., Tacconi, L.J., (1999), "The Molecular Gas in the
   Circumnuclear Region of Seyfert Galaxies", Ap.J. 524, L5
2. Schinnerer, E. , Eckart, A., Tacconi, L.J., (2000), "Distribution and Kinematics
   of the Circumnuclear Molecular Gas in the Seyfert 1 Galaxy NGC 3227", Ap.J.
   accepted
3. Schinnerer, E. , Eckart, A., Tacconi, L.J., Genzel, R., Downes, D. (2000), "Bars
   and Warps traced by the Molecular Gas in the Seyfert 2 Galaxy NGC 1068",
   Ap.J. accepted

Centaurus A: The Supermassive Black Hole in the Nearest AGN

Ethan J. Schreier[1] and Alessandro Marconi[2]

[1] Space Telescope Science Institute, Baltimore MD 21218, USA
[2] Osservatorio Astronomico di Arcetri, Firenze I-50125, Italia

Abstract. Centaurus A (NGC5128) is the nearest active galaxy and one of the first discovered, via its radio and X-ray emission. Although its proximity should make it an ideal test case of the standard model invoking massive black holes, the heavy obscuration due to the dust lane makes it very difficult to study the AGN at high spatial resolution in the optical. Near-infrared observations with HST and with the VLT are now providing high-resolution images and spectra and new insights into the region around the black hole. We summarize what is known about Cen A, discuss the implications of recent HST optical and infrared imaging and polarimetry, and present some preliminary results from VLT/ISAAC.

1 Historical Preface

At this symposium honoring Riccardo Giacconi, it is interesting to recall that Riccardo was a key mover in several of the observatories from which we have gained our current understanding of Centaurus A. The initial extra-galactic observations reported from Riccardo's first major observatory, UHURU, identified the source of Cen A's X-rays with the galaxy itself, not the outer radio lobes, and showed that the X-rays were highly absorbed at low energies. Riccardo was a co-author of that paper, despite his professed lack of interest in jets and nearby AGN. Riccardo's Einstein Observatory enabled the discovery of an X-ray jet in Cen A and revealed the unresolved X-ray source at the galaxy's core. HST, which Riccardo helped make into the forefront space observatory it is, mapped the central region of the galaxy in the near-IR with a resolution of a few parsecs. And the VLT has now provided, for the first time, high spatial resolution near-IR spectroscopy of the nuclear region of Cen A, enabling kinematic mapping of the mass distribution at the center of this nearest active galaxy. We look forward to what ALMA will teach us.

2 Background

Centaurus A, at a distance of 3.5Mpc (Hui et al. 1993), is the nearest active galaxy, and one of the first identified via its radio and X-ray emissions. It was one of first powerful radio galaxies found (Slee 1949, Baade & Minkowski 1954). Amusingly, early optical images from Cerro Tololo identified NGC 5128 as a "source of radio noise". Bowyer (1970) identified a hard X-ray source with Cen A, and then the higher resolution UHURU observations (Kellogg et al. 1971) suggested that the source lay in the galaxy NGC5128 itself. Of course, it was the Einstein Observatory that found the point-like source at the nucleus of the galaxy, along with the first X-ray jet (Schreier et al. 1979).

As more and more active galaxies were discovered via their radio, X-ray, and optical characteristics, mass accretion onto a super-massive black hole became by far the most commonly accepted mechanism for powering AGN. It provided a feasible and efficient way to produce the large amounts of energy necessary to power the AGN themselves and their associated jets and large radio sources. While a unified model explaining the various types of AGN via orientation effects is being actively tested, establishing the existence of super-massive black holes and measuring their masses is still a crucial element in understanding the nature of AGNs.

The advent of high spatial resolution optical and UV spectroscopy with HST revolutionized the field over the last few years, enabling direct dynamical mass determinations of BHs in nearby galaxies in sufficient numbers and to sufficient accuracy to enable demographic studies; many of these are being presented at this meeting. The possible correlation between the mass of the BH and that of the host elliptical galaxy or spiral galaxy bulge is prompting serious studies of the relationship between BH growth and host-galaxy evolution (cf. van der Marel 1999).

Unfortunately, the nuclei of many AGN as well as normal galaxies are heavily obscured by dust, making it difficult to study kinematics at visible wavelengths. Dust obscuration can totally mask visible emission, or distort both absorption and emission-line velocity-field measurements to an extent that makes reliable BH mass estimates difficult using optical data alone. Unfortunately, Centaurus A is not only the nearest example of an AGN but, with its famous dust lane across the center of the galaxy, the nearest example of a heavily obscured AGN. Near-IR observations are essential for its study. HST observations, especially those with NICMOS, were instrumental in revealing the morphology of the central region, and VLT with ISAAC is now providing high resolution near-IR spectroscopy.

3   HST Observations

HST observed Cen A with WFPC in R and I (including polarimetry), with WFPC2 in U, V and I, and with NICMOS in H, K, K-band polarimetry, and narrow band filters on Paschen α and [FeII]$\lambda 1.643 \mu$m. The latter observations showed an elongated ionized structure around the nucleus in addition to the expected galaxy contribution. The visible light images revealed not only new detail of the morphology within the dust lane, but also, finally, the unresolved nucleus. The results are summarized in a series of papers (Schreier et al. 1996, Schreier et al. 1998, Marconi et al. 1999, Capetti et al. 2000).

HST detected the active nucleus in the near–IR (H, K) and, for the first time, in I and V (Marconi et al. 1999). Fig. 1 shows the HST photometry of the nucleus of Centaurus together with the 2.5–240 keV X-ray power law observed by RXTE, with photon index ~1.86, absorbed by a column density of 9.42×10^{22} cm^{-2} (Rothschild et al. 1998), the ISO photometric observations and SCUBA data of Mirabel et al. (1999), and the VLA radio fluxes for the nucleus of Burns, Feigelson and Schreier (1983). Note that the mid-IR/far-IR emission observed from IRAS is completely dominated by the galaxy since, as shown by ISO, the nucleus contributes at most 10% of the total flux (Mirabel et al. 1999). Indeed, the ISO points themselves overestimate the contribution from the nucleus, since even the smallest 7″ ISO aperture includes significant extended galaxy emission. The sharp break between 2 and 3μm combined with the absorbed non-thermal spectrum that fits the X-ray data shows that no single spectrum can simultaneously explain both the IR and the optical. We concluded (Marconi et al. 1999) that at least two components are present – the near-IR data are explained by thermal emission from dust with a temperature of 500 - 1300 K, while the optical emission is consistent with an extrapolation of the X-ray power law, reddened by $A_V \sim 14$.

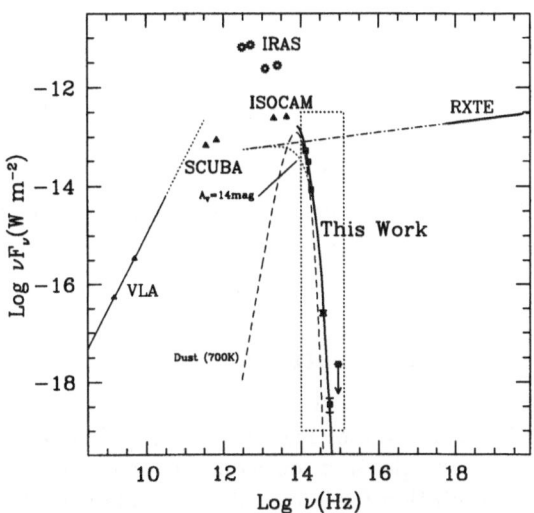

Fig. 1. HST photometry ("This Work") and the overall Cen A SED.

The NICMOS observations in Paschen α and [Fe II] revealed an elongated ionized structure (cf. Fig. 2), interpreted as a disk of gas around the nucleus (Schreier et al. 1998). This inclined disk is approximately 20pc in radius and roughly perpendicular to the larger dust lane, on another principal plane of the galaxy. Notably, the disk is not perpendicular to the jet, despite the fact that the standard model suggests that the central torus around a black hole would be perpendicular to a jet. The ionized feature was interpreted as an extended accretion disk, aligned with the gravitational potential of the galaxy, perhaps transferring material in from the dust lane and barred structure seen in the mid-IR to the unseen dense torus around the black hole at the center. It was the detection of this ionized disk that argued for the VLT/ISAAC observations to measure the velocity of the gas.

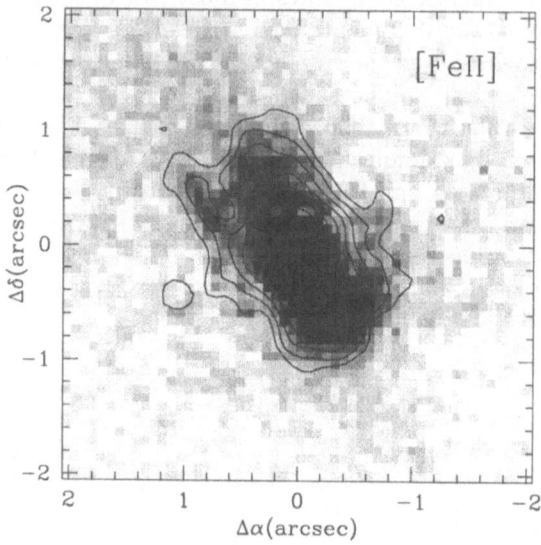

Fig. 2. Ionized gas emission from the disk around the AGN.

4   VLT/ISAAC Observations

We used the near-IR spectrometer ISAAC on VLT/UT1 to map the velocity field at several position angles and locations in the central 2" of Cen A and obtain high spatial resolution (~ 0.5") velocity fields from both ionized and molecular gas (Paβ, FeII, H$_2$ and Brγ). The observations were performed in service mode in May, June and July 1999, and consisted of medium resolution spectra obtained with the 0.3" slit and the grating centered at 1.27 (J) and 2.15 μm (K). The dispersion was 0.6 Å/pix and 1.2 Å/pix respectively, yielding a resolving power of 10,000 in J and 9,000 in K. The slit positions are shown in Fig. 3, superimposed on HST/NIC2 and 3 images. Only some of the data have been analysed, but preliminary results are very interesting.

The spectra are very high quality and show Keplerian velocities within a few arcseconds of the nucleus, suggesting that the ionized gas is indeed part

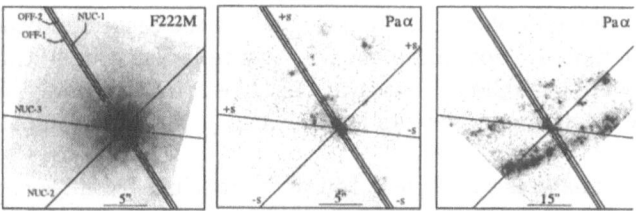

Fig. 3. ISAAC slits superimposed on HST images (NIC2 K and Paα, NIC3 Paα).

of a thin disk rotating around a large central mass condensation. Fig. 4 shows a sample of the ISAAC spectra.

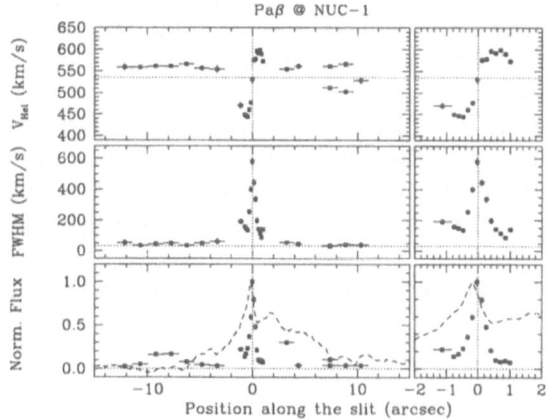

Fig. 4. Paβ velocity, FWHM, and relative flux along the disk. The right panel enlarges the central section.

Initial estimates of the possible black-hole mass may be $\sim 10^8$ $M_\odot$ – smaller than we might expect for a very active galaxy. However, these numbers are very preliminary. There are several other potential surprises suggested by the data. First, the velocities for the ionized gas taken at the different position angles do not appear consistent with the line of nodes being along the principal axis of the disk seen with NICMOS. Second, the velocities at larger distances (more than a few arcseconds) from the nucleus are not consistent with the velocity curves seen close to the nucleus. Third, the molecular (H_2) rotation curves are not consistent with those of the ionized gas. Detailed analyses will be presented in a forthcoming paper.

5 Summary

In summary, we have confirmed there is ionized gas in Keplerian rotation around the nucleus of Cen A. The observed velocities suggest a mass consistent with the range of black hole masses seen at the centers of other galaxies. However, the data are not consistent with a simple disk around the nucleus as suggested by HST/NICMOS observations. It is very well possible that there are multiple velocity components in the vicinity of the nucleus of Centaurus A.

NOTE: In analyses subsequent to the Symposium, we have confirmed the presence of multiple velocity components. One, at $|r| > 2''$, is characterized by small velocity gradients and a FWHM close to the instrumental resolution (30 km s^{-1} in J, 33 km s^{-1} in K). A second is confined to $|r| < 2''$, with large velocity amplitudes and gradients, and FWHM rising sharply at the nucleus, peaking at ~ 600 km s^{-1}. We believe there are three kinematical systems:

1) A thin ionized disk in Keplerian rotation, the illuminated part of which represents the NICMOS Paα feature. The disk has a low inclination ($i < 60°$), i.e. not edge-on. The implied central mass is $M_{BH} > 5 \times 10^7 M_\odot$, unresolved at our resolution ($R < 0.25'' \sim 4$ pc). The corresponding M/L $\sim 100 \, M_\odot/L_{V\odot}$ suggests the existence of a supermassive black hole.

2) An edge on ($i > 80°$) ring with an inner radius $\sim 6''$, detected only in H$_2$ the counterpart of the 100 pc scale "torus" detected in CO by other authors. The mass within the 100 pc molecular ring is $M = 0.6 - 2.5 \times 10^9 M_\odot$. It could be accounted for by stars. This ring may constitute the inner part of the dust lane.

3) A normal extended component of gas rotating in the galactic potential.

References

1. Baade W., Minkowski R., 1954 ApJ, 119, 215
2. Bowyer C.S., Lampton M., Mack J., 1971, ApJ, 161, L1
3. Burns J.O., Feigelson E.D., Schreier E.J., 1983 ApJ, 273, 128
4. Hui X., Ford H.C., Ciardullo R., Jacoby G.H., 1993 ApJ, 414, 463
5. Capetti A., Schreier E.J., Axon D., Young S., Hough J., Clark S., Marconji A., Macchetto D., and Packham C. (submitted)
6. Marconi A., Schreier E.J. Koekemoer A., Capetti A., Axon D., Macchetto D., and Caon N., 2000 ApJ, 528, 276.
7. Kellogg E., Gursky H., Leong C., Schreier E., Tananbaum H., Giacconi R., 1971, ApJ, 165, L49
8. Mirabel I.F., Laurent O., Sanders D.B., et al., 1999 A&A, 341, 667
9. Rothschild R.E., Band D.L., Blanco P.R., et al., 1999 ApJ, 510, 651
10. Schreier E.J., Feigelson E., Delvaille J., et al., 1979 ApJ, 234, L39
11. Schreier E.J., Capetti A., Macchetto F., Sparks W.B., Ford H.J., 1996 ApJ, 459, 535 (Paper I)
12. Schreier E.J., Marconi A., Axon D.J., Caon N., Macchetto D., Capetti A., Hough J.H., Young S., Packham C., 1998, ApJ, 499, L143
13. van der Marel R.P., 1999 AJ, 117, 744

SINFONI – Galaxy Dynamics at 0.″05 Resolution with the VLT

Niranjan Thatte[1], Frank Eisenhauer[1], Matthias Tecza[1], Sabine Mengel[1], Reinhard Genzel[1], Guy Monnet[2], Domenico Bonaccini[2], and Eric Emsellem[3]

[1] Max-Planck-Institut für extraterrestrische Physik, Giessenbachstraße, D-85740 Garching, Germany
[2] European Southern Observatory, Karl-Schwarzschild-Straße 2, D-85748 Garching, Germany
[3] CRAL – Observatoire de Lyon, 9 Avenue Charles Andre, F-69561 Saint Genis Laval, France

Abstract. The SINFONI integral field spectrometer for the VLT will provide near-infrared spatially resolved spectra at spatial resolutions close to the diffraction limit of the telescope (0.″05 at 2 μm). 1024 spectra can be simultaneously obtained, covering a 32 × 32 pixel field of view with ~100% filling factor. The spectral resolution is R ~ 4500, corresponding to a kinematic resolution of 67 km s^{-1}. SINFONI is ideally suited to study stellar kinematics in the nuclear regions of normal spiral galaxies, using the near-infrared H and K band CO stellar absorption features. Integral field data from SINFONI will provide high-resolution two-dimensional maps of nuclear velocity dispersion and rotation, which in turn will constrain the anisotropy parameter and yield robust estimates of the central dark mass.

1 The SINFONI Integral Field Spectrometer

Several integral field instruments are planned for the 8 meter class telescopes which have recently been or will soon be put into operation. However, even a smaller fraction of these integral field spectrographs plan to operate at near-IR wavelengths, and of those, only a fraction will be optimized for use with an adaptive optics system. SINFONI will be unique in combining the good intrinsic seeing of the VLT site, the excellent correction provided by adaptive optics in the near infrared, and the light gathering capability of the VLT to provide near-infrared high-resolution spectroscopy at better than 0.″1 resolution.

SINFONI (Thatte et al. 1998) is a scientific collaboration between the Max-Planck-Institut für Extraterrestrische Physik (MPE) and the European Southern Observatory (ESO) covering the field of high spatial resolution studies of compact objects (e.g. star forming regions, centers of galaxies, cosmologically distant galaxies). SINFONI is a combination of two instruments, a near-infrared integral field spectrometer, SPIFFI II (Tecza et al. 1998), and a curvature sensor based adaptive optics system MACAO (Bonaccini et al. 1998).

1.1   SPIFFI II

The near-infrared integral field spectrometer, SPIFFI II, which is the responsibility of the MPE group, obtains 1024 simultaneous spectra covering a 32 × 32 pixel contiguous square field of view, with a filling factor close to 100%. The pixel size may be chosen to be $0\rlap{.}''25$ per pixel for seeing limited observations, or $0\rlap{.}''025$ per pixel for diffraction limited observations, or $0\rlap{.}''1$ for intermediate observing conditions. The corresponding fields of view are $8''$ × $8''$, $1''$ × $1''$, and $3\rlap{.}''2$ × $3\rlap{.}''2$ respectively. Four different gratings are available, providing spectral resolutions of ∼4500 covering the J, H or K band, or both H and K bands simultaneously at a spectral resolution of 2000. The high spectral resolution settings would allow an *OH avoidance* technique to be implemented. Since 98% of the night sky emission flux in the wavelength range from 1.0 to 2.2 μm is contained in OH lines, a substantial reduction in sky background flux may be achieved by simply blanking out those pixels in the spectrum which are contaminated by OH line emission.

In addition, SPIFFI II incorporates a *Himmels-spinne*, which allows light from one of four pre-selected offset positions to be redirected to the image slicer, and then onto the detector. This precludes the need to nod the telescope every couple of minutes in order to measure the variable sky background accurately. Combined with the OH avoidance capability, this would allow integrations of roughly 1 hour duration. A pupil imaging system is also included, so as to ensure accurate alignment of the telescope and adaptive optics system pupils with that of the spectrometer.

1.2   MACAO

The 60 element curvature sensor based adaptive optics system MACAO (Bonaccini et al. 1998) is the responsibility of ESO. MACAO uses photon counting Avalanche Photodiodes (APDs) for wavefront sensing, providing extremely high sensitivity. The system is specifically optimized to provide low-order correction using very faint guide stars. Its performance is well adapted to spectroscopic applications, where energy concentration is more important than achieving diffraction limited image quality. MACAO can use a guide star within a field of $2'$ diameter centered on the scientific target. Its novel design implies only four warm reflections (apart from the primary and secondary) before the light enters the cryogenic spectrometer. The dewar window itself serves as the dichroic, reflecting the visible light to the wavefront sensor. The extra thermal background added by the relay optics is thus kept to an absolute minimum, dramatically improving the sensitivity of SINFONI in the K band, where thermal emission from warm telescope optics and surroundings can dominate the background flux seen by the spectrometer. MACAO can also be used with a laser guide star, which is foreseen to be a standard facility for the VLT, formerly UT3, telescope *Melipal*.

Stellar kinematics with SINFONI

A galactic nuclei at 10 Mpc.   Image quality FWHM = 0."2

Average surface brightness (r < 1"): K = 14.5 mag.arcsec⁻²

T = 3600s, R = 4500

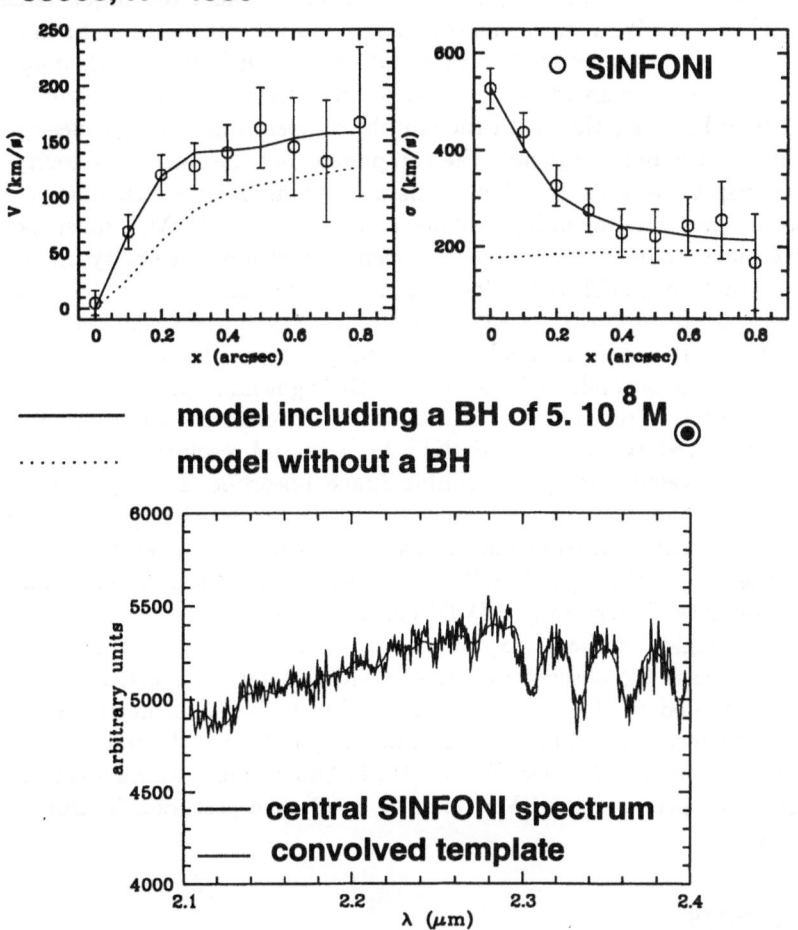

Fig. 1. Simulated observations of the nucleus of a normal galaxy harbouring a central black hole made with SINFONI. The top right panel shows the tell tale signature of a central dark mass concentration – a rapidly increasing velocity dispersion toward the center. High spatial resolution ($\sim0.''1$) is essential to probe within the radius of influence of the black hole

2   Stellar Kinematics with SINFONI

Figure 1 shows simulated SINFONI observations of the nuclear region of a normal galaxy at a distance of 10 Mpc. For each spatial pixel within the field of view, the line of sight velocity and the velocity dispersion can be determined by comparing the observed spectrum with the spectrum of a template star using a Fourier cross correlation quotient technique (Bender 1989). Alternatively, a least squares fitting routine can be used to find the optimal gaussian broadening function and velocity shift, which, when applied to the template star spectrum best reproduces the data. Such a broadened, shifted template, together with the simulated spectrum of the nuclear pixel, are shown in the bottom panel. The top panels plot the rotation velocity and velocity dispersion as a function of distance from the nucleus. Also plotted are two models, one including a nuclear dark mass of 10^8 $M_\odot$, most likely a massive black hole. As can be clearly seen, the observed velocity dispersion is only fit with a model with a large central dark mass.

High spatial resolution spectroscopy is the key to detecting central dark mass concentrations in the nuclei of normal galaxies. The radius of influence of the black hole is small, as illustrated in the right hand plot in Figure 1. An observation with a resolution of 1″ would fail to detect the central dark mass in this simulated example. SINFONI's high spatial resolution spectroscopic capability is rivaled only by the Hubble Space Telescope. The high sensitivity of the MACAO adaptive optics system will allow the use of bright star forming regions in the spiral arms of normal galaxies as wavefront reference objects. Another strong advantage of SINFONI is that the rotation velocity and line-of-sight velocity dispersion (LOSVD) can be measured for every point within the nuclear region of the galaxy, not only along one or two slit positions, as is often the case for observations made with a long-slit spectrometer. This allows a detailed model of the nuclear region to be constructed, including anisotropy effects. The latter can dramatically influence the inferred central dark mass, as seen in the case of NGC 3115 (Anders 1999) for data taken with the MPE 3D spectrometer (Weitzel et al. 1996), the precursor instrument to SINFONI.

References

1. Anders, S. 1999, Ph.D. thesis, LMU München
2. Bender, R. 1989, A&A, **229**, 441,
3. Bonaccini, D., Rigaut, F., Dudziak, G., and Monnet, G. 1998, Proc. SPIE, **3353**, 553
4. Tecza, M., Thatte, N., Krabbe, A., and Tacconi-Garman, L. E. 1998, Proc. SPIE, **3354**, 394
5. Thatte, N. et al. 1998, Proc. SPIE, **3353**, 704
6. Weitzel, L., Krabbe, A., Kroker, H., Thatte, N., Tacconi-Garman, L. E., Cameron, M., and Genzel, R. 1996, A&ASupp, **116**, 531

Part 3

BLACK-HOLE PHENOMENOLOGY: VARIABILITY, JETS, DISKS AND ACCRETION TORI

Accretion onto Black Holes and Neutron Stars: Differences and Similarities

Rashid Sunyaev

[1] Max-Planck-Institut für Astrophysik, Garching, Germany
[2] Space Research Institute, Moscow, Russia

Abstract. Accreting black holes and neutron stars at luminosities above 0.01 of the critical Eddington luminosity have a lot of similarities, but also drastic differences in their radiation and power density spectra. The efficiency of energy release due to accretion onto a rotating neutron star usually is higher than in the case of a black hole. The theory of the spreading layer on the surface of an accreting neutron star is discussed. It predicts the appearance of two bright belts equidistant from the equator. This layer is unstable and its radiation flux must vary with high frequencies.

1 Introduction

One of the most important properties of accreting black holes in our Galaxy was discovered by Riccardo Giacconi and the *Uhuru* Team in 1971, when they discovered the spectral transition of Cyg X-1 from the soft to the hard state (Tananbaum et al. 1972). Simultaneously, a radio source appeared in the vicinity of Cyg X-1. Radio observations permitted its localization with high accuracy and the identification of the X-ray source with a bright star of the 9th magnitude. Immediately thereafter, measurements of its optical spectrum showed that this star is member of a 5.6-day non-eclipsing binary with an optically invisible companion (Bolton 1972). Lyuty et al. (1973) interpreted the observed ellipsoidal variations in the brightness of the optical star as a result of the gravitational influence of a nearby black hole invisible in optical light. Today Cyg X-1 is the best-known steadily accreting black hole in our Galaxy. Now we have a list with more than 12 excellent black-hole candidates and many of them show similar soft- to hard state transitions (Tanaka & Shibazaki 1996). Recently, Cyg X-1 experienced the third transition from a hard to a soft state in 18 years (Fig. 1).

Such transitions became a signature of black holes. Today we know that all galactic black-hole candidates show a very soft X-ray spectrum. As predicted by standard accretion theory, this is a multicolor disk spectrum (cf. Shakura & Sunyaev 1973) or a power-law hard X-ray spectrum with a Wien-type decay at high energies formed due to comptonization (Sunyaev & Trümper 1979, Sunyaev & Titarchuk 1980). Sometimes we do not even see the high frequency decay yet. Therefore, usually when a newly discovered X-ray transient shows an extremely hot tail in its X-ray spectrum, we immediately refer to it as a black-hole candidate.

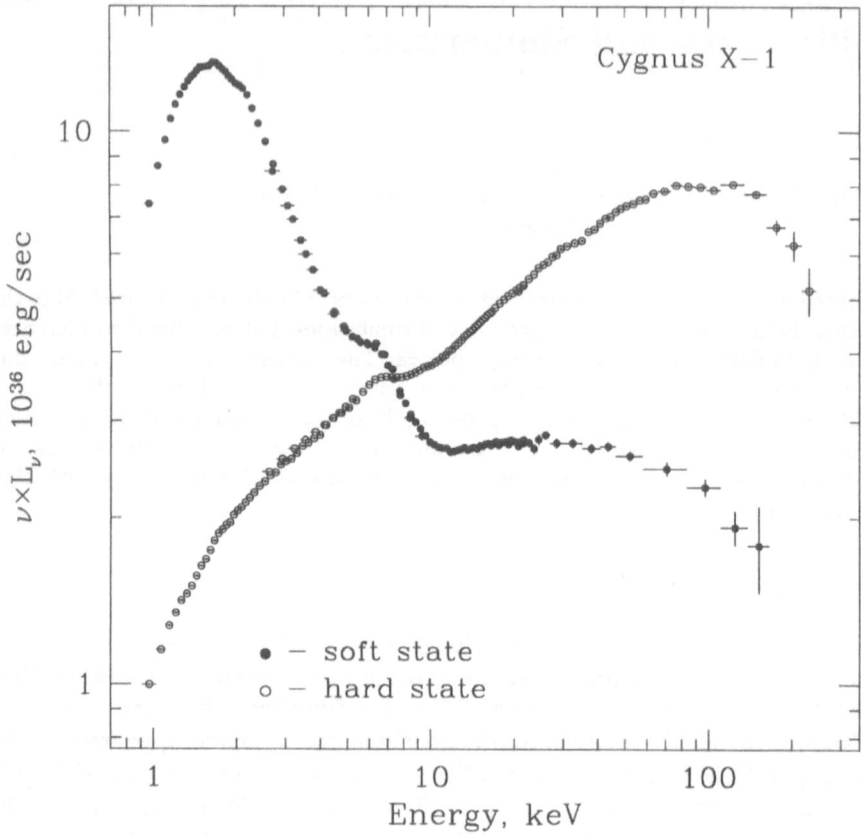

Fig. 1. The spectral energy distribution of Cyg X-1 in the soft (filled circles) and hard (open circles) spectral state. Data are shown of nearly simultaneous ASCA and RXTE observations on March 26, 1996 (hard state) and May 30, 1996 (soft state); cf. Gilfanov, Churazov & Revnivtsev (2000).

Neutron stars without magnetic fields and black holes have practically the same gravitational potential and must show many similarities. Nevertheless, we know now that they have very different X-ray spectra and variability characteristics. One of the great surprises of the last 15 years of observations is the discovery that neutron stars also exhibit soft- to hard-state transitions (Fig. 2).

Neutron stars with small magnetic fields usually have spectra which are significantly harder than the spectra of multicolor accretion disks around black-hole candidates in a high/soft state. But their spectra are usually much softer than the spectra of black-hole candidates in the hard/low state. Sometimes we observe hot tails in the persistent flux of X-ray bursters. However,

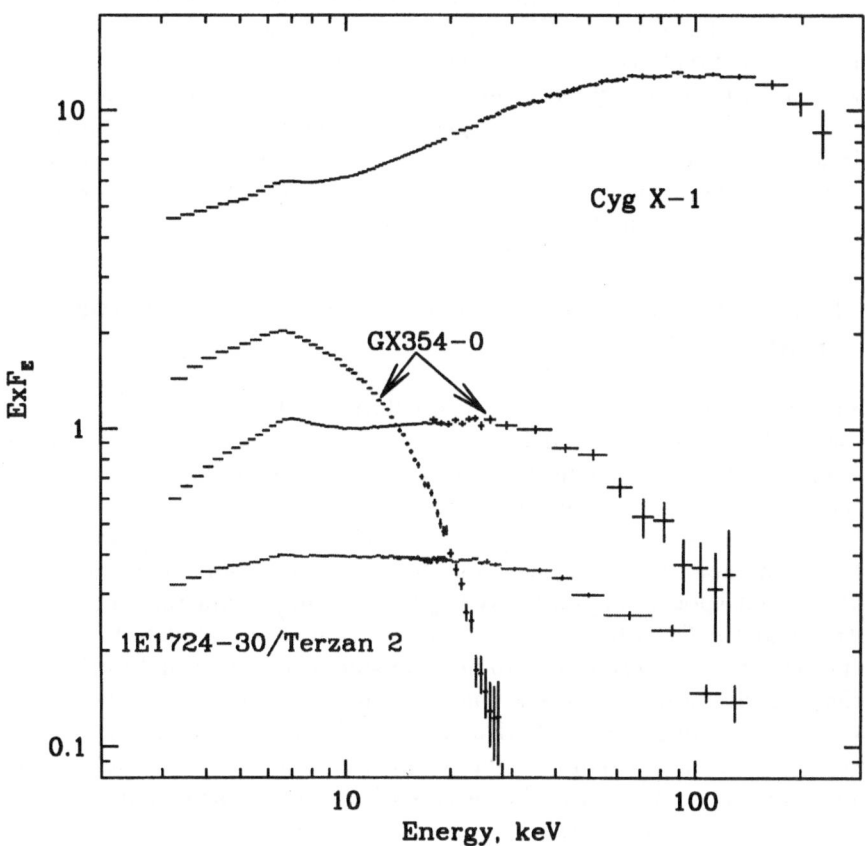

Fig. 2. The energy spectra of three X-ray binaries. The neutron star in GX354-0 (4U1728-34) is shown in two spectral states – low/hard and high/soft. Even in the hard state the neutron-star spectra are much softer than the spectrum of Cyg X-1 (accreting black hole). Adapted from Revnivtsev & Sunyaev (2000).

spectra of these hot tails from neutron stars are much steeper than in the case of black holes and contain a smaller fraction of the source luminosity. It seems that now we know the reason. In the case of black-hole accretion we only see the radiation of accretion disk – plus, maybe, the corona above it (Galeev et al. 1979) or the advection flow with even smaller accretion efficiency (Narayan & Yi 1995). In the case of neutron stars we have an object with a solid surface. Therefore, part of the gravitational energy of the accreting matter must be released in an extended accretion disk, and another part in the narrow boundary layer in the vicinity of the neutron star where accreting matter is decelerating from the Keplerian velocity (of the order of half the velocity of light) to the velocity of rotation at the equator of the neutron star. The surface of the star is able to produce enough soft protons for comptonization to cool down the hot parts of the disk and boundary layer to temperatures below 20 keV (Sunyaev & Titarchuk 1989). The physics of the boundary layer permits us to explain the strong differences between the radiation spectra of accreting black holes and neutron stars. It also predicts a strong difference in the characteristic variability timescales of the X-ray flux from black holes and neutron stars (see below).

2   Efficiency of Accretion onto a Rapidly Rotating Neutron Star

The recent discovery of quasi-periodic oscillations (QPO) with frequencies of the order of 500-600 Hz during the nuclear bursts on the surface of a neutron star appears to be very strong evidence of neutron-star rotation with the same frequency, or with periods of the order of 1.6-2 ms (Strohmayer et al. 1998). This interpretation is natural for a nuclear burning front propagating on the surface of a rapidly rotating neutron star. A bright front region manifests itself as a hot spot giving rise to the QPO. It is important that for a given neutron star the QPO frequency remains the same from burst to burst.

The efficiency of accretion onto neutron stars is higher (usually) than the efficiency of accretion onto black holes. The reason is obvious: in the case of a black hole we have an event horizon and an effective energy release and the release of the observed radiation flux might occur only in the accretion flow well beyond the event horizon. In the case of a neutron star without a strong magnetic field part of the energy is released in the extended accretion disk and another part is liberated in the narrow boundary layer near the surface of the neutron star. In Newtonian mechanics energy release in the boundary layer is equal to

$$L_{\bullet} = \frac{1}{2}\frac{GM\dot{M}}{R_*}(1 - \frac{f}{f_k})^2 ,$$

or is equal to the energy liberated in the disk

$$L_d = \frac{1}{2}\frac{GM\dot{M}}{R_*}$$

in the case of a slowly rotating compact star. Here and below M is the gravitational mass of the star, R_* is its radius, $f_* = \frac{1}{2\pi}\sqrt{\frac{GM}{R_*^3}}$ the cyclic keplerian frequency near the its surface, f is the frequency of stellar rotation and $\dot{M}$ is the accretion rate.

The problem becomes much more complicated in the case of General Relativity. Kerr metrics is not applicable to the case of rapidly rotating neutron star because the mass distribution within the star is no longer spherically symmetric. There is a strong quadrupole component in the mass distribution. Fortunately, there is an exact solution of the GR equations for the case when the mass distribution has a quadrupole component. Using this solution, Sibgatullin & Sunyaev (2000) plotted the dependence of the energy release due to the accretion onto a neutron star as a function of the rotation frequency of that star (Fig. 3). The existing GR solution permits us to find the efficiency of the energy release only in the case when the spin directions of the neutron star and accretion disk are parallel or anti-parallel. Unfortunately, the problem with an arbitrary angle between the axes of rotation of the neutron star and the accretion disk is much more complicated. In Figure 3 the positive values of the rotational frequency f correspond to the case of corotation and negative values describe the case of counterrotation. The figure presents the computations for the equation of state (EOS) FPS in the classification of Lorenz et al. (1993) for a gravitational mass of the neutron star $M = 1.4\ M_\odot$.

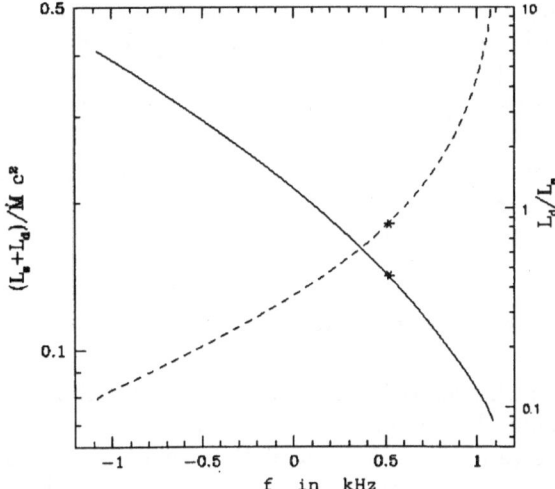

Fig. 3. Efficiency of the energy release in the disk (L_d) and on the stellar surface (L_s) as a function of the stellar rotation frequency f. Negative f values correspond to the case of counter rotation of disk and star. The solid line and the numbers on the left axis give the value $(L_s + L_d)/\dot{M}c^2$. The dashed line and the values on the right axis give the ratio L_d/L_s. A gap between the marginally stable orbit and the stellar surface exists in the region leftside of the two asterisks on the solid and dashed curves. There is no such gap in the case of rapid corotation of star and disk ($f > 550$ Hz). Adapted from Sibgatullin & Sunyaev (2000).

We see that the energy release efficiency drops rapidly with increasing frequency in the case of corotation and increases rapidly towards high frequencies of counter rotation. The ratio of the disk luminosity to the luminosity in the boundary layer or in the spreading layer near the surface of the star also strongly depends on the frequency of rotation. It is close to 1 for the case of corotation with $f = 600$ Hz and decreases up to 0.2 in the case of counter rotation with the same frequency.

The asterisks in Figure 3 give information on the existence of a gap between the marginally stable orbit in the accretion disk and the radius of the star. For frequencies of corotation higher than 550 Hz such a gap does not exist; then the disk is in contact with the surface of the neutron star. For lower frequencies of corotation and in the case of counter rotation for the EOS FPS and $M = 1.4$ M$_\odot$ there is a gap $R_m - R_* \approx$ $[1.44 - 3.06(f/\text{kHz}) + 0.843(f/\text{kHz})^2 + 0.6(f/\text{kHz})^3 - 0.22(f/\text{kHz})^4]$ km. In the most interesting case of corotation the gap is very narrow and the thickness of the boundary layer or the hight of the spreading layer usually exceeds the dimension of the gap. However, in the case of counter rotation (negative values of f) the gap could be sufficiently large that it has to be taken into account.

The energy release efficiency due to accretion onto a counter-rotating neutron star may reach very large values up to 0.67 $\dot{M}c^2$ for the case of a neutron star with baryonic mass $m = 2.1$ M$_\odot$ for $f = 1.5$ kHz and the EOS FPS. Obviously, such a high energy release efficiency is connected with the spin down of the rapidly (counter) rotating star. This efficiency is much higher than that of disk accretion onto a Kerr black hole. In the case of corotation the energy release efficiency, due to accretion onto a Kerr black hole, is higher than in the case of counterrotation. This is reversed in the case of accretion onto a neutron star .

3   Structure of the Boundary Layer

The problem of disk accretion onto a neutron star without a magnetic field is two-dimensional. The height of an accretion disk at low accretion rates and luminosities $(0.01 < L/L_{\text{Edd}} < 0.3)$ is small in comparison with the radius of the neutron star. Here and below $L_{\text{Edd}} = \frac{4\pi GM m_p}{\sigma_T}$ is the critical Eddington luminosity. The angular rotation frequency Ω in the disk is close to keplerian and increases when matter approaches the neutron star. In the boundary layer the matter velocity must decrease to the velocity of rotation at the neutron-star surface and then matter must be redistributed over its equipotential surface. This surface is defined by the common influence of gravity and centrifugal forces. It is obvious that there must be a ring where Ω reaches its maximum, $d\Omega/dR = 0$. There are two possible approaches to consider the matter flow beyond this point. We could assume that the boundary layer is described by the same equations as those valid for the accretion disk or we could consider the motion of matter in the spreading layer as belonging to the surface of the neutron star. We tried to investigate both of these approaches in one-dimensional approximations. In the paper by Popham & Sunyaev (2000) we computed the structure and properties of the boundary layer considering it as a part of the disk. Figure 4 shows how the height of the boundary layer depends on the distance from the stellar surface for different accretion rates. In the case of a low accretion rate or $L \sim 0.01\,L_{\mathrm{Edd}}$, the height of the disk in the "neck" between the accretion disk and the boundary layer is close to only 40 meters and the extension of the boundary layer about 1.5 km. The situation drastically changes when we go to the case of high accretion rates with a luminosity close to the critical Eddington luminosity. The height of the neck between the boundary layer and the accretion disk in this case exceeds 2 km and the boundary layer extends up to 2 neutron-star radii.

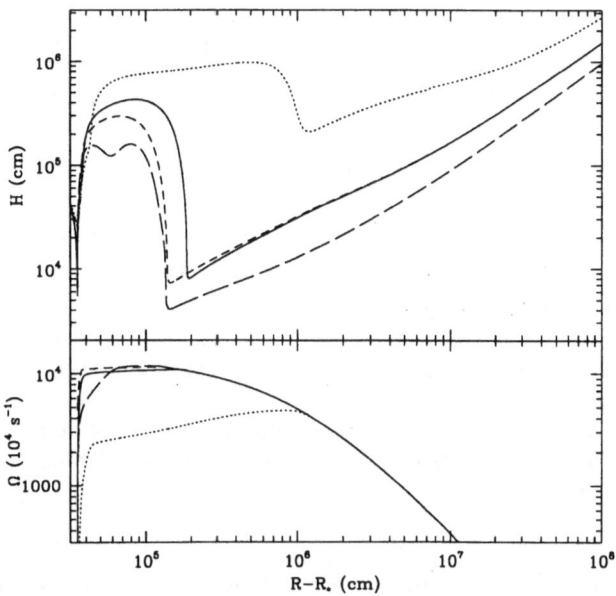

Fig. 4. The vertical pressure scale height H (top) and the angular velocity Ω (bottom), for solutions with $\dot{M} = 10^{-10}$ (long dash), 10^{-9} (solid) and $10^{-8}\,M_{\odot}yr^{-1}$ (dotted), all for a non-rotating neutron star, and for $\dot{M} = 10^{-9}\,M_{\odot}yr^{-1}$ and a neutron star rotation frequency $f_{*} = 636$ Hz (dashed), all with standard viscosity and $\alpha = 0.1$. Note the very small values of H at the "neck" between the disk and the boundary layer in the lower $\dot{M}$ solutions, and the rapid increase in H in the boundary layer. Adapted from Popham & Sunyaev (2000).

A more natural approach was considered by Inogamov & Sunyaev (1999). This approach uses the shallow water or hydraulic approximation. It assumes that the thickness of the spreading layer on the surface of the neutron star is less than the circumference of the neutron-star equator $H \ll 2\pi R_*$. This approach assumes that matter entering the equatorial ring with a very high rotational velocity of the order of $0.5c$, where c is the velocity of light. Then the matter begins to spiral slowly towards the poles losing its kinetic rotation energy due to turbulent friction with the dense underlying layer (Fig. 5 and Fig. 6).

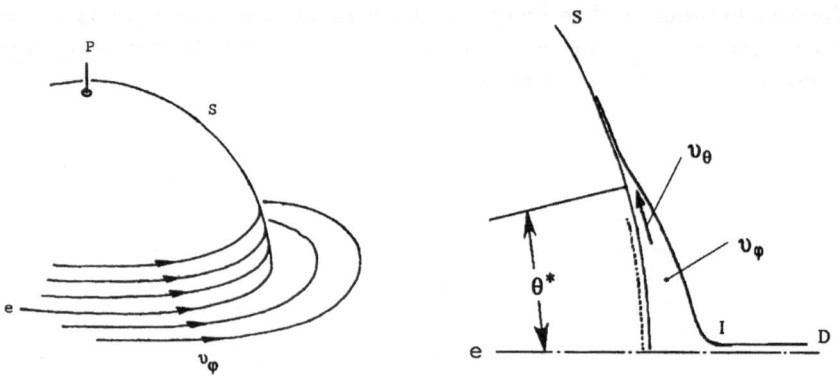

Fig. 5. Rotation of matter in the disk and on the stellar surface in the spreading layer model; S is the stellar surface, P is the pole, and e is the equator.

Fig. 6. Spreading of the rotating plasma from the disk, D, over the neutron star surface, S. Here, I is the intermediate zone near the disk neck, θ^* corresponds to the position of the hot belt, and $\theta > \theta^*$ is the cold part of the spreading layer. The rotation velocity v_ϕ (filled circle) is directed along the normal to the plane of the figure. The slowly circulating dense underlying layers of matter beneath the spreading layer are indicated by the dashes. Both figures are adapted from Inogamov & Sunyaev (1999).

The thickness of the spreading layer is highest in the vicinity of the equator and decreases towards the poles. This means that matter is moving down the hill under the influence of gravity, the centrifugal force and the light pressure force. The problem is extremely interesting. We are dealing with radiation dominated plasma when the radiation pressure strongly exceeds the matter pressure. The sound speed is close to $0.1 - 0.15c$. Radiative viscosity is also much stronger than the viscosity of plasma. The solution of the set of hydrodynamic equations results in the following picture (see Inogamov & Sunyaev 1999 for details). Two bright belts equidistant from the equator appear on the surface of the neutron star due to disk accretion. Figure 7 gives the distance of the bright belts from the equator for 4 luminosities of the neutron star: 0.01, 0.04, 0.20 and 0.80 L_{Edd}. The distance from the equator increases with luminosity.

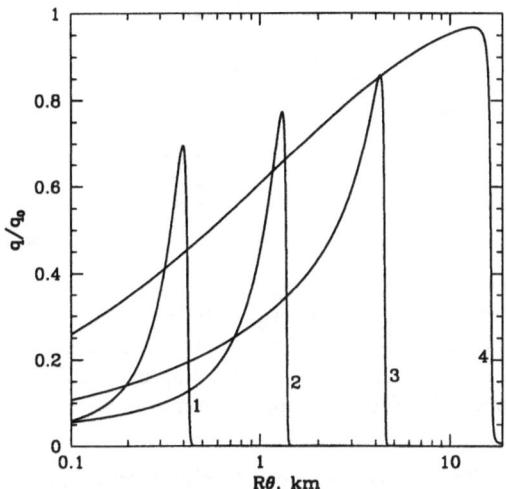

Fig. 7. Dependence of the surface brightness as a function of distance from the equator. Labels 1, 2, 3 and 4 refer to $L_{\mathrm{SL}}/L_{\mathrm{Edd}} = 0.01$, 0.04, 0.2, and 0.8, respectively; q_o is the critical Eddington flux. Adapted from Inogamov & and Sunyaev (1999).

The energy release in the vicinity of the equator is very low because there centrifugal forces compensate gravity with high precision. Therefore, any substantial radiation flux could destroy the structure of the thin spreading layer. Fortunately, advection takes the radiation energy density and transports it to the bright belts above and below the equator (Fig. 8).

In these bright belts the rotational velocity of the spreading matter becomes low enough to permit the existence of a large radiation flux comparable to the critical Eddington flux $q_0 = \frac{m_p c^3}{2\sigma_T R_g}(\frac{R_g}{R_*})^2 = 10^{22} \frac{\mathrm{W}}{\mathrm{m}^2}$, where R_g is the gravitational radius. This flux value is comparable to radiation fluxes achieved in the most intense petawatt laser facilities (Perry 1996, Budil et al. 2000).

We are dealing here with a critical Eddington flux even in the case of a low luminosity of the neutron star ($0.01 < L/L_{\mathrm{Edd}} < 1$). The surface of the bright belts is small and the high radiation flux from the narrow belts is consistent with the low luminosity of the star.

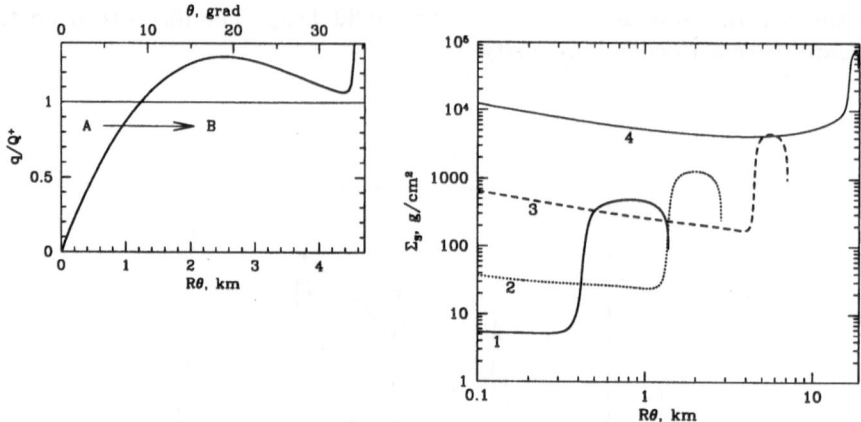

Fig. 8. The dynamics of the spreading layer is determined in many ways by the hydrodynamic transfer of radiative energy. The ratio of the energy flux, q, emitted per unit area of the radiating layer to the frictional energy release Q^+ per unit area of the contact surface between the spreading layer and the star. The energy is transferred from zone A into zone B through meridional advection. This calculation is for $L_{SL}/L_{Edd} = 0.2$.

Fig. 9. Column density of matter in the spreading layer as a function of the distance from the equator. A strong increase of Σ occurs when the flow cools down and begins to move very slowly. Bright belts correspond to the regions of minimal Σ. Labels 1, 2, 3 and 4 refer to $L_{SL}/L_{Edd} = 0.01$, 0.04, 0.2 and 0.8, respectively. Both figures are adapted from Inogamov & Sunyaev (1999).

The matter in the spreading layer is practically levitating. The difference between the gravitational force and the centrifugal- and radiation pressure force is close to $(1-3) \times 10^{-3}$ of gravity. At higher longitudes the rotational velocity of matter and the velocity of the flow along the meridian decreases and the flow becomes subsonic, cool, dense and very slow.

One of the most interesting predictions of the theory of the spreading layer is the strong dependence of the matter column density in the spreading layer on the accretion rate or the luminosity of the neutron star (see Fig. 9). In the case of a low luminosity the levitating layer in the bright belts is optically thin against Thompson scattering $\tau_T \sim 2$. Under these circumstances it is impossible to radiate the energy released due to accretion at low temperatures. Comptonization forms hard tails. In the case of a high luminosity the bright belt has a large column density (up to 10 kg/cm^2). Then free-free processes and comptonization form Bose-Einstein type spectra inside the spreading layer and the resulting spectrum is much softer than in the case of low luminosity.

4   Time Variability in the Accretion Disk and in the Boundary Layer

All instabilities existing in the accretion disk modulate the flow of matter onto the neutron-star surface. Therefore, we could expect that the majority of the types of variability we observe in accreting black holes must manifest themselves in accreting neutron stars with characteristic timescales proportional to the mass of the accreting object (see e.g. Shakura & Sunyaev 1976, Wijnands & van der Klis 1999). The spreading layer on the surface of the neutron star is the source of additional high-frequency instabilities (see the discussion in Sunyaev & Revnivtsev 2000). Their origin is obvious – the matter in the bright belts is radiation dominated, levitating, the height is smaller than in the region of the main energy release in the accretion disk, the sound velocity is huge and corresponding sound frequencies are very high.

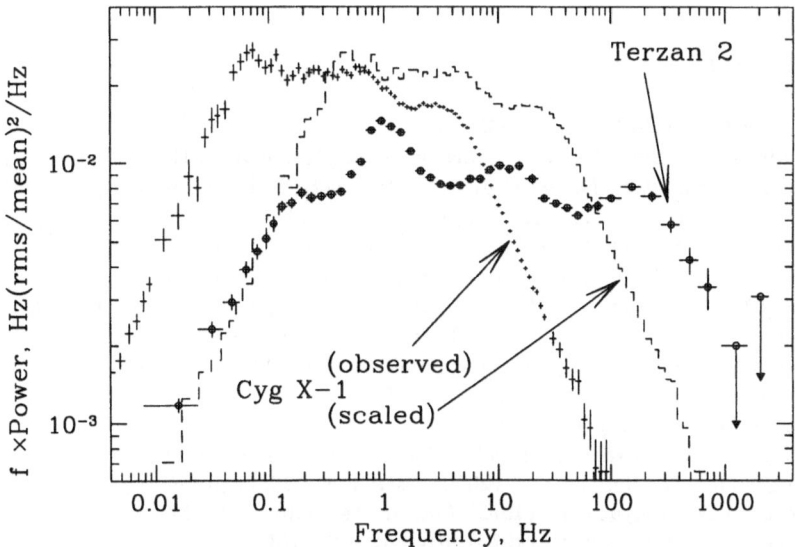

Fig. 10. Comparison of the power spectra of a black hole (Cyg X-1) and a neutron star (Terzan 2). The dashed line shows the power spectrum of Cyg X-1 scaled according to the mass ratio with Terzan 2 ($f_{\text{CygX}-1} \times 7 \to f^{\text{scaled}}_{\text{CygX}-1}$. This simple scaling is important but insufficient to explain fully the difference in the high frequency variability of Terzan'2 and Cyg X-1. Note that the slopes of the power density spectra of Terzan 2 and Cyg X-1 in the high and low frequency limits are similar, but that the power spectrum of Terzan 2 is significantly broader than that of Cyg X-1. Adapted from Sunyaev & Revnivtsev (2000).

Sunyaev & Revnivtsev (2000) compared the power density spectra of 9 black holes and 9 neutron stars observed by RXTE in their low/hard state.

There is a very strong difference. In the power density spectra of accreting neutron stars with a weak magnetic field significant power is contained at frequencies close to one kHz. At the same time, most Galactic accreting black holes demonstrate a strong decline in the power spectra at the frequencies higher than 10-50 Hz. In principle this might open an additional way to distinguish the accreting neutron stars from black holes in X-ray transients (we do not mention in this paper the well-known differences: X-ray bursts or X-ray pulsations). Fig. 10 compares the power density spectrum of Cyg X-1 with the power density spectrum of the X-ray burster in Terzan 2.

The simplest assumption is that the characteristic frequencies in the power spectra of the sources scale as M^{-1} (Shakura & Sunyaev 1976). This scaling law is valid for e.g. the keplerian frequency in the vicinity of the marginally stable orbit, the thermal and secular instabilities of the accretion disk in the region of main energy release, and the Balbus-Hawley instability. However, this assumption does not account for the observed difference in the high frequency variability between neutron stars and black holes.

References

1. Bolton C.T. (1972) Nature 235, 271
2. Budil, K. S., Gold, D. M., Estabrook, K. G., Remington, B. A., Kane, J., Bell, P. M., Pennington, D. M., Brown, C., Hatchett, S. P., Koch, J. A., Key, M. H. and Perry, M. D. (2000) ApJ Suppl 127, 261-265
3. Galeev, A., Rosner, R., Vaiana, G. (1979) ApJ, 229, 318
4. Gilfanov, M., Churazov, E., Revnivtsev, M. (2000) MNRAS 316, 923
5. Inogamov, N., Sunyaev, R., (1999) Astr. Letts 25, 269, astro-ph 9904333
6. Lorenz, C. P., Ravenhall, D. G., Pethick, C. J. (1993) Phys. Rev. Lett. 70, 379
7. Lyuty, V., Sunyaev, R., Cherepashchuk, A. (1973) Soviet Astronomy 50, 3
8. Narayan, R., Yi, I. (1995) ApJ 452, 710
9. Perry, M. D. (1996) Sci. Tech. Rev. 12, 4
10. Popham, R., Sunyaev, R. (2000) ApJ (in press), astro-ph 0004017
11. Sibgatullin, N., Sunyaev, R. (2000) Astr. Letts 26, in press
12. Shakura N., Sunyaev, R. (1973) Astron. Astrophys. 24, 337
13. Shakura N., Sunyaev, R. (1976) MNRAS 175, 613
14. Strohmayer, T., Zhank, J. H., White, N. E., Lapidus, I. (1998) Astrophys.J. 498, 1358
15. Sunyaev, R., Revnivtsev, M. (2000) Astronomy and Astrophysics 358, 617
16. Sunyaev, R., Titarchuk, L. (1980) Astron. Astrophys 86, 121
17. Sunyaev, R., Titarchuk, L. (1989) Proceedings of the 23rd ESLAB Symposium 1, 627
18. Sunyaev, R, Trümper, J. (1979) Nature 279, 506
19. Tanaka, Y., Shibazaki, N. (1996) ARA&A 34, 607
20. Tananbaum, H., Gursky, H., Giacconi, R., Jones, C. (1972) ApJ 177, L5
21. Wijands, R., van der Klis, M. (1999) ApJ 514, 939

X-Ray Variability of Neutron Stars versus Black Holes

Tomaso Belloni

Osservatorio Astronomico di Brera, Via E. Bianchi 46,
I-23807 Merate, Italy

Abstract. In the recent years, thanks to the RXTE satellite, our knowledge of the properties of aperiodic time variability of compact binaries increased enormously. Here I briefly summarize the current status both for neutron star and black hole systems and concentrate on the similarities between these two classes. I show that there is increasing evidence that the main processes that produce the variability are the same for black holes and neutron stars.

1 Introduction

The analysis of aperiodic variability in the X-ray emission from X-ray binaries is a powerful model-independent tool to probe the accretion disk in its innermost regions, close to the compact object. Excluding the case of X-ray pulsars, where the magnetic field dominates the flow, we would expect little difference in the X-ray properties of neutron star and black hole binaries. In both types of systems, the accretion disk extends up to very close to the compact object; moreover, the radius of a neutron star and the innermost stable orbit of a 10 solar mass black hole are comparable in size. Until the launch of the *Rossi X-ray Timing Explorer* (RXTE) at the end of 1995, all timing features observed were much slower than the millisecond regime characteristic of the expected dynamical time scales close to the compact object, where they would be expected. With RXTE, we were able for the first time to observe millisecond phenomena. In the following, I will summarize the current observational picture for both types of systems (which for reasons of space will not be complete) and discuss the similarities between their properties.

2 Neutron-Star Binaries

Since the discovery of 20-50 Hz QPOs in GX 5-1 (van der Klis et al. 1985), a number of additional QPOs and broad noise components have been observed (for a complete picture see van der Klis 1995, 1998, 1999). Traditionally, the sources are divided into two classes: the *Z sources*, which have a high X-ray luminosity (and possibly a stronger magnetic field) and the *atoll sources*, with a lower luminosity. Sources in both classes show broad-band noise components and QPOs. A QPO common to both classes is what is called HBO (in the case of the Z sources), with a central frequency in the range 15-60 Hz, positively correlated with the accretion rate.

With RXTE the so-called *kilohertz QPOs* have been discovered. Their main properties are:

- Their frequency range is 200-1200 Hz
- They often appear in pairs
- Their frequency increases with accretion rate
- The difference frequency remains *almost* constant

During thermonuclear bursts from a number of sources, an additional oscillation has often been observed, whose nature is compatible with being coherent. When these oscillations (between 330 and 590 Hz) are seen, their frequency is close to the difference between the frequency of the two kHz QPOs.

3   Black-Hole Candidates

Unlike the case of neutron star systems, the timing properties of black-hole candidates have so far escaped a tight classification. In the recent years, thanks to the observation of a number of black-hole transients, it has been possible to isolate four characteristic (or "canonical") states of BHCs as a function of accretion rate. As a function of decreasing accretion rate, a "canonical" BHC is expected to go through: a Very High State (VHS), a High State (HS), an Intermediate State (IS) and a Low State (LS) (see van der Klis 1995; Méndez & van der Klis 1997 for a more detailed description of the four states). Unfortunately, not all sources are strictly "canonical", in the sense that some objects do not change state despite obvious large changes in luminosity (and therefore presumably in accretion rate, see Tanaka & Lewin 1995), and some others show mixed or anomalous features that make it difficult to clearly classify them (see Oosterbroek et al. 1998, Tanaka & Lewin 1995, Morgan et al. 1997).

All the states besides the HS are characterized by a band-limited noise component. 1-10 Hz QPOs are observed in the IS and the VHS, usually with a positive correlation with accretion rate. In the LS, a broad bump is often observed on the declining part of the noise, and sometimes a low-frequency (below 1 Hz) QPO is observed. For a more detailed description and references see van der Klis (1995). In three sources, RXTE detected a high-frequency QPO: GRS 1915+105 (67 Hz, Morgan et al. 1997), GRO J1655-40 (300 Hz, Remillard et al 1999a) and XTE J1550-564 (102-280 Hz, Remillard et al. 1999b, Homan et al. 2000).

4   Theoretical Models

A number of models have been proposed for the different QPOs and noise components in these systems. These models attribute the oscillations to Keplerian orbital frequencies in the disk, to the beat of these frequencies with the stellar spin, to radiation-hydrodynamic or oscillatory disk and stellar modes, to general relativistic effects and so on (see Psaltis, Belloni & van der Klis 1999 for a list of references). Some of these models involve the presence of a magnetic field and are therefore limited to neutron star systems, while others can in principle be applied to both classes. So far none of the models has been able to explain all the observed features.

5   Comparing the Systems

Since years it was noticed that there are similarities between the aperiodic variability originating from matter around neutron stars and black holes. In particular, comparing the power spectral distributions of black hole candidates in their Low State, i.e. at low accretion rates, and atoll sources in their *island* state, also at low accretion rates, the similarities look indeed strong (van der Klis 1994a,b). The power spectra of the black hole candidate GRO J0422+32 and the burster 1E 1724-3045 look so similar even in the smallest details that it would be difficult to distinguish them if it wasn't for the label in the figure (Olive et al. 1998). Wijnands & van der Klis (1999) compared the common features from a number of sources from both classes, i.e. the characteristic frequency of the broad-band noise and the QPO frequency. They found that these two quantities in atoll sources and black hole candidates correlate remarkably well over roughly three orders of magnitude.

The Wijnands & van der Klis correlation is based on the presence of similar features both in black hole and neutron star systems, namely a characteristic noise frequency and a low-frequency QPO. Besides the three QPOs mentioned in the previous section, which however are all below 300 Hz, there are no high-frequency QPOs known for black hole candidates, so that further comparisons with neutron star systems seem to be excluded. However, if the power spectra of different systems are shifted so that the low-frequency QPOs are aligned (see Fig. 1, from Psaltis, Belloni & van der Klis 1999), it appears that the lower of the two kHz QPO frequencies aligns with the bump often observed in black hole candidates and bursters. This is also true for the peculiar NS system Cir X-1 (Shirey et al. 1996, 1998). On this basis, Psaltis, Belloni & van der Klis (1999) produced the plot shown in Fig. 2. Different things are plotted here. For Z and atoll sources, the X axis is the lower of the two kHz peaks, the Y axis the HBO QPO. For Cir X-1, low-luminosity bursters and black hole candidates, the X axis is the central frequency of the "bump" in their power spectra, and the Y axis their low frequency QPO. The upper kHz peak (when observed) and the NBO QPOs (van der Klis 1995) are included for completeness. A remarkable correlation over three orders of magnitude is evident, with a possible second, steeper, branch to its right. A detailed description of the points included in the correlation and the appropriate references would not fit this space and can be found in Psaltis, Belloni & van der Klis (1999).

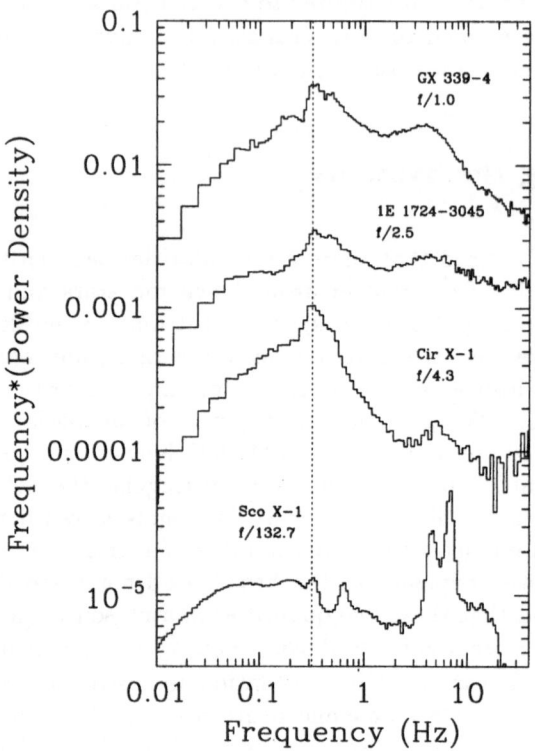

Fig. 1. Example of power spectra (in νP_ν representation) for the main classes of sources discussed in the text (from Psaltis, Belloni & van der Klis 1999). The spectra are shifted in frequency to coalign the low-frequency QPO.

The correlation suggests that the bumps observed in the power spectrum of bursters and BHCs are the low-frequency version of the kHz QPOs observed in Z and atoll sources. This suggestion is strengthened by one source, the BHC XTE J1550-564, which in the early phases of its outburst is observed to show a broad bump (lower open circles, Cui et al. 1999), but later a 100-280 Hz QPO peak (Remillard et al. 1999b, Homan et al. 2000). Therefore, this correlation indicates that, contrary to what is thought, kHz QPOs have already been observed in black hole candidates. They have not been recognized as such only because they are not coherent enough to be called QPOs and are observed at frequencies much lower than the kHz regime. Despite the fact that they do not fulfill the two defining characteristics of kHz QPOs, these QPOs are simply an extreme version of them.

6   Theoretical Models II

As I have shown above, there is increasing evidence that at least some of
the basic physical processes that produce the aperiodic variability in neutron
star and black hole systems are the same. If this is the case, it is difficult to
see how theoretical models based on the presence of a magnetic field, like the
beat-frequency model (see e.g. Miller, Lamb & Psaltis 1998), can account for
it. Interestingly, there is at least one model which can reproduce the observed
correlation without any additional assumption. In the general-relativistic pre-
cession/apsidal motion model discussed by Stella & Vietri (1998,1999), the
kHz QPOs are due to the orbital motion around the compact object and
to the general relativistic apsidal motion of the slightly eccentric orbit. The
low frequency QPO is the Lense-Thirring precession of the same orbit. This
model does not assume the presence of a magnetic field and can therefore
be applied both to the neutron star and the black hole case. Stella, Vietri
& Morsink (1999) have shown that the model naturally predicts the QPO
correlation shown in Fig. 2.

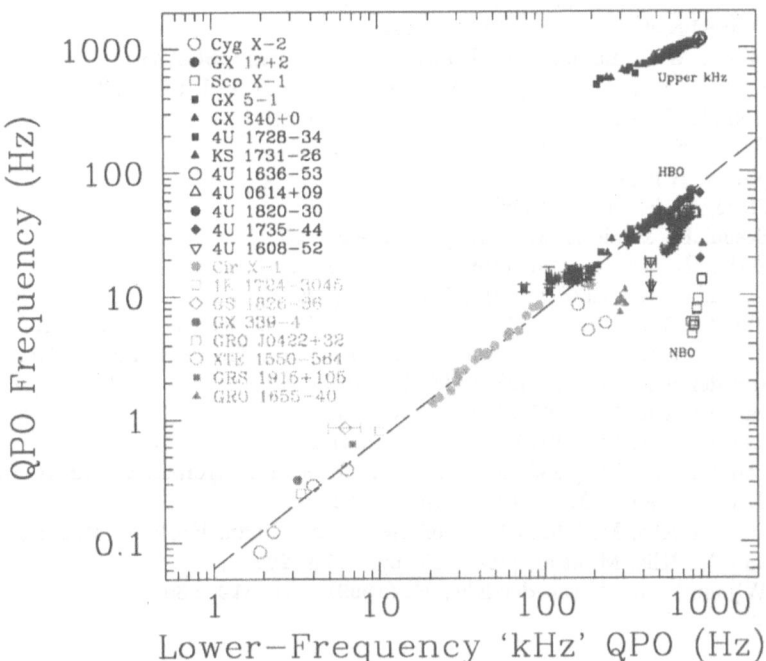

Fig. 2. The QPO correlation from Psaltis, Belloni & van der Klis (1999).

7   Conclusions

By putting together the results obtained with RXTE, two important facts are becoming clear. The first is that the phenomenology of QPOs and noise in neutron star systems, however complex, is much better defined that in the case of black hole candidates, where only a much rougher classification is possible. The second is that comparing the two classes of sources, strong similarities appear, which cannot be ignored. As the results by Wijnands & van der Klis (1999) and Psaltis, Belloni & van der Klis (1999) have shown, the phenomenological differences between the two classes might be only secondary and the basic processes that are behind the production of X-ray variability in these systems could be the same.

References

1. Cui, W., et al. (1999) ApJ, **512**, L43
2. Homan, J. et al. (2000) in preparation
3. Méndez, M., & van der Klis, M. (1997) A&A, **479**, 926
4. Morgan, E.H. et al. (1997) ApJ, **482**, 993
5. Olive, J.F., et al. (1998) A&A, **333**, 942
6. Oosterbroek, T., et al. (1998) A&A, **340**, 431
7. Miller, M.C., Lamb, F.K., Psaltis, D. (1998) ApJ, **508**, 791
8. Psaltis, D., Belloni, T., van der Klis, M. (1999) ApJ, **520**, 262
9. Remillard, R.A. et al. (1999a) ApJ, **517**, L127
10. Remillard, R.A. et al. (1999b) ApJ, **522**, 397
11. Shirey, R.E., et al. (1996) ApJ, **469**, L21
12. Shirey, R.E., et al. (1998) ApJ, **506**, 374
13. Stella, L., & Vietri, M. (1998), ApJ, **492**, L59
14. Stella, L., & Vietri, M. (1999), Phys. Rev. Letters, **82**, 17
15. Stella, L., & Vietri, M., Morsink, S.M. (1999), ApJ, **524**, L63
16. Tanaka, Y., & Lewin, W.H.G. (1995) in "X-ray binaries", ed. W.H.G. Lewin, J. van Paradijs, E.P.J. van den Heuvel, Cambridge Univ. Press, 126
17. van der Klis, M., (1994a) ApJS, **92**, 511
18. van der Klis, M., (1994b) A&A, **283**, 469
19. van der Klis, M. (1995) in "X-ray binaries", (op. cit.), 252
20. van der Klis, M. (1998) in "The many faces of neutron stars", ed. R. Buccheri, J. van Paradijs, M.A. Alpar, Kluwer, 337
21. van der Klis, M. (1999) Proc. of the Third William Fairbank Meeting, in press
22. van der Klis, M. et al. (1985) Nature, **316**, 225
23. Wijnands, R., & van der Klis, M. (1999) ApJ, **514**, 939

RXTE Monitoring of LMC X-3

Jörn Wilms[1], Michael A. Nowak[2], Katja Pottschmidt[1], William A. Heindl[3], James B. Dove[4], Mitchell C. Begelman[2,5], and Rüdiger Staubert[1]

[1] Institut für Astronomie und Astrophysik, Waldhäuser Str. 64,
    D-72076 Tübingen, Germany
[2] JILA, University of Colorado, Boulder, CO 80309-440, USA
[3] CASS, University of California San Diego, La Jolla, CA 92093, USA
[4] Dept. of Physics, Metropolitan State College of Denver,
    Denver, CO 80217-3362, USA
[5] APS, University of Colorado, Boulder 80309, U.S.A.

Abstract. We present results from an RXTE monitoring campaign of the black hole candidate LMC X-3 since 1996 December. The spectral properties of this canonical soft state source show considerable variability on timescales of ~ 100 d. We report on these changes, including on two or possibly three instances where we found that the soft spectral component vanishes and where the X-ray spectrum of LMC X-3 resembles that of a hard state source such as Cyg X-1.

1 The RXTE Monitoring Campaign

Although the two persistent black hole candidates in the LMC, LMC X-1 and LMC X-3, have rather similar companion stars, X-ray spectra, and X-ray luminosities, their long term behavior is quite different: LMC X-1 does not exhibit any long term variability, while LMC X-3 was known to be variable on a ~ 100 d timescale (Cowley et al., 1991, 1994). Prompted by this difference and in the light of theoretical modeling of the long term variability of X-ray binaries, we initiated an X-ray monitoring campaign of these sources with RXTE in 1996. From 1997 throughout 1998 (for LMC X-1) resp. 1999 (LMC X-3), RXTE observed LMC X-1 and LMC X-3 for ~ 10 ksec every three to four weeks. This interval was chosen to provide enough data points to sample the ~ 100 d period of LMC X-3. Detailed results of the campaign will be presented elsewhere (Wilms et al., 1999b, Nowak et al., 1999).

The data analysis was performed using the standard RXTE ftools, using the methods described by Wilms et al. (1999a). Due to the low source intensity we increased the signal to noise level by working with data from the top layers of the detectors only. The spectral data accumulated over each observation was then modeled using the standard spectrum for a soft-state black hole candidate: the sum of a multi-temperature disk black body (Mitsuda et al., 1984, Makishima et al., 1986) plus a power-law component. Spectral parameters were the inner temperature of the accretion disk, kT_{in}, the disk normalization, A_{disk}, the normalization of the power-law component, and the power-law index, Γ.

2 Results

Our observations in 1997 and 1998 indicate that there was a very significant variation of the spectral parameters of LMC X-3 that is mainly correlated with the total soft X-ray luminosity as measured by the All Sky Monitor on RXTE. During episodes of high ASM flux, i.e., at times where $F_{\mathrm{ASM}} > 1\,\mathrm{cps}$, the X-ray spectrum is a classical soft state spectrum. Similar to Ebisawa et al. (1993) we find that A_{disk} is only slightly variable around its average value of $A_{\mathrm{disk}} \sim 27$, while the disk temperature varies significantly. Our results indicate that the 2–10 keV X-ray luminosity variation during the soft state is mainly caused by these variations of kT_{in} (over our observations, kT_{in} varied from 0.8 keV to as high as 1.25 keV). During the soft state, the power-law is weak and soft, with $\Gamma \sim 3$ or even softer.

On the other hand, episodes where $F_{\mathrm{ASM}} < 1\,\mathrm{cps}$ are characterized by a strong decrease of kT_{in} while the relative flux in the power-law component increases and the photon index hardens. During at least two episodes in 1997 and 1998 we are unable to detect any curvature in the X-ray spectrum of LMC X-3. The X-ray spectrum during these times can be described as a pure power-law with $\Gamma \sim 0.7$ – in other words, as a classical hard state spectrum. These state switches occur on the lows of the $\sim 150\,\mathrm{d}$ quasi-periodic oscillations seen in the long term RXTE ASM lightcurve of LMC X-3.

The data of the monitoring observations of LMC X-1 were treated in the same way as those for LMC X-3. In contrast to LMC X-3, LMC X-1 stays in the soft state throughout our monitoring campaign, exhibiting the same qualitative behavior as that outlined for the soft state of LMC X-3 above. In passing we note that we did not detect any low-frequency quasi-periodic oscillations for LMC X-1 in the $\sim$30 pointings analyzed so far.

Since the X-ray luminosity of LMC X-1 and LMC X-3 is comparable, our results might indicate that the variation in spectral shape and X-ray luminosity for LMC X-3 is not due to partial obscuration, e.g., by a precessing accretion disk, but rather due to changes in the accretion disk geometry. The origin for these changes might be changes in the mass accretion rate.

This work has been partly financed by NASA grants NAG5-3225, NAG5-4621, NAG5-4737 and DFG grant Sta 173/22.

References

Cowley, A. P., et al., 1991, ApJ, 381, 526
Cowley, A. P., et al., 1994, ApJ, 429, 826
Ebisawa, K., et al., 1993, ApJ, 403, 684
Makishima, K., et al., 1986, ApJ, 308, 635
Mitsuda, K., et al., 1984, PASJ, 36, 741
Nowak, M. A., et al., 1999, ApJ, submitted
Wilms, J., et al. 1999a, ApJ, 522, 460
Wilms, J., et al. 1999b, ApJ, submitted

Monitoring the Short-Term Variability of Cyg X-1

Katja Pottschmidt[1], Jörn Wilms[1], Rüdiger Staubert[1], Michael A. Nowak[2], James B. Dove[3,4], William A. Heindl[5], and David M. Smith[6]

[1] Institut für Astronomie und Astrophysik, Waldhäuser Str. 64, D-72076 Tübingen, Germany
[2] JILA, University of Colorado, Boulder, CO 80309-440, USA
[3] CASA, University of Colorado, Boulder, CO 80309-389, USA
[4] Dept. of Physics, Metropolitan State College of Denver, Denver, CO 80217-3362, USA
[5] University of California San Diego, La Jolla, CA 92093, USA
[6] SSL, University of California Berkeley, Berkeley, CA 94720, USA

Abstract. We present first results from an RXTE monitoring campaign of the black hole candidate Cygnus X-1 in 1999. The spectral and temporal properties of this hard state black hole show considerable variability, even though the state does not change. We find that a shortening of the time lags is accompanied by a hardening of the X-ray spectrum, as well as by an increase of the characteristic "shot time scale". We briefly discuss possible physical/geometrical reasons for this variability of the hard state properties.

1 Stability of the Hard State Properties

With broad band instruments like RXTE it is now possible to study the canonical black hole states, i.e., their associated accretion characteristics, with high time resolution over a time base of up to several years. Using the ftools 4.2, we extracted RXTE/PCA spectra and high (2 ms) time resolution lightcurves for four hard state observations of our 1999 Cyg X-1 monitoring campaign. The observations are separated by 2 weeks from each other and were performed before the PCA gain change in 1999 March.

Fourier frequency dependent time lags have been proven to be very helpful in constraining models for the accretion process [4,2]. Fig. 1 shows that the typical time lag in the hard state can vary by at least a factor three over the course of weeks: the first three observations show a gradual decrease in the time lags, while the fourth observation has intermediate values. In addition, we used the linear state space model (LSSM) to model the lightcurves in the time domain. This method allows to derive a characteristic τ, that can be interpreted in terms of a shot noise relaxation time scale [5]. We find that τ gets larger for observations with smaller time lags (Fig. 1).

Spectral fitting of black hole candidate spectra with the PCA is severely affected by the uncertainty of the PCA response matrix. In order to characterize the spectral variability of Cyg X-1 during these observations, we resorted

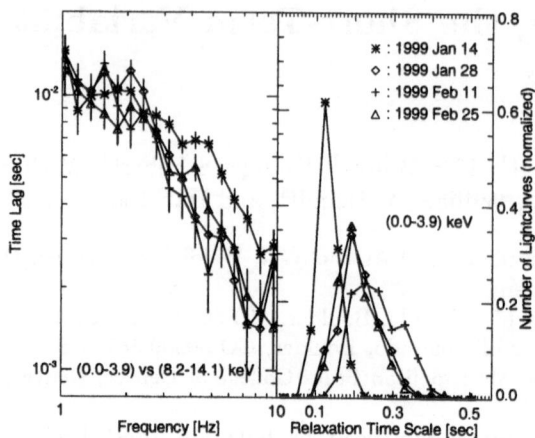

Fig. 1. (Left:) Time lags as a function of Fourier frequency. **(Right:)** Distribution of the relaxation time scale τ found from short (32 s long) time segments.

to directly comparing the data in the detector space [6]. We not only find that Cyg X-1 is clearly spectrally variable on a time basis of 14 d but also that a decrease of the time lags correlates with a spectral hardening of the source.

2 Discussion

First results from our systematic analysis of RXTE data of Cyg X-1 in its canonical hard state show possible evidence for harder spectra to be associated with shorter time lags (similar to the hard state of GX 339−4 , see [3]). One possible interpretation would be that the accretion disk penetrates to smaller disk radii at times of harder flux, thereby increasing the reflection fraction of the Comptonized radiation (see also [1]). Alternatively, our results might indicate that coronae with larger optical depth and/or temperature are physically smaller. This is also consistent with the development of the shot time scale in the sense that more scattering events lead to longer relaxation times. — This work has been partially financed by DFG grant Sta 173/22.

References

1. Gilfanov, M., Churazov, E., & Revnivtsev, M., 1999, A&A, in press
2. Hua, X.-M., Kazanas, D., & Titarchuk, L., 1997, ApJ, 482, L57
3. Nowak, M. A., Wilms, J., & Dove, J. B., 1999a, ApJ, 517, 355
4. Nowak, M. A., Wilms, J., Vaughan, B. A., Dove, J. B., & Begelman, M. C., 1999b, ApJ, 515, 726
5. Pottschmidt, K., König, M., Wilms, J., & Staubert, R., 1998, A&A, 334, 201
6. Pottschmidt, K., Wilms, J., Staubert, R., Nowak, M. A., Dove, J. B., Heindl, W. A., & Smith, D. M., 2000, in Proc. 5th Compton Symposium, in press

Simultaneous Infrared and mm–Wave Observations of Disc Ejection Events in GRS 1915+105

Guy Pooley[1] and Rob Fender[2]

[1] Mullard Radio Astronomy Observatory, Cavendish Laboratory, Cambridge, UK
[2] Astronomical Institute 'Anton Pannekoek'/Center for High-Energy Astrophysics
University of Amsterdam, Kruislaan 403, 1098 SJ Amsterdam, The Netherlands

Abstract. We have observed in the Galactic X-ray source GRS 1915+105 some of the brightest 20–min quasi-periodic oscillations yet seen, simultaneously at 2.2 μm (with *UKIRT*) and at 1.3 mm (with *JCMT*). These oscillations are associated with the disappearance and probable ejection of substantial parts of the inner accretion disc. The emission shows remarkable similarities at the two wavebands, both in timing and intensity. The mechanism for the emission is still unclear.

1 Introduction

GRS 1915+105 is one of the more extreme Galactic objects. It is most probably an X-ray binary, and it is frequently suggested that it contains a black hole. This object is one of a small number known to make repeated ejections of radio-emitting plasma. It is also very unusual in that it often shows wild, quasi-repetitive changes in its X-ray emission: the variations in intensity and spectrum have been modelled in terms of the disappearance (and ejection, in whole or in part?) of the inner few hundred km of the accretion disc. We have reported the discovery of changes in the radio emission with typical "periods" from 20 to 40 min, and which are related in phase to the X-ray oscillations. Similar light-curves in the infrared were reported soon after this, and it was shown that the infrared oscillations have flux densities and pulse-shapes which have a remarkable similarity to those in the 2-cm radio [1, and references therein]. The X-ray, infrared and radio are all related in phase though the relative time-delays may not be constant.

2 Simultaneous IR and mm Observations

As part of a program to investigate the emission process further, we were granted target-of-opportunity observations on the UK Infrared Telescope (*UKIRT*), using *IRCAM3*, and on the James Clerk Maxwell Telescope (*JCMT*) using *SCUBA*.

After our monitoring observations at 2 cm with the Ryle Telescope in Cambridge had shown several days of QPO with increasing amplitudes, we asked for these programs to be triggered. The results were spectacular. Fig. 1 shows the light-curves for the overlapping part of the two datasets.

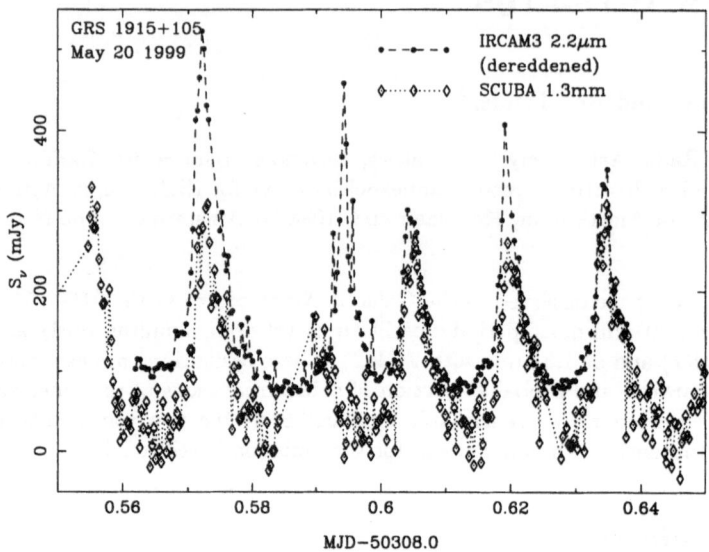

Fig. 1. Simultaneous observations of GRS 1915+105 with IRCAM3 at 2.2 μm and SCUBA at 1.3 mm

3 Interpretation

The spectrum of these oscillatory emissions is remarkably flat across the band from 2 cm to 2.2 μm. The emission is non-thermal (we suppose synchrotron) in origin, from the inferred high brightness temperatures. For a synchrotron source, the consequences include the following:

- a total energy of the order of 5×10^{33} J (5×10^{40} erg) is required to provide the magnetic and particle energy for each pulse, corresponding to a mean power output of 10^{30} W (10^{37} erg s^{-1});
- it is difficult to fit a traditional "expanding synchrotron cloud" model to such a system, since although the delays may be reproduced via optical-depth effects, the relative intensities are more troublesome.

We also note, without quantitative explanation at present, that one pulse failed to appear in the 1.3 mm data.

References

1. Fender, R.P., Pooley, G.G. (1998) MNRAS **300** 573–576.

Observations of Cygnus X-3 in Quiescence in the 2.4 – 12 μm Range with ISO

Lydie Koch Miramond[1], Jean-Marc Bonnet-Bidaud[1], Peter Abraham[2], and Arnaud Claret[1]

[1] DAPNIA/Astrophysics, CEA Saclay, France
[2] Max Planck Institute for Astronomy, Heidelberg, Germany

Abstract. We present mid-infrared spectrophotometric and imaging photometric results obtained with ISO on the peculiar X-ray binary Cygnus X-3. The observations covered the orbital phases 0.83 to 1.04, during quiescence. The flux observed with ISOCAM at 11.5 μm is 3.51±0.73 mJy, in good agreement with that observed in the K and mm ranges, during quiescence (Fender et al, 1999; Ogley et al, 1999). Spectrophotometry with ISOPHOT-S in the range 2.4 – 12 μm shows an unresolved line at about 4 μm, peaking at 61±10 mJy above a mean flux in the 3.5 – 4.5 μm range of 9.0 ±3.2 mJy; it is interpreted as the expected HeII(10-8) line at 4.05 μm.

1 Objectives

Cygnus X-3 is known as a binary system, since its discovery (Giacconi et al, 1967), but there is still debate about the masses of the two stars in the system. The main objective of the ISOPHOT observations was to constrain further the nature of the companion star to the compact object, suggested to be a Wolf-Rayet star (van Kerkwijk et al, 1992). The objectives of the ISOCAM mapping were to deduce the characteristics of the surrounding medium and to find evidence of acceleration of high energy particles.

2 Results

Cygnus X-3 was observed by ISO on 1996, April 7. The subsequent observing modes were: ISOPHOT-S spectrophotometry in the range 2.4 – 12 μm , during 4096s, covering the orbital phases 0.83 to 1.04 (according to the parabolic ephemeris of Kitamoto et al, 1995); ISOPHOT multi-filter photometry at 3.6, 10, 25 and 60 μm and ISOCAM imaging photometry at 11.5 μm (bandwidth 8 to 15 μm). The flux observed at 3.6 μm in the ISOPHOT multi-filter mode was 8.1±3.3 mJy (about 20 mJy dereddened). The flux observed on the ISOCAM map at 11.5 μm was 3.51±0.7 mJy (about 4 mJy dereddened). No detection above the galactic noise was obtained at 10, 25 and 60 μm in the ISOPHOT mulfilter mode. No extended emission was observed on the ISOCAM image at 11.5 μm but this conclusion is still preliminary.

The Cygnus X-3 spectrum in the 2.4 – 12 μm range was obtained after applying a Fast Fourier Transform analysis to the corrected ISOPHOT-S data. The observed continuum flux is about 9.7±3.3 mJy in the range 2.4 – 4.5 μm, in good agreement with that observed in the K and mm ranges during quiescence (Fender et al, 1999; Ogley et al, 1999). An unresolved line at 4 μm peaking at 61 ± 10 mJy (127 mJy dereddened) is observed at a confidence level of 3 sigma. The linewidth is 0.04 μm, consistent with the instrumental response and corresponding to 2494 km/s. This line is interpreted as the HeII(10-8) line at 4.05 μm. Strong He II lines have been previously observed in the K-range during quiescence (van Kerkwijk et al, 1992; Fender et al, 1996, 1999). These lines have been interpreted (van Kerkwijk et al, 1996; Cherepashchuk and Moffat, 1994), as emission from the wind of a massive companion star (possibly a Wolf-Rayet) to the compact object. Note that the expected He II lines at 3.09 and 3.54 μm are not observed in our spectrum. Since the length of this spectrophotometric measurement was comparable to the 4.8 h modulation period seen in both the K-band and the radio (Fender et al, 1995), we attempted to detect this modulation in our data set also. Although the measurement uncertainty of the orbitally phase-resolved spectra was relatively high, the data clearly exclude periodic variations of amplitude higher than 15 per cent.

The quiescent state observed in the mid-infrared range with ISO was also seen during radio observations of Cygnus X-3, shortly before and after the ISO observations, the mean flux at 2 cm being about 120 mJy (Pooley, 1999). In the X-ray range, at the same epoch, the XTE/All Sky Monitor observed a mean flux of about 1 mJy from 2 to 12 keV (XTE team, 1999). This quiescent state was still present in 1996 May, June and July (Fender, 1999).

3   Conclusions

Cygnus X-3 was in quiescence during the ISO observing period. The mean continuum flux in the range $2.6 - 12$ μm was about 10 mJy with an almost flat spectrum. An unresolved line at about 4 μm was observed at a 3 sigma confidence level. If real, this line can be interpreted as the HeII(10-8) emission line at 4.05 μm from the wind of the companion He star to the compact object, the upper limit of the line width being 2494 km/s. Cygnus X-3 was also observed in quiescence in the radio and X-ray ranges at the same epoch.

References

1. Cherepashchuk A., Moffat A., 1994, ApJ, 243, L53
2. Fender R. et al, 1995, MNRAS 274, 633
3. Fender R. et al, 1996, MNRAS 283, 798
4. Fender R. et al, 1999, MNRAS 308, 473
5. Fender R., 1999, private communication
6. Giacconi R. et al, 1967, ApJ 148, L119
7. Kitamoto S. et al, 1995, PASJ 47, 233
8. Ogley R. et al, 1999, MNRAS submitted
9. Pooley G., 1999, private communication
10. van Kerkwijk M. et al, 1992, Nature 355, 703
11. van Kerkwijk M. et al, 1996, A & A 314, 521

Determination of Mass Limits Around Pulsars at 10 and 90 μm with ISO

Lydie Koch Miramond[1], Philipp Podsiadlowski[2], Martin Haas[3],
Tim Naylor[4], and Marc Sauvage[1]

[1] DAPNIA/Astrophysics, CEA Saclay, France
[2] Oxford University, UK
[3] Max Planck Institute for Astronomy, Heidelberg, Germany
[4] Keele University, UK

Abstract. We present mid-infrared photometric results obtained with ISOCAM and ISOPHOT on 3 millisecond pulsars and 3 ordinary radio pulsars. No detections have been obtained for the three ms pulsars nor the two more distant radio pulsars. A faint enhancement in the brightness map at 90 μm is seen at about 5 arcsec from the radio position of PSR J0108-1431, the nearest radio pulsar (Tauris et al, 1994), located at 85 pc from us. We conclude that this 90 μm emission, amounting to about 12 mJy, originates either from material orbiting the pulsar or from cirrus on the line of sight. We deduce the upper limits on mass of dust orbiting this pulsar and on the mean temperature of grains.

1 Objectives

1) Search for thermal dust emission from circumstellar disks or clouds (by-products or progenitors of planet formation) around pulsars
2) Discover intermediate stages of evolution between evaporating binary pulsars and isolated millisecond pulsars with planets
3) Discover residual material from the envelope of the progenitor, that was not ejected in the supernova explosion and has settled in a post- supernova disk. On planet formation scenarios around pulsars see (Podsiadlowski, 1993). We also aim to deduce the mass of radiating dust and compare its physical properties to that of disks around main-sequence and post main-sequence stars seen by space and ground-based infra-red and millimeter observations.

2 Results

The ISO observing modes used were ISOCAM imaging photometry at 15 μm and ISOPHOT photometry at 90 μm. We assume that the disk of dust has similar physical properties to the disk around the star Beta Pictoris (Pantin et al, 1997), where the grains are made of 'cosmic silicates' as defined in Draine et al, 1985. The upper limits on the mass of radiating dust and the mean temperature of grains are constrained by the measurements at 15 and 90μm, but the mass of gas is not constrained. If the mass ratio of gas to dust is similar to that in T Tauri disks, the upper limits given below should be 100 times higher. Table 1 presents the 6 pulsars observed by ISO, the upper limits of radiating mass around them and the mean temperature of the grains.

An upper limit on the total mass of circumstellar material of $1.6 \times 10^{-2}\,M_\odot$ around PSR B1534+12 has been obtained by Philips and Chandler (1994), in the sub-mm and mm ranges, using the Beckwith and Sargent (1990) results on circumstellar disks of T Tauri stars.

Table 1. Upper limits on the mass and temperature of emitting dust around Pulsars

Pulsar	P [s]	d [pc]	M [g]	M/M$_\odot$	T [K]
B1534 + 12	0.038	430	2.5×10^{26}	1.2×10^{-7}	134
J2322 + 2057	0.0048	780	3.4×10^{26}	1.7×10^{-7}	127
J2019 + 2425	0.0039	850	8.5×10^{26}	4.2×10^{-7}	117
B0149 − 16	0.8	790	3.8×10^{26}	2×10^{-7}	124
B1604 − 00	0.42	590	2.6×10^{26}	1.3×10^{-7}	123
J0108 − 1431	0.85	85	3.4×10^{23}	1.7×10^{-10}	194

Stars, dust and molecular clouds in the direction of PSR J0108-1431: in the IRAS All Sky Survey a flux of 12 ± 1 mJy is observed at 90μm in an aperture of 46x46 arcsec, equal to the ISOPHOT beam size. There is no visible object in the Digitized Sky Survey within 10 arcsec of the pulsar's position and only two very faint stars at 10.2 and 14.5 arcsec from the pulsar, in the ISOPHOT beam. The CO survey at high galactic latitude (Magnani et al, 1998) found a mass surface density of molecular gas of $0.2\,M_\odot/pc^2$.

3   Conclusions

The flux enhancement measured in the direction of PSR J0108 − 1431 can be interpreted either as the emission from material around the pulsar or as the emission from cirrus or molecular clouds on the line of sight. We aim to remove this ambiguity by observations in infra-red, sub-mm and millimeter ranges, that will allow the measurement of the mass and mean temperature of the dust, thus constraining the distribution of grain sizes and the physical parameters of the grains.

References

1. Beckwith S., Sargent A. (1990) AJ 99, 924
2. Draine B., Lee H. (1984) ApJ 285, 89
3. Magnani L., et al (1998) AAS 193, 6505
4. Pantin E., Lagage P.O., Artymowicz P. (1997) A&A 327, 1123
5. Philips J.A., Chandler C.J. (1994) ApJ 420, L83
6. Podsiadlowski P. (1993) in ASP Conf.Ser. 36, 149
7. Tauris T.M., et al (1994) ApJ 428, L53

X-Ray Spectra
of Neutron Stars vs. Black Holes

Yasuo Tanaka

[1] Max-Planck-Institut für Extraterrestrische Physik, D-85748 Garching, Germany
[2] Institute of Space and Astronautical Science, Sagamihara, Kanagawa-ken, Japan

Abstract. This paper gives an overview of the observed X-ray spectra of low-mass X-ray binaries containing a weakly-magnetized neutron star (NS-LMXRB) and those of X-ray binaries containing a black hole (BH-XRB). The X-ray spectra of these systems change with luminosity (the mass-accretion rate). At high X-ray luminosities ($L_X > 10^{37}$ erg s^{-1}), while soft thermal emission from the accretion disk dominates, there is a distinct difference in the X-ray spectrum between NS-LMXRBs and BH-XRBs. At a lower luminosity (typically $L_X \sim 10^{37}$ erg s^{-1}), both systems undergo transitions between a soft thermal state and a hard power-law state. The characteristics of these X-ray spectra are discussed.

1 Introduction

A wealth of observational results on X-ray binaries has become available from various X-ray astronomy satellites. Among them, luminous X-ray binaries are those containing either a neutron star (NS-XRB) or a black hole (BH-XRB). This paper gives a brief review of X-ray spectra of these systems. For NS-XRB, we deal here with those containing a weakly magnetized neutron star, for which the magnetic fields has little influence on the accretion flow. They are all low-mass X-ray binaries (NS-LMXRB).

In the following sections, it is shown that there is a distinct difference in the spectral shape between NS-LMXRB and BH-XRB when the luminosity is well above 10^{37} erg s^{-1}. Whereas, at lower luminosities, both systems undergo a radical change in the spectral shape, and the key features that distinguish these two systems go away. The important properties of the X-ray spectra of these systems are discussed. For previous reviews on the subject, see e.g. Tanaka & Lewin 1995; Tanaka & Shibazaki 1996.

In addition, some other related subjects of interest are also discussed.

2 X-Ray Spectrum at High Luminosities

NS-LMXRB at high luminosities ($> 10^{37}$ erg s^{-1}) show a common spectral shape, as shown in Fig. 1. It actually consists of two separate components: a soft component and a hard component. Each of the two components can be individually determined from the analysis of changes of the spectral shape with intensity (see Tanaka 1997).

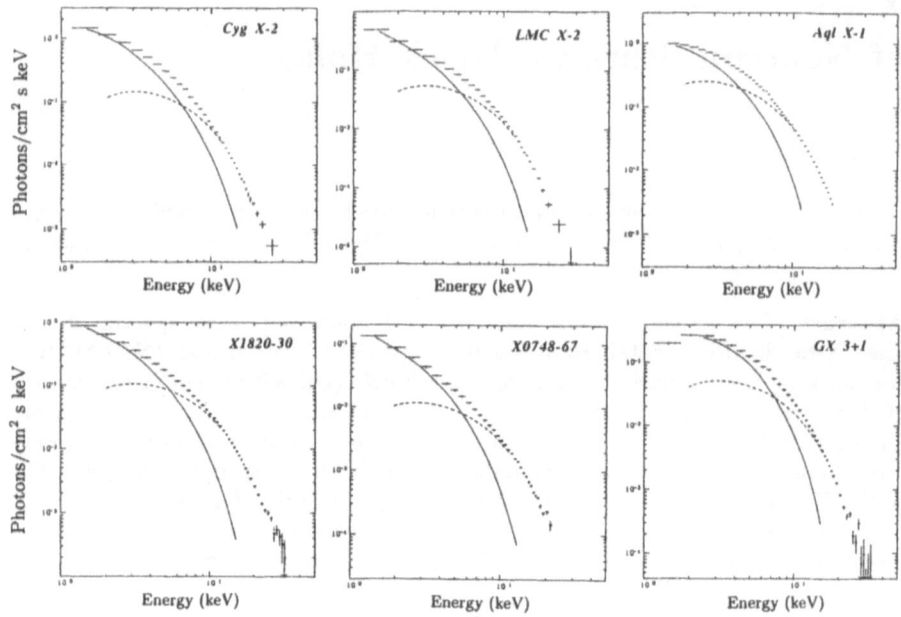

Fig. 1. X-ray photon spectra of NS-LMXRB at high luminosities, each consisting of a soft component (solid curve) and a blackbody component (dashed curve).

The soft component is well expressed by the multicolor blackbody disk model for an optically-thick accretion disk (Mitsuda et al. 1984) based on the standard Shakura-Sunyaev disk model (Shakura & Sunyaev 1973). This multicolor blackbody disc model includes only two free parameters, i.e. r_{in} and kT_{in}, where r_{in} is the innermost disk radius and kT_{in} is the color temperature at r_{in}. The excellent agreement with this model makes it certain that the soft component is the emission from an optically-thick accretion disc. The absence of emission lines that are expected for thin thermal plasma also supports that the disk is indeed optically thick. The observed color temperature kT_{in} is typically $1.4 - 1.5$ keV at $L_X \sim 10^{38}$ erg s^{-1} and goes down as luminosity decreases.

The hard component is best expressed by a modified blackbody spectrum (color temperature $kT_c \sim 2.3 - 2.5$ keV). This component is most probably the emission from the neutron star surface (or an optically-thick boundary layer) where the kinetic energy of accreting matter is eventually thermalized. The fact that this spectrum is very similar to that of X-ray bursts (thermonuclear flashes on the neutron star surface) also supports this interpretation. This blackbody component varies irregularly by as large as a factor of 5 (its maximum luminosity being comparable to the soft component) without changing the shape (see Tanaka 1997). The reason for this variation is unknown.

Table 1. Black-hole binaries established from the mass functions

Source name		Spectrum	Companion	$F(M)$ $(M_\odot)$	BH mass $(M_\odot)$
Cyg X-1	persistent	S+PL	O 9.7 Iab	0.241±0.013	~16 (>7)
LMC X-3	persistent	S+PL	B 3 V	2.3±0.3	>7
LMC X-1	persistent	S+PL	O 7−9 III	0.14±0.05	~6(?)
J0422+32	XNova Per	PL	M 2 V	1.21±0.06	>3.2
0620−003	XNova Mon	S+PL	K 5 V	3.18±0.16	>7.3
1124−684	XNova Mus	S+PL	K 0−4 V	3.1±0.4	~6
1543−475	XN '71,'83,'92	S+PL	A 2 V	0.22±0.02	2.7−7.5
J1655−40	XNova Sco	S+PL	F 3−6	3.24±0.09	7.02±0.22
1705−250	XNova Oph'77	S+PL	K ~3 V	4.0±0.8	~6
2000+251	XNova Vul	S+PL	early K	4.97±0.10	6−7.5
2023+338	XNova Cyg	PL	K 0 IV	6.26±0.31	8−15.5

S+PL: soft + power-law, PL: power law.

For references, see Tanaka & Shibazaki (1996) except for 1705−250 (Remillard et al. 1996) and 1543−475 (Orosz et al. 1998)

There appears to be a distinct difference in the X-ray spectrum between BH-XRB and NS-LMXRB when the X-ray luminosity is well above 10^{37} erg s^{-1}. So far, eleven X-ray binaries including two in the Large Magellanic Cloud have been shown to contain a compact object whose mass lower limit is greater than 3 $M_\odot$, as listed in Table 1. Hence, they are considered to be "reliable" black holes. Among these eleven sources, only three (Cyg X-1, LMC X-1 and LMC X-3) are persistent sources, and they are all high-mass systems. The other eight are all transient sources, and are all low-mass binaries. Of the eleven reliable BH-XRB (Table 1), nine show X-ray spectra of a common characteristic shape, as shown in Fig. 2. It consists of a soft thermal component and a hard power-law tail.

The soft component of BH-XRB is also well expressed by a multicolor blackbody disk model of the same functional form as that of NS-LMXRB, hence identified to be the emission from an optically-thick disk. The observed color temperature kT_{in} is typically ~ 1.2 keV at $L_X \sim 10^{38}$ erg s^{-1}, significantly lower than that for NS-LMXRB of the same luminosity level.

On the other hand, a blackbody component that is always present in the spectra of luminous ($L_X > 10^{37}$ erg s^{-1}) NS-LMXRB is absent in the BH-XRB spectra. A fundamental difference between a neutron star and a black hole is the presence or absence of a solid surface. We consider the absence of a blackbody component to be a strong indication of an accreting black hole.

Furthermore, for a standard accretion disk model, kT_{in} scales as $\propto M_X^{-1/4}$ for a given accretion rate, where M_X is the compact object mass. That the observed kT_{in} for BH-XRB is significantly lower than that for NS-LMXRB, is consistent with a larger M_X. For a non- or slowly spinning black hole, r_{in}

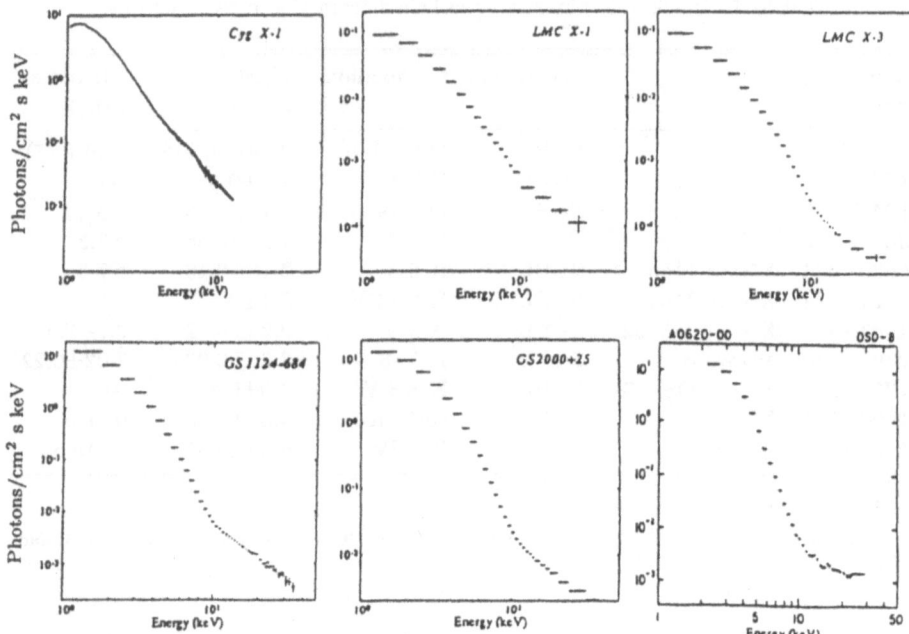

Fig. 2. X-ray photon spectra of reliable BH-XRB at high luminosities.

may represent three Schwarzschild radii, hence proportional to the compact object mass. If the source distance is known, r_{in} is obtained from the observed L_X and kT_{in} of the soft component. The estimated values of r_{in} for most BH-XRB turned out to be larger by a factor of 3 to 4 than those for NS-LMXRB, implying that the compact objects are more massive by this factor than a neutron star (see Tanaka & Lewin 1995; Tanaka 1997 for more detail). These results make it convincing that such a "soft + hard-tail" spectrum is a signature of an accreting black hole. Note that an estimate of the actual mass requires the consideration of general relativistic effects and a correction of the color temperature to the effective temperature (see e.g. Hanawa 1989; Ebisawa, Mitsuda & Hanawa 1991; Zhang, Cui & Chen 1997).

It is of interest that two systems GRO J1655–40 and GRS 1915+105 (black-hole candidate, see below), both superluminal jet sources, show significantly higher kT_{in} than other BH-XRB. Zhang, Cui & Chen (1997) suggest that the black holes in these systems rapidly spin in the same direction as the accretion disk. This could be related to the formation of relativistic jets.

In addition to these reliable BH-XRB, so far about fourteen more low-mass transients (including GRS 1915+105) are found to show the characteristic spectral shape (soft + power-law tail) when $L_X > 10^{37}$ erg s^{-1}. They are also believed to be BH-LMXRB. In fact, X-ray bursts (a definitive signature of an accreting neutron star) have never been detected in any of them.

The hard power-law tail that characterizes the BH-XRB spectrum extends well over 100 keV without a cut-off, sometimes observed up to ~ 1 MeV (e.g. Grove et al. 1998). The luminosity of the hard component relative to the soft component varies irregularly by an extremely large factor, from a comparable luminosity down to $1 \sim 2$ orders of magnitude less (see Fig. 3a). The photon index of the hard tail remains fixed at $2.0 - 2.5$ against these changes. The origin of the power-law tail is still unclear. It is generally considered to be produced by Comptonization of soft photons by high-energy electrons. Yet, how to accelerate electrons to such high energies and to maintain their energy against Compton cooling are serious problems. The fact that such a power-law tail is absent in the luminous NS-LMXRB might suggest that it is formed in the gap between the innermost stable orbit and the Schwarzschild radius. (In the case of a neutron star, the gap is presumably small.) Such a model has been proposed (see e.g. Laurent & Titarchuk 1999 and references therein), though it contains the above-mentioned problems as well.

It is to be remarked that among eleven reliable BH-XRB, GS 2023+338 and GRO J0422+32, are exceptions. Both sources showed an approximately single power-law spectrum even at high luminosities. The reason why they did not show the soft + hard-tail spectrum is still unknown.

3   X-Ray Spectrum at Lower Luminosities

Both NS-LMXRB and BH-XRB undergo a dramatic change in the spectral shape around a certain luminosity level between a soft thermal state (high state) as described in Section 3.1 and a hard state (low state) with an approximately single power-law form, as shown in Fig. 3. This transition between the two spectral states is considered to be a fundamental property of accretion disks regardless of whether the compact object is a neutron star or a black hole. The transition occurs at about $L_X \sim 10^{37}$ erg s^{-1} or a mass-accretion rate around 10^{17} g s^{-1}, but it might vary from source to source and even from one transition to another (see Tanaka & Shibazaki 1996, and references therein). Associated with the spectral change, the time variability also changes. As sources enter into the hard state, rapid large-amplitude intensity fluctuations (flickering) build up on all time scales down to milliseconds.

The power-law spectrum in the hard state is substantially harder than the power-law tail of BH-XRB in the soft state, and shows a high-energy cut-off. The observed photon indices in the hard state are in the range $1.7-1.9$ for both BH-XRB and NS-LMXRB. Hence, once they go into the hard state, the distinction in spectral shape between these two systems is lost. The photon index in this state also remains remarkably constant against large luminosity changes. It is important to note that the power law emission in the hard state and that of the BH-XRB in the soft state are qualitatively different with respect to the presence or absence (at least up to the highest energy observed) of a cut-off and the variations with time.

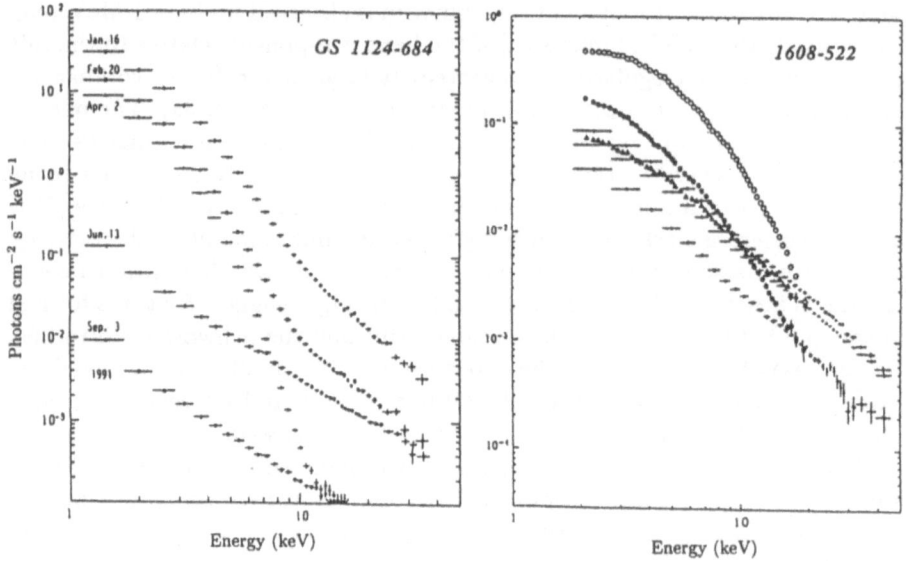

Fig. 3. Changes in the spectral shape with luminosity. (a) The spectra of the BH-LMXRB GS 1124–684, and (b) the spectra of the NS-LMXRB 4U 1608–522.

It is important to note that, despite a radical change in the spectral shape, the transition between the two spectral states does not cause a big jump in the total luminosity before and after transition (see e.g. Zhang et al. 1997 for the 1996 transition of Cyg X-1). Hence, it is not due to switching between a radiation-dominated accretion flow and an advection-dominated accretion flow (ADAF).

Transition from a soft state to a hard state has been considered as due to a change in the disk structure. There is evidence that a quick build-up of an optically-thin hot plasma occurs when a source goes into the hard state. For instance, when X 1608–522 was about to go into the hard state, the spectra of X-ray bursts (blackbody emission from the neutron star surface) also began to show a significant hard tail (Nakumura et al. 1989). Note that, for this effect, NS-LMXRB spectrum may mimic "soft + hard tail" near the critical mass-accretion rate. Therefore, it is important that the spectral distinction between BH-XRB and NS-LMXRB holds only when L_X is well above 10^{37} erg s^{-1}. The observed power-law spectra with a cut-off can be reproduced by thermal Comptonization of soft photons (e.g. Sunyaev & Titarchuk 1980). Yet, the mechanism of the transition and other striking properties (constancy of the photon index against large intensity changes, different photon-index values before and after transition, and flickering) remain to be explained.

It is worth noting that the properties of X-ray binaries in the hard state are strikingly similar to those of AGN, i.e. the same power-law index, and strong variations on time scales down to the shortest Keplerian periods. These similarities suggest that the basic process of accretion is essentially the same in both systems, despite huge differences in scale and power.

3.1 Reflection Component

When an optically-thick disk is illuminated with X-rays, part of the X-rays are reflected by Thomson scattering. This reflection component, predicted by Lightman & White (1988), was first discovered from AGN (Pounds et al. 1990). Since photoabsorption dominates Thomson scattering at low energies, the reflected component is characterized by a hard continuum, much harder than the incident spectrum, and a K-absorption edge of iron accompanied by a fluorescent emission line. In particular, the iron lines from Seyfert galaxies are found to be relativistically broadened as described by Fabian et al. (1989), which provides clear evidence for the general relativistic effect unique to the close vicinity of a massive black hole (see e.g. Tanaka et al. 1995).

Since the reflection component provides a useful diagnostic tool of an accretion disk, it has been studied also for X-ray binaries, in particular Cyg X-1. The ASCA spectrum (the best spectral resolution so far available) of Cyg X-1 in the hard state shows a weak narrow line and a shallow K-edge of iron

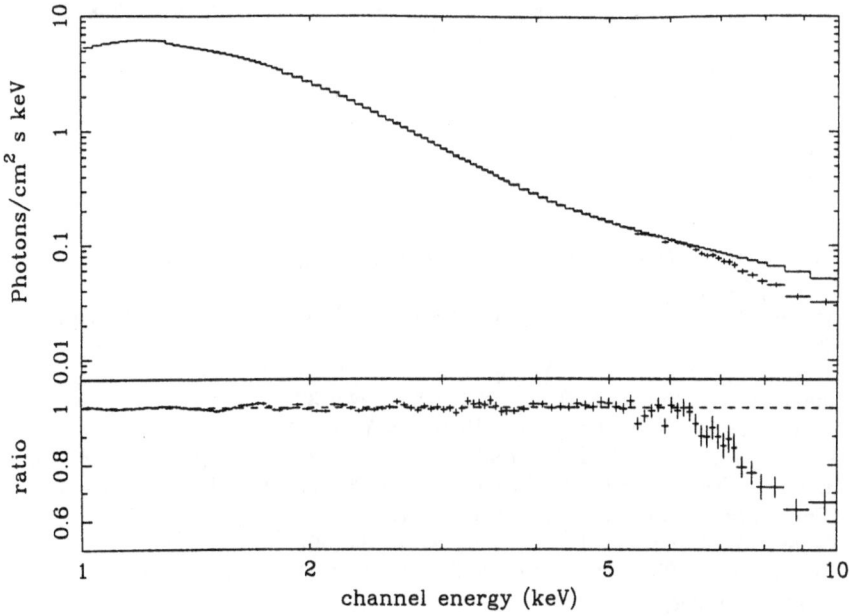

Fig. 4. Cyg X-1 spectrum in the soft state (upper panel) observed with ASCA SIS, and the ratio to the best-fit model determined below 6 keV (lower panel).

(Ebisawa et al. 1996; Done & Życki 1999), indicating a relatively minor contribution of the reflection component. In the soft state, the effect of reflection appears significantly different. As shown in Fig. 4, intensity decreases substantially above 7 keV due to iron K-absorption, but the structure is quite broad (smeared edge). On the other hand, the fluorescence line that must be emitted is hardly visible.

The situation in X-ray binaries seems to be more complicated than in Seyfert galaxies. Because of an orders of magnitude smaller size scale, the effect of photoionization, particularly at high luminosities, becomes much more important. Ross, Fabian & Young (1999) demonstrated a pronounced dependence of the line intensity and relative contribution of the reflected component on the ionization degree of the disk. The line profile and the edge structure are degraded by Compton broadening and also relativistic blurring (if near the black hole). Thus, detailed investigation of the properties of reflection will lead to understanding of the fundamental questions on the disk structures in the two states and where and how the illuminating X-rays are generated.

References

1. Done, C., Życki, P.T., 1999, MNRAS, 305, 457
2. Ebisawa, K., Mitsuda, K., Hanawa, T., 1991, ApJ, 367, 213
3. Ebisawa, K., Ueda, Y., Inoue, H., Tanaka, Y., White, N.E., 1996, ApJ, 467, 419
4. Fabian, A.C., Rees, M.J., Stella, L., White, N.E., 1989, MNRAS, 238, 729
5. Grove, J.E. et al., 1998, ApJ, 500, 899
6. Hanawa, T., 1989, ApJ, 341, 948
7. Laurent, P., Titarchuk, L., 1999, ApJ, 511, 289
8. Lightman, A.P., White, T.R., 1988, ApJ, 335, 57
9. Mitsuda, K. et al., 1984, PASJ, 36, 741
10. Nakamura, N. et al., 1989, PASJ, 41, 617
11. Orosz, J.A., Jain, R.K., Baylin, C.D., McClintock, J.E., Remillard, R.A., 1998, ApJ, 499, 375
12. Pounds, K.A., Nandra, K., Stewart, G.C., George, I.M., Fabian, A.C., 1990, Nature, 344, 132
13. Remillard, R.A., Orosz, J.A., McClintock, J.E., Baylin, C.D., 1996, ApJ, 459, 226
14. Ross, R.R., Fabian, A.C., Young, A.J., 1999, MNRAS, 306, 461
15. Shakura, N.I., Sunyaev, R.A., 1973, A&A, 24, 337
16. Sunyaev, R.A., Titarchuk, L.G., 1980, A&A, 86, 121
17. Tanaka, Y., Lewin, W.H.G., 1995, in Cambridge Astrophys. Ser. Vol. 26, X-Ray Binaries, eds. W.H.G. Lewin, J. van Paradijs & E.P.J. van den Heuvel (Cambridge: Cambridge Univ. Press), 126
18. Tanaka, Y., Shibazaki, N., 1996, ARA&A, 34, 607
19. Tanaka, Y., 1997, in LNP Vol. 487, Accretion Disks - New Aspects (Berlin: Springer Verlag) 1
20. Tanaka, Y. et al., 1995, Nature, 375, 659
21. Zhang, S.N., Cui, W., Chen, W., 1997, ApJ, 482, L155
22. Zhang, S.N. et al., 1997, ApJ, 477, L95

Why Do Black-Hole X-Ray Binaries Tend to Be Transient?

Jean-Pierre Lasota

Institut d'Astrophysique de Paris
98bis Bd Arago, 75014 Paris, France

Abstract. Black hole X–ray binaries are transient probably because their discs are subject to the same thermal–viscous instability which is present in dwarf nova binary systems. I discuss applications of the dwarf–nova instability model to transient, low–mass, X–ray binary systems. When disc truncation and X–ray irradiation are taken into account this model is capable of reproducing the basic properties of X–ray binary outbursts [3].

1 The Thermal-Viscous Instability

All Low Mass X–ray Binaries (LMXBs) in which the presence of a black hole is inferred from dynamical mass determination are transient. This could mean that all Black Hole LMXBs (BHLMXBs) are transient, but there could be non–transient systems in which black holes have masses in the range allowed for neutron stars. For this reason it is more prudent to say that BHLMXBs 'tend' to be transient (as suggested by the organizers of this meeting) rather than to affirm that all of them are transient.

The responsibility for outbursts in low mass, close binary systems is commonly imputed (e.g. [22]) to the usual suspect: the thermal–viscous instability present in discs of a subclass of cataclysmic variable stars: the dwarf–nova systems. There, for temperatures corresponding to hydrogen recombination, changes in opacity (emissivity) strongly affect cooling mechanism dependence on temperature and lead to a thermal instability under the standard assumption about viscous heating. This instability is intimately related to a viscous instability. A disc's thermal equilibria, at a given distance from the center, form a characteristic S shape in the surface density – effective temperature plane. The middle branch of the S corresponds to unstable solutions, the upper one to hot, almost fully ionized configurations, while the lower branch contains cold, quasi–neutral states. (The existence of the cold branch does not depend on the presence of convection, contrary to erroneous assertions in recent lectures on the subject [18]. In fact some attempts to modify the viscosity prescription required switching off the convection in order to get a sufficient amplitude of the jump between the cold and hot branches [2]). If the rate at which the mass is transferred from the black–hole's stellar companion is such that the disc is (somewhere) unstable, a limit cycle behaviour will follow taking the system through a cycle of oscillations between hot and cold states. Such a cycle is supposed to explain dwarf nova and transient LMXB outburst cycles.

The instability and the limit cycle do not, by themselves, produce outbursts similar to the ones observed in dwarf novae and transient systems. The thermal instability creates temperature gradients which propagate through the disc in the form of heating (transition to the hot branch) and cooling (transition to the cold branch) fronts. The form of the resulting luminosity variations depends on what is assumed about the viscosity or, more precisely, what is assumed about the viscosity parameter α appearing in the kinematic viscosity prescription $\nu = \alpha c_s^2 / \Omega_K$, where c_s is the speed of sound and Ω_K is the Keplerian frequency. Is α is assumed to be constant, then, because of insufficient temperature and density contrasts between the cold and hot states, the resulting light–curve has the form of regular, small amplitude outbursts with no resemblance to dwarf–nova outbursts or X–ray binary transient events. In order to obtain dwarf–nova type outbursts the viscosity parameter α must be assumed to be larger in the hot than in the cold state [24]. This *einsatz* has not been, for the moment, based on any model of viscosity, although a recent argument in favour of different values of α in hot and cold states [8] is based on such a model (beware, however of the erroneous interpretation in [18]: it applies hot equilibrium disc solutions to describe the properties of a cold non–equilibrium disc!).

The assumption of a jump in the α value, however, does not suffice to reproduce observed properties of outbursts in close binary systems and additional physical effects have to be added to the 'pure' disc instability model (see e.g. [26][10]). Some of these effects, such as inner disc truncation [19][20] and disc X–ray irradiation [3][4], which are important for low mass X–ray transients, will be discussed in this article.

Figure 1 shows that the parameters of all BHLMXBs put them well below the limit for the disc thermal stability. Here I used the stability criterion for X–ray irradiated discs as proposed by Jan van Paradijs [27]. Irradiation (discussed in Sect. 3) of the outer disc by accretion–produced X–rays from the inner disc gives a stability criterion which can be expressed in the form [4], [16]:

$$\dot{M}_{transf} > \dot{M}_{crit}^{irr} \approx 2.0 \times 10^{15} \left(\frac{M_1}{M_\odot}\right)^{0.5} \left(\frac{M_2}{M_\odot}\right)^{-0.2} P_{hr}^{1.4}$$

$$\times \left(\frac{C}{5 \times 10^{-4}}\right)^{-0.5} \text{ g s}^{-1} \qquad (1)$$

where $\dot{M}_{transf}$ is the mass–transfer rate, M_1 is the black hole mass, M_2 is the mass of the companion, P_{hr} the orbital period in hours. C is a parameter describing effects of irradiation (see [4]). Systems satisfying condition (1) are stable.

However, systems with $\dot{M}_{transf} < \dot{M}_{crit}^{irr}$ are not necessarily unstable. In fact BHLMXBs are well below the stability limit (lower than one would expect from evolution models [19]). They are close to the stability limit for cold discs if their discs are truncated in the inner regions. There are several arguments in favour of such a truncation.

2   The Necessity of Truncation

According to the model the quiescent accretion rate must *everywhere* satisfy the condition [12]:

$$\dot{M}(r) < \dot{M}_{\text{cold}} = 9.5 \ 10^{15} \ \alpha^{0.01} \left(\frac{M_1}{M_\odot}\right)^{-0.89} \left(\frac{R}{10^{10} \ \text{cm}}\right)^{2.68} \ \text{g s}^{-1} \qquad (2)$$

For a disc extending down to the last stable orbit and typical parameters, this predicts quiescent accretion several orders of magnitude lower than observed [15]. A disc truncated at $10^3 - 10^4 R_S$ (where R_S is the Schwarzschild radius) will solve this problem, in particular if the inner 'hole' is filled by an Advection Dominated Accretion Flow (ADAF) (e.g. [23], [11], [6]). For such truncation radii, what is left of the disc could be cold and stable as shown in Fig. 1 (where the truncation radius is scaled by the circularization radius). Also, if we take into account irradiation, disc truncation is necessary in order to obtain the light–curves required by observations [3].

3   Effects of Irradiation

There is overwhelming evidence that discs in LMXBs are strongly X–ray irradiated [28]. In outburst, a disc's optical emission is due to X-ray reprocessing. As mentioned above, irradiation modifies the stability criterion. It also affects all the properties of the outburst cycle. Its main effect is to retard the propagation of the cooling front, leading to an exponential decay in the light–curves [14]. An example of this effect is shown in Fig. 2. A systematic study of irradiation effects will be presented in Dubus et al. [3]. It appears that most of the fundamental properties of at least one class of LMXB transient systems can be reproduced by the model if both truncation and irradiation are taken into account. The transient systems concerned are those with a fast-rise exponential decay (FRED) light curves. (See next section for the discussion of recurrence times).

In systems with more complex forms of light–curves one should also take into account (the observed) irradiation of the secondary [7],[5].

4   Recurrence Times

Compared with most dwarf novae, black-hole X–ray transients have very long recurrence times (> 30 years). These also seem to be longer than the

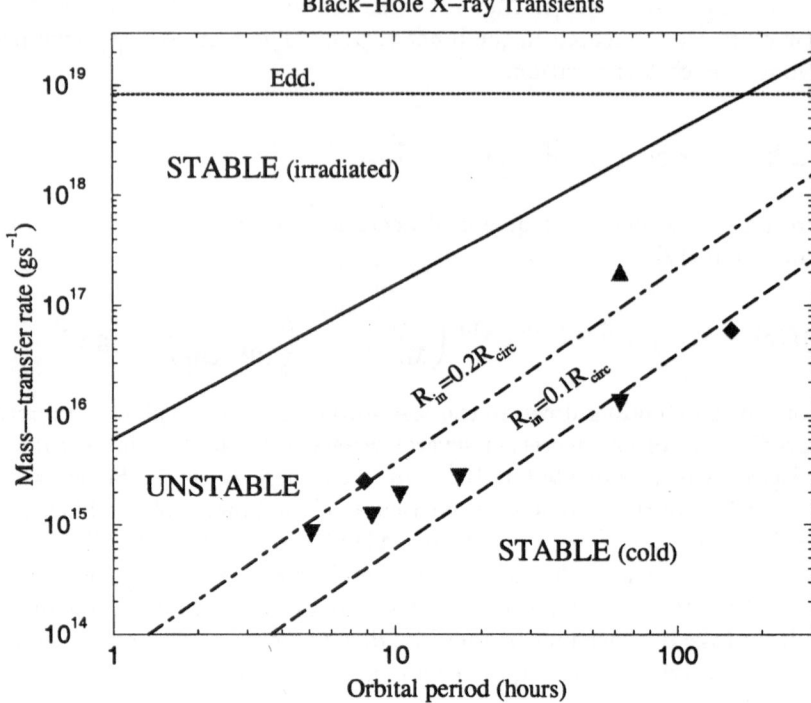

Fig. 1. Stability criteria for accretion discs in black–hole low–mass X–ray binaries. The mass–transfer rate is estimated by dividing the mass accreted during outburst by the recurrence time. This mass accumulation rate might be slightly lower than the actual transfer rate [19]. Since, except for two systems, only the lower limits on recurrence times are known, we have, for most systems, only the upper limits on the mass accumulation rate. All these systems are well below the upper line corresponding to stable irradiated discs but are rather close to the stability limit for truncated discs, represented here by two lines corresponding to truncation radii $R_{in} = 0.1$ and $0.2R_{circ}$, where R_{circ} is the circularization radius. Data is taken from ref. [19]. Two values are given for GRO J1655-40: the higher one corresponds to the 1996 outburst (see [7]). The dotted line marked 'Edd' corresponds to the Eddington accretion rate $\dot{M}_{Edd} = L_{Edd}/0.1c^2$ for $6M_\odot$.

recurrence times of neutron-star transients. Very long recurrence times can be due to several effects. First, truncated discs in these systems could be cold and stable (Fig. 1) and outbursts due to (upward) fluctuations of mass transfer. The recurrence time then would correspond to some cycle in the secondary star. Such a model was proposed for the dwarf nova WZ Sge, whose recurrence time is ~ 30 years [17][9].

Second, the inner truncation radius can be such that the disc is marginally stable (see Fig. 1). Since for a stable disc the recurrence time is infinite, for a given mass-transfer rate one can always fine-tune the inner disc radius so

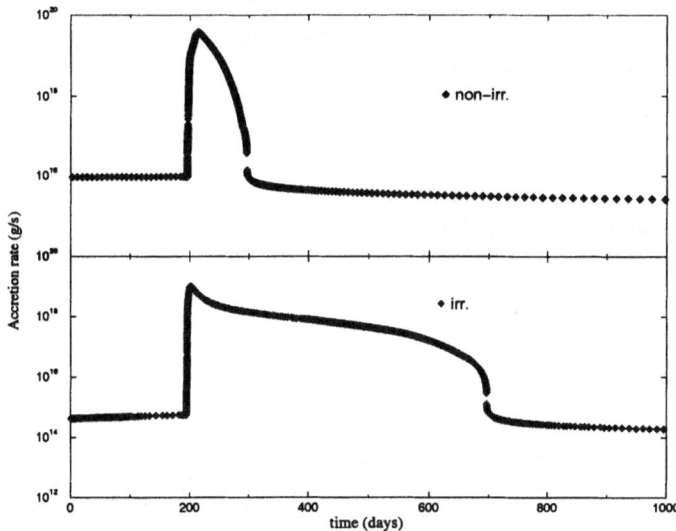

Fig. 2. Examples of lightcurves obtained in the disc instability model for truncated, non–irradiated (upper panel) and irradiated (lower panel) discs in black–hole low-mass X-ray binaries. In the non–irradiated case the cooling front rapidly propagates inwards, whereas irradiation postpones propagation of this front as suggested in [14]. As a result the light curve is close to exponential, outbursts last longer and more mass is drained from the disc.

that the recurrence time is arbitrarily long (see e.g. [20]). This is for example the case of the model proposed in [21]. Such an assumption is, however, equivalent to assuming global stability, because the fine tuning would be at the mercy of even very small fluctuations of mass transfer.

Two other possibilities can be described by using an approximate formula for the recurrence time (see [25][20])

$$t_{\rm rec} \approx 3 \left(\frac{\xi}{3}\right) \left(\frac{M_1}{M_\odot}\right)^{0.62} \left(\frac{R}{10^{10}\ \rm cm}\right)^{0.14} \left(\frac{\alpha_{\rm cold}}{0.02}\right)^{-0.83} \left(\frac{T_{\rm eff}}{3000\ \rm K}\right)^{-4} \quad {\rm yr} \quad (3)$$

where $T_{\rm eff}$ is the disc effective temperature, $\alpha_{\rm cold}$ is the viscosity parameter in quiescence and ξ is a factor (3 - 5) taking into account the fact that the quiescent disc is not in *viscous* equilibrium [13]. In a quiescent disc $T_{\rm eff}$ is almost independent of the radius (this is *not* the result of low viscosity in quiescence, contrary to the erroneous assertions in [18]).

We can see that if we assume $\alpha_{\rm cold} \approx 10^{-3}$, Eq. 3 gives recurrence times longer than tens of years. This assumes that the quiescent effective temperature is not lower than 3000 K. In *non- irradiated* discs this is indeed the case (note that irradiation is important only during the outburst) and only by lowering $\alpha_{\rm cold}$ can one obtain long recurrence times [20]. Irradiation during outburst, however, allows one to drain more mass before the cooling front brings the disc into the cold state. As a result quiescent temperatures are lower than in the non– irradiated case: $T_{\mathrm{eff}} \lesssim 2000$ K and accordingly recurrence times are longer [3].

Acknowledgments

I am grateful to Guillaume Dubus, Anya Esin, Jean-Marie Hameury and Kristen Menou for many enlightening discussions. I thank J.-M. Hameury for data used in Fig. 2. This work was supported in part by an ASPS/CNRS grant.

References

1. Cannizzo, J.K. (1998) ApJ, 494, 318
2. Cannizzo, J.K., Chen, W., Livio, M. (1995), 454, 880
3. Dubus, G., Hameury, J.-M., Lasota, J.-P. (2000) to be submitted to A&A
4. Dubus, G., Lasota, J.-P., Hameury, J.-M., Charles, P. (1999) MNRAS, 303, 139
5. Esin, A.A. (2000) these proceedings
6. Esin, A.A., McClintock, J.E., Narayan, R. (1997) ApJ, 489, 865
7. Esin, A.A., Lasota, J.-P., Hynes, R.I. (2000) A&A, submitted
8. Gammie, C.F., Menou, K. (1998) ApJ, 492, L75
9. Hameury, J.-M., Lasota, J.-P., Huré, J.-M. (1997) MNRAS, 287, 937
10. Hameury, J.-M., Lasota, J.-P., Warner, B. (2000) A&A, in press
11. Hameury, J.-M., Lasota, J.-P., McClintock, J.E., Narayan, R. (1997) ApJ, 489, 234
12. Hameury, J.-M., Menou, K., Dubus, G., Lasota, J.-P., Huré, J.-M. (1998) MNRAS, 298, 1048
13. Idan, I., Lasota, J.-P., Hameury, J.-M., Shaviv, G., (1999) Phys. Rep. 311, 213
14. King, A.R., Ritter, H. (1998) MNRAS, 293, 42
15. Lasota, J.-P. (1999) in van Paradijs, J., van den Heuvel, E.P.J, Kuulkers, E. eds., Compact Stars in Binaries, IAU Symp. 165, Kluwer, Dordrecht, p. 43
16. Lasota, J.-P. (1999) in S. Mineshige, J.C. Wheeler eds. Disk Instabilities in Close Binaries - 25 years of the Disk Instability Model, Universal Academy Press, Tokyo, p. 191
17. Lasota, J.-P., Hameury, J.-M., Huré, J.-M. (1995) A&A, 302, 29
18. Livio, M. (1999) in J.A. Sellwood, J. Goodman eds., Astrophysical Discs - An EC Summer School, ASP Conference Series Vol. 160, p. 33
19. Menou, K., Narayan, R., Lasota, J.-P. (1999) ApJ, 513, 811
20. Menou, K., Hameury, J.-M., Lasota, J.-P., Narayan, R. (2000) MNRAS, in press
21. Meyer-Hofmeister, E., Meyer, F. (1999) A&A, 348, 154
22. Mineshige, S., Wheeler, J.C. (1989) ApJ, 343, 241
23. Narayan, R., Barret, D., McClintock, J.E. (1997) ApJ, 482, 448
24. Smak, J. (1984) Acta Astron., 32, 101
25. Smak, J. (1993) Acta Astron., 43, 161
26. Smak, J. (1999) Acta Astron., 49, 391
27. van Paradijs, J. (1996) ApJ, 464, L139
28. van Paradijs, J., McClintock, J.E. (1995), in Lewin, W.H.G., van Paradijs, J., van den Heuvel, E.P.J. eds., X-ray Binaries, CUP, Cambridge, p. 58

Black-Hole Transients
and the Eddington Limit

A.R. King

Astronomy Group, University of Leicester, Leicester LE1 7RH, U.K

Abstract. I show that the Eddington limit implies $P_{\rm crit}({\rm BH}) \simeq 3.3$ d, i.e. a critical orbital period beyond which black–hole LMXBs cannot appear as persistent systems. The unusual behaviour of GRO J1655-40 may result from its location close to $P_{\rm crit}({\rm BH})$.

1 Introduction

It is now well understood that the accretion discs in low–mass X-ray binaries (LMXBs) are strongly irradiated by the central X-ray source, and that this has a decisive effect on their thermal stability (van Paradijs 1996; King, Kolb & Burderi 1996). Irradiation stabilizes LMXB discs compared with the otherwise similar ones in cataclysmic variables (CVs) by removing their hydrogen ionization zones. In CVs this instability causes dwarf nova outbursts, and in LMXBs it produces transient outbursts rather than persistent accretion. The irradiation effect appears to be weaker if the accretor is a black hole rather than a neutron star, possibly because of the lack of a hard surface (King, Kolb & Szuszkiewicz 1997). The result is that neutron–star LMXBs with short ($\sim$ hours) orbital periods tend to be persistent, while similar black–hole binaries are largely transient. Both types of LMXBs must be transient at sufficiently long orbital periods, since a long period implies a large disc, so that a large X–ray luminosity would be needed to keep the disc edge ionized and thus suppress outbursts. We can write this stability requirement as

$$\dot{M}_{\rm crit}^{\rm irr} \sim R_{\rm disc}^2 \sim P^{4/3}, \tag{1}$$

where $\dot{M}_{\rm crit}^{\rm irr}$ is the minimum central accretion rate required to keep the disc stable, $R_{\rm disc}$ is the outer disc radius, and P is the orbital period, and we have used Kepler's law. The precise coefficient in Eq. (1) depends on uncertainties in the vertical disc structure (see the discussion in Dubus et al. 1999). Thus for large P, $\dot{M}_{\rm crit}^{\rm irr}$ will rise above any likely steady accretion rate, making long–period systems transient. This simple prediction seems to be borne out by the available evidence (see King, Frank, et al. 1997).

2 The Eddington Limit

Here I concentrate on another aspect of Eq. (1) which does not seem to have received much attention. Namely, for large enough P, $\dot{M}_{\rm crit}^{\rm irr}$ must exceed the

Eddington accretion rate

$$\dot{M}_{\text{Edd}} \simeq 1 \times 10^{-8} m_1 \text{M}_\odot \text{ yr}^{-1}, \tag{2}$$

where m_1 is the mass of the accretor in $\text{M}_\odot$. The obvious consequence of eqs. (1, 2) is that for sufficiently long orbital periods irradiation will be unable to suppress outbursts, as the required central luminosity exceeds the Eddington limit, and the system presumably cannot be both super–Eddington and persistent. Note that this conclusion holds whatever the *actual* value of the mass transfer rate in the particular binary happens to be. Thus we should expect to find no persistent LMXBs above a certain critical orbital period P_{crit}.

For the neutron–star case this was recognised by Li & Wang (1998), who found $P_{\text{crit}}(\text{NS}) \simeq 20$ d, in agreement with observation. For the black–hole case, combining (1, 2) (the latter with the normalization suggested by King, Kolb & Szuszkiewicz; 1997) gives

$$P_{\text{crit}}(\text{BH}) \simeq 3.3 \left(\frac{f_{\text{disc}}}{0.7}\right)^{-1.5} \left(\frac{\dot{M}}{0.5\dot{M}_{\text{Edd}}}\right)^{0.75} \left(\frac{m_1}{m_2}\right)^{1/8} \text{d}, \tag{3}$$

where $f_{\text{disc}} = R_{\text{disc}}/R_{\text{lobe}}$ is the disc filling fraction, i.e. its size relative to the accretor's Roche lobe, and m_2 is the companion star mass in $\text{M}_\odot$. We thus expect to find no persistent black–hole LMXBs above this period. This is indeed supported by the available data, but hardly surprising in view of the difficulty in identifying black holes in persistent systems. Note that in *high-mass* black–hole systems such as Cygnus X–1, the powerful UV luminosity of the companion star, as well as the small disc size expected in a wind–fed system, are both likely to keep the disc hot and therefore give a persistent system.

3   GRO J1655-40

The value of $P_{\text{crit}}(\text{BH})$ found above is close to the observed period $P = 2.62$ d of the soft X–ray transient GRO J1655-40. Indeed Kolb et al (1997) pointed out the system's proximity to the Eddington limit, and Hynes et al. (1998) explicitly suggested that no globally steady disc solution might be possible for this system with $\dot{M} < \dot{M}_{\text{Edd}}$. GRO J1655-40 is unusual in at least two respects:

1. The companion star has spectral type F3 – F6IV and mass $M_2 \simeq 2.3\text{M}_\odot$. On a conventional view, this places it in the Hertzsprung gap. The companion star should therefore be expanding on a thermal timescale and thus driving a mass transfer rate $-\dot{M}_2 \sim 10^{-7}\text{M}_\odot \text{ yr}^{-1}$ (Kolb et al., 1997). This is well above the appropriate value of $\dot{M}_{\text{crit}}^{\text{irr}}$, making it puzzling that the system is nevertheless transient. Regös, Tout & Wickramasinghe (1998) appeal to convective overshooting to increase the main–sequence radius of stars of $\sim 2\text{M}_\odot$. The companion might then be on the main sequence rather than

in the Hertzsprung gap. This implies a slower evolutionary radius expansion, reducing the discrepancy. However the predicted $-\dot{M}_2$ is still uncomfortably far above $\dot{M}_{\text{crit}}^{\text{irr}}$.

2. The system was first detected in an outburst in 1994, and had probably been quiescent for at least 30 yr before that. Yet two more outbursts followed in the next two years.

The considerations given here offer explanations for both of these unusual features. First, if $P > P_{\text{crit}}(\text{BH})$, the system must be transient in some sense, regardless of the actual mass transfer rate (cf. Hynes et al. 1998). Second, we see from (3) that the value of $P_{\text{crit}}(\text{BH})$ is sensitive to the filling factor f_{disc}. Thus if f_{disc} decreases, $P_{\text{crit}}(\text{BH})$ can increase above the actual orbital period, allowing irradiation to keep the disc in the high state (prolong an outburst) for as long as f_{disc} remains sufficiently small. Thus the unusual outburst behaviour of GRO J1655-40 may be explicable in terms of the time evolution of the disc size. Encouragingly there is some observational evidence (see the discussion in Orosz & Bailyn 1997) that the grazing eclipses seen in the optical are time–dependent, just as expected if the disc size varies. In fact we do expect f_{disc} to evolve systematically: in the early part of an outburst, the central accretion of low angular–momentum material will raise the average disc angular momentum and thus cause f_{disc} to increase, hence lowering $P_{\text{crit}}(\text{BH})$ and making the system more vulnerable to a return to quiescence. However at some stage matter transferred from the companion will tend to reduce the angular momentum of the outer disc, thus decreasing f_{disc}, raising $P_{\text{crit}}(\text{BH})$ and allowing irradiation to stabilize the disc in the high state. But eventually the disc must grow towards its tidal limit, increasing f_{disc} and thus lowering $P_{\text{crit}}(\text{BH})$ again, finally enforcing a return to quiescence. Obviously a full disc code is required to follow this sequence in detail and to check if it can account qualitatively for the unusual outburst behaviour of GRO J1655–40.

4   Conclusions

I have shown that the Eddington limit implies a critical orbital period $P_{\text{crit}}(\text{BH})$ beyond which black–hole LMXBs cannot appear as persistent systems. The unusual behaviour of GRO J1655-40 may result from its location very close to $P_{\text{crit}}(\text{BH})$. This system, and those at longer orbital periods, probably have central accretion rates which are highly super–Eddington during outbursts. Since observed radiative luminosities are mildly sub–Eddington, most of this mass must be expelled. Strong support for this comes from the observation of P Cygni profiles in GRO J1655–40 (Hynes et al. 1998). The superluminal jets observed (Hjellming & Rupen 1995) in an outburst of this system may therefore simply represent the most dramatic part of this outflow.

Acknowledgment

I gratefully acknowledge the support of a PPARC Senior Fellowship.

References

1. Dubus, G., Lasota J.-P., Hameury, J.-M., Charles, P.A., 1996, MNRAS, 303, 139
2. Hjellming, R.M., Rupen, M.P., 1995, Nat, 375, 464
3. Hynes, R.I., Haswell, C.A., Shrader, C.R., Chen, W., Horne, K., Harlaftis, E.T., O'Brien, K., Hellier, C., Fender, R.P., 1998, MNRAS, 300, 64
4. King, A.R., Frank, J., Kolb, U., Ritter, H., 1997, ApJ, 484, 844
5. King, A.R., Kolb, U., Burderi, L., 1996, ApJ, 464, L127
6. King, A.R., Kolb, U., Szuszkiewicz, E., 1997, ApJ, 488, 89
7. Kolb, U., King, A.R., Ritter, H., Frank, J., 1997, ApJ 485 L33
8. Li, X.-D., Wang, Z.-R., 1998, ApJ, 500, 935
9. Orosz, J.A., Bailyn, C.D., 1997, ApJ, 477, 876
10. Regös, E., Tout, C.A., Wickramasinghe, D., 1998, ApJ, 509, 362

Limits on Accretion Rates
in Black Hole X-Ray Transients

Emmi Meyer-Hofmeister and Friedrich Meyer

Max-Planck-Institut für Astrophysik,
Karl-Schwarzschild-Str. 1, D-85740 Garching, Germany

Abstract. Several black hole X-ray transients have a very long recurrence time, decades of years [2]. We interpret this as a marginal occurrence of a dwarf nova type disk instability in the cool outer accretion disk. We compute how the recurrence time varies with the mass overflow rate from the companion star. Evaporation of the inner accretion disk is an essential feature of disk evolution which depends on the black hole mass. For accretion rates only slightly lower than those in the known black hole X-ray transients the disk does not reach the instability. We argue that many of those stationary, optically faint black hole X-ray binaries exist.

1 Evolution of the Accretion Disk in BHSXTs During Quiescence

We compute the evolution of the cool geometrically thin outer disk in equilibrium with a hot corona above. The inner edge of the thin disk is reached where the coronal flow via evaporation has picked up all of the mass flow brought in in the cool disk. The mass flow rate in the cool disk and the mass of the black hole determine this location.

Parameters chosen are:

- mass of the black hole $M_1 = $ 4, 6, 8, 12 $M_\odot$
- binary orbital period $P = $ 4, 8, 16 hours (enters in determination of size of the accretion disk)
- mass of companion star according to orbital period (assumed: Roche lobe-filling main sequence).
- rates of mass transferred from the companion star
  1 - 15 $10^{-10} M_\odot$/yr
- viscosity in the cool state $\alpha_{cool} = 0.05$

2 Results of Computations

During quiescence matter accumulates in the disk until the surface density reaches a critical value to trigger the thermal-viscous instability in the thin disk. The inner edge of the disk inside of which all matter flows in form of a coronal flow depends on the mass flow rate and the black hole mass. It is farther in for higher rates and farther out for larger black hole mass.

2.1   Outburst Recurrence Time

In Fig. 1 we show how the recurrence time depends on black hole mass, the orbital period and mass transfer rate.

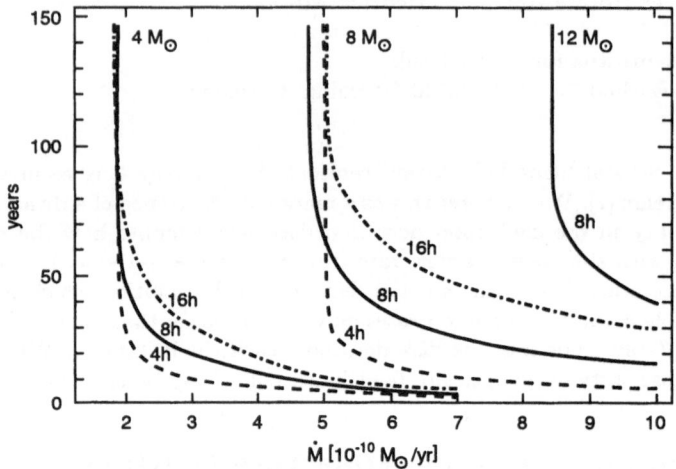

Fig. 1. Outburst recurrence time as a function of mass transfer rate in $10^{-10} M_\odot/\mathrm{yr}$ for different black hole masses and orbital periods

For the black hole transients with long recurrence time a small difference in the mass transfer rate causes a large difference in recurrence time. This explains the observed wide range of recurrence times of black hole X-ray transients.

2.2   Accumulated Matter in the Disk

The computed amount of matter in the disk lies in the range $3\text{-}10 \times 10^{24}\mathrm{g}$ for $M_1 = 4 M_\odot$ and about a factor 3 higher for 8 $M_\odot$. This agrees with the amount estimated from the observed outbursts of transients, for example for A0620-00 [1].

The fraction of matter accumulated during quiescence to matter transferred from the companion star decreases when the outbursts occur more rarely. For a recurrence time of about 50 years our computations yield the value 0.35 to 0.40, the same for 4 and $8 M_\odot$ black holes.

References

1. McClintock J.E., Horne K., Remillard R.A. (1983) The dim accretion disk of the quiescent black hole A0620-00. ApJ **442**, 358–365
2. Tanaka Y., Shibazaki N., (1996) X-ray novae. ARA&A **34**, 607–644

Supermassive Black Holes Can Hardly Be "Silent"

Alvio Renzini

European Southern Observatory, Garching b. München, Germany

Abstract. There is now ample evidence that most - perhaps all - galactic spheroids host a supermassive BH at their center. This has been assessed using a variety of observational techniques, from stellar and/or gas dynamics to megamasers. Yet another promising technique is offered by the case of the Virgo elliptical NGC 4552, in which early HST/FOC observations revealed a central low-luminosity flare. Subsequent HST/FOS observations with a 0.21 arcsec aperture have revealed a rich emission-line spectrum with broad and narrow components with FWHM of 3000 and 700 km/s, respectively. This variable, mini-AGN at the center of NGC 4552 is most naturally the result of a sporadic accretion event on a central BH. It has a Hα luminosity of only $\sim 10,000\ L_\odot$, making it the likely, intrinsically faintest AGN known today. Only thanks to the superior resolution of HST such a faint object has been discovered and studied in detail, but adaptive optics systems on large ground-based telescopes may reveal in the future that a low level of accretion onto central massive BHs is an ubiquitous phenomenon among galactic spheroids. FOC/FOS observations of a central spike in NGC 2681 reveal several analogies with the case of NGC 4552, while yet another example is offered by a recent exciting finding with STIS by R.W. O'Connell in NGC 1399, the third galaxy in our original program.

1 Discovery of a Central Flare in NGC 4552

Sophisticated techniques have been used to infer the presence of supermassive black holes (BH) at the center of galactic spheroids: from detailed stellar dynamical modeling fitting 2D spectroscopy data cubes to megamaser observations, all aspects widely discussed at this meeting. Here I will present observational evidence concerning three randomly-selected galactic spheroids, showing that high spatial resolution imaging and/or spectroscopy can reveal sporadic mini-AGN activity likely related to the presence of a central BH.

To investigate the UV upturn of ellipticals, in 1993 HST/FOC images in several UV bands were obtained for the central regions of the elliptical galaxies NGC 1399 and NGC 4552 and of the bulge of the S0/a galaxy NGC 2681. A point-like source – or *spike* – was evident at the center of both NGC 4552 and NGC 2681, with their photometric profile being indistinguishable from the PSF of the pre-COSTAR HST (Paper I [1]).

Comparison with another FOC image of NGC 4552 taken in 1991 [2] showed that this spike had increased its luminosity in the U band (F342W) by a factor $\sim 7 \pm 1.5$ between the two epochs, reaching $\sim 10^6 L_\odot$ (Paper I). A second point-like source is also present in the 1991 image, $\sim 0''.14$ from the central spike and with nearly the same luminosity. While both sources were detected at the $\sim 4\sigma$ level in the 1991 image, the off-center source was not detectable in the 1993 image at the $\sim 2.5\sigma$ level.

In Paper I several possible interpretations were discussed for the origin of this *flare* at the center of NGC 4552, favoring an accretion event onto a central massive black hole (BH). The accreted material could have been tidally stripped from a star in a close fly-by with the BH, though alternative ways of feeding the BH were also considered. The low luminosity of the spike was nevertheless at variance with the predicted luminosity in the case of a total stellar disruption [3,4], hence the case of partial stripping was favored, being also such an event much more frequent than total disruptions.

2   More HST Follow Up Reveals a Mini AGN in NGC 4552

If the flaring spike was due to accretion onto a BH, its spectrum should show prominent, rotationally broadened emission lines, typical of an accretion disk. To test this expectation further HST observations were obtained in 1996, both in imaging with FOC and in spectroscopy with FOS. The results of these later HST observations are extensively reported in Paper II [5].

2.1   FOC Imaging

The 1991, 1993, and 1996 observations were made with different telescope-instrument configurations and therefore differences in detector efficiency and non-linearity effects were carefully treated along with an adequate modeling of the PSF. The analysis revealed that the center spike in NGC 4552 increased its U-band luminosity by a factor of 4.5 from 1991 to 1993 and faded from 1993 to 1996 by a factor ~ 2 at all observed wavelengths (1700–3500 Å) (Paper II). In 1996 the UV fluxes indicated a black-body temperature of $T \sim 15,000$ K (assuming the emission to be thermal) implying a spike bolometric luminosity of $\sim 3 \times 10^5 \, L_\odot$ (at a distance of 15.3 Mpc). The off-center spike that was present in 1991 did not appear in any of the later images.

2.2   FOS Spectroscopy

The 1996 FOS spectra were obtained through a $0''.21 \times 0''.21$ aperture centered on the spike and covering the range from ~ 2200 to ~ 8500 Å. Fig. 1 compares the 1996 FOS spectrum of this central region with a composite spectrum meant to represent the spectrum of the inner $r \leq 7''$ of this galaxy. This composite spectrum is a good match to the spectral energy distribution (SED) of the FOS spectrum, with two notable exceptions: 1. the FOS spectrum shows strong emission lines that are absent in the composite spectrum,

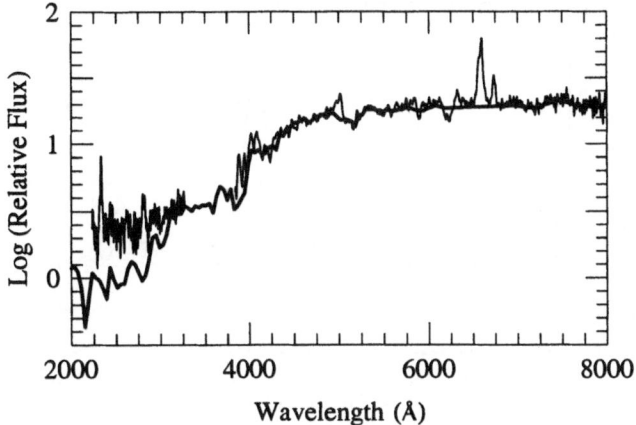

Fig. 1. The overall 1996 FOS spectrum of NGC 4552 within the $0''.21 \times 0''.21$ aperture centered on the spike (thin line), is superimposed to a scaled combination of the IUE $10'' \times 20''$ aperture of NGC 4552 [6] matched to ground-based optical spectrum of NGC 4649, a giant elliptical whose SED is virtually the same as that of NGC 4552 [7] (thick line). The spectra have been normalized to the visual region. The FOS spectrum shows a strong UV excess and prominent emission lines.

and 2. shortward of 3200 Å the FOS spectrum is far stronger than the IUE SED.

The most prominent emission lines include C II] $\lambda2326$, MgII $\lambda2800$, [O II] $\lambda3727$, [S II] $\lambda4072$, Hβ, [O III] $\lambda\lambda4959, 5007$, [N I] $\lambda5700$, [O I] $\lambda6300$, [N II] $\lambda\lambda6548, 6583$, Hα and [S II] $\lambda\lambda6717, 6731$. The emission line ratios of the narrow components place the spike among extreme AGNs, while the [O III]/Hβ ratio falls just on the borderline between Seyferts and LINERs.

The emission line profiles were fitted with gaussian components, reaching the following main conclusions: 1. Good fits of the emission lines can be obtained only with a combination of broad and narrow components for *both* the permitted as well as the forbidden lines. 2. The emission lines are very broad, with very large velocity widths for both the broad (FWHM $\simeq$ 3000 km s^{-1}) and narrow components (FWHM $\simeq$ 700 km s^{-1}). 3. The shape of the Hα+[N II] complex has definitely changed from the 1996 spectrum to a FOS spectrum taken 8 months later, indicating a shift to the blue of $\sim$ 230 km s^{-1} of the whole (narrow + broad) Hα line.

2.3 Interpretation

This complex phenomenology clearly points towards the presence of a very weak AGN at the center of this galaxy, most likely powered by a low level of accretion onto a central BH. Other interpretations were already considered unlikely, given the early evidence (Paper I). The additional evidence gathered in 1996 was conclusive in this respect. The mass of the supermassive BH at

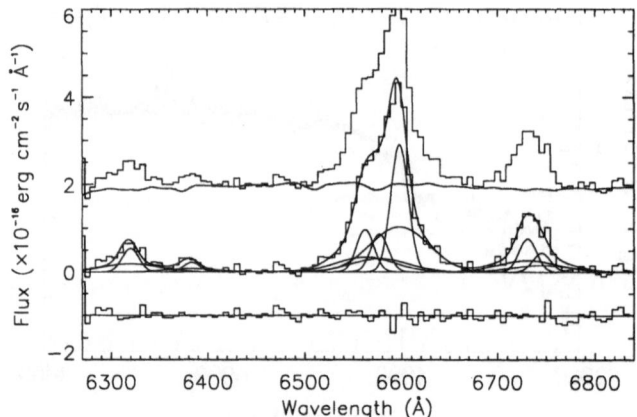

Fig. 2. The observed [O I], [N II], Hα, and [S II] lines in the red region of the FOS G780H spectrum (top) together with their gaussian decomposition into narrow and broad components (middle) and the corresponding residuals (bottom). The narrow and broad components have FWHM≃ 700 and ≃ 3000 km s^{-1}, respectively.

the center of NGC 4552 was then estimated to be between 3×10^8 to 2×10^9 $M_\odot$, consistent with other ground-based estimates as well as with the BH mass–bulge mass (or -bulge luminosity) relation [8,9].

With the adopted distance of 15.3 Mpc the broad Hα luminosity is $\sim$ 5.6×10^{37} erg s^{-1}, a factor of two less than the broad Hα luminosity of the (previously) faintest known AGN, the Seyfert 1 nucleus of NGC 4395, and a $\sim$ 20 times fainter than M81, the next faintest Seyfert 1 nucleus [10].

As suggested in Paper I, this phenomenology is generically consistent with a scenario in which the flare is caused by the tidal stripping of a star in a close flyby with a central supermassive BH. From the theoretical side, only the extreme case of the total disruption of a $\sim 1\, M_\odot$ main sequence star has been widely investigated so far (e.g. [3,4,11]). The frequency of such events is estimated to be of one every $\sim 10^3 - 10^4$ yr in a giant elliptical galaxy [3]. The flare is predicted to be very bright for several years ($\sim 10^{10}L_\odot$), much brighter than the observed flare. This indicates that if the flare in NGC 4552 was caused by a tidal stripping in a BH-star flyby, then this flyby was rather wide and led to only partial stripping. One expects wider flybys to be vastly more frequent than the hard ones causing total disruption. To be consistent with the observed luminosity, only $\sim 10^{-3}M_\odot$ should have been stripped (Paper I), perhaps more if an ADAF is established.

Tidal stripping of a star is but one possible way to feed matter to a massive central BH at a low rate. Other mechanisms mentioned in Paper II include: a) Roche lobe overflow from a star in bound orbit around the BH; b) accretion from a clumpy interstellar medium which is actually seen near the nucleus (see Fig. 12 in Paper II); or c) gas fed to the BH via a cooling flow within the X-ray emitting hot interstellar medium. Concerning alternative c), occasional cooling catastrophes lead to a transient major cooling flow that can feed the central BH, while mini-inflows may be active most of the time. Such flows can lead to sizable excursions in the (mini-)AGN luminosity (flickering), reminiscent of the behavior of the spike in NGC 4552 [12,13].

It may well be that each of these mechanisms operates from time to time in the central regions of elliptical galaxies like NGC 4552, and in all these options an accretion disk is established, hence no clear cut discrimination can easily be achieved. However, some evidence favoring the tidal stripping option comes from the noticed shift in the broad Hα emission between 1996 and 1997. Tidal stripping/disruption is indeed predicted to produce an *elliptical* accretion disk which precession results in Hα line profile variations [14]. Finally, in Paper II it was speculated that the off-center spike seen only in the 1991 image could be due to a relativistic jet emerging from the central mini-AGN having shocked the dense dusty ring seen at nearly the same distance, with the jet having been possibly produced by a previous accretion event.

3   Also NGC 2681 and NGC 1399 Display Central Activity

In 1997 FOC and FOS observations were secured also for the central spike in NGC 2681, with a virtually identical strategy to that used for NGC 4552 (Paper III [15]). The main results can be summarized as follows.

The photometric profile is well represented by a *Nuker-law* of the *power-law* type, from the innermost $0''.005$ up to $\sim 100''$ from the center. Given the very high surface brightness of the central regions, a transient UV spike such as that in NGC 4552 would have not been revealed in either the FOC images (see Fig. 3) or in the FOS UV continuum. The UV continuum and the presence of Balmer absorptions indicate that the central region of NGC 2681 is dominated by the light of a relatively young stellar population (1–2 Gyr).

Contrary to the case of NGC 4552, the FOS spectrum of the innermost $0''.21 \times 0''.21$ region matches very well the IUE UV spectrum, in spite of the much larger (a factor of $\sim$4000) area sampled by the IUE aperture. Together with the absence of gradients in the (Far-UV)−(Near-UV) and UV−IR colors, this implies the homogeneity of the stellar population within $r \lesssim 10''$. The FOS shows emission lines whose ratios indicate that the nucleus is a LINER. The emission lines are well-modeled by a single Gaussian with FWHM$\simeq$ 480 km s^{-1}, which is a factor ~ 2 higher than that measured from the ground, within a $2'' \times 4''$ aperture, indicating the presence of a central mass concentration.

This steepening of the (gas) velocity dispersion is not accounted for by a spherical isotropic dynamical model with constant M/L, derived by deprojecting the Nuker-law. The same kind of model gives a good fit to the FOS data when assuming a central dark mass (BH) with $M_{\bullet} \lesssim 6 \times 10^7 M_{\odot}$, consistent with the BH mass–bulge luminosity relation [9]. This holds under the assumptions that the emitting gas has an isotropic velocity-dispersion tensor and that its density is proportional to the stellar density. Models without a

BH can also fit the data if these assumptions are relaxed, e. g. either the nuclear gas is in a disk, or gas clouds are on radially-anisotropic orbits close to the nucleus. In summary, there are quite many analogies between the central spikes in NGC 4552 and NGC 2681. The possibility that the line emission in the NGC 2681 spike originates in an accretion disk seen more face on than in NGC 4552 cannot be ruled out.

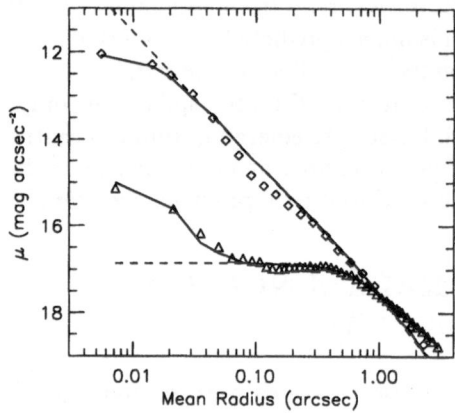

Fig. 3. The FOC U-band surface brightness profiles of the innermost $3''$ of NGC 4552 (lower lines) and NGC 2681 (upper lines). The dashed lines represent the best fit Nuker laws and the solid lines their convolution with the PSF of the FOC, with a point-like source being added in the case of NGC 4552.

On 1993 FOC images the central $\sim 20''$ of NGC 1399 appear *as smooth as silk*, i.e. no spike similar to either that of NGC 4552 or that of NGC 2681 is present. Indeed, NGC 1399 is well fit by a Nuker law of the core type, and its central surface brightness is much lower than that of NGC 2681. However, a recently obtained STIS far UV spectrum of NGC 1399 (R.W. O'Connell, private communication) appears to contain a distinct nuclear point source –definitely above the background light of the galaxy's core– which may be even brighter than the spike in NGC 4552. A very nice surprise indeed!

4   No Supermassive Black Hole Can Be Totally Silent

The case of the flare in NGC 4552 adds yet another option at our disposal to gather evidence for supermassive BHs lurking at the center of apparently *in-active* galactic nuclei. Supermassive BHs in galactic spheroids sit at the bottom of their gravitational potential well, where stellar and ISM densities reach their maximum, and where any cannibalized material tends to converge. As such, one is tempted to say that the real problem is how to *avoid* a low, fluctuating level of accretion onto a massive BH — hence of low level AGN activity — rather than how to produce it. After all, it must be hard for guests as bulky as 10^8 or 10^9 $M_\odot$ to really hide at the center of galaxies: wherever a massive BH exists, its presence is likely to be betrayed by at least a low level of AGN-like *activity*.

The case of NGC 4552 offers a lesson in this respect. Thanks to its angular resolution, HST observations in either UV or optical imaging or narrow aperture spectroscopy allow to reveal a mini-AGN activity which would be essentially invisible to similar ground-based observations. However, a high spatial resolution comparable to that of HST is now possible also from the ground at near-IR wavelength thanks to adaptive optics (AO). AO-fed, near-IR, 1D and 2D spectroscopy of the center of galactic spheroids (e.g., with SINFONI at the VLT) might reveal broad emission lines of the Paschen and Brackett series, hence potentially offering additional opportunities for the demography of central BHs in galactic spheroids. The generic prediction is that virtually all galactic spheroids (and especially those for which dynamical evidence exists) should show a low-level AGN activity similar to that so clearly detected by HST in NGC 4552.

I would like to thank Bob O'Connell for his permission to mention his discovery of a newly appeared point-like source in NGC 1399. It is also a pleasure to thank the Paper I-III team, and especially Michele Cappellari who made an effort to conclude the analysis of NGC 2681 in time for this meeting.

References

1. Renzini, A., Greggio, L., Di Serego Alighieri, S., Cappellari, M., Burstein, D., & Bertola, F. 1995, Nature, 378, 39 (Paper I)
2. Crane, P., Stiavelli, M., King, I.R., Deharveng, J.M., Albrecht, R., Barbieri, C., Blades, J.C., Boksenberg, A., et al. 1993, AJ, 106, 1371
3. Rees, M.J. 1990, Science, 247, 817
4. Kochanek, C.S. 1994, ApJ, 422, 508
5. Cappellari, M., Renzini, A., Greggio, L., Di Serego Alighieri, S., Buson, L.M., Burstein, D., & Bertola, F. 1999, ApJ, 519, 117 (Paper II)
6. Burstein, D., Bertola, F., Buson, L.M., Faber, S.M., & Lauer, T.R. 1988, ApJ, 328, 440
7. Oke, J.B., Bertola, F., & Capaccioli, M. 1981, ApJ, 243, 453
8. Magorrian, J., Tremaine, S., Richstone, D., Bender, R., Bower, G., Dressler, A., Faber, S.M., Gebhardt, K., et al. 1998, AJ, 115, 2285
9. van der Marel 1999, in Galaxy Interactions at Low and High Redshifts (IAU Symp. 186), eds. J. E. Barnes & . B. Sanders (Dordrecht: Kluwer), in press
10. Filippenko, A.V., Ho, L.C., & Sargent, W.L.W. 1993, ApJ, 410, L75
11. Ulmer, A. 1999, ApJ, 514, 180
12. Ciotti, L., Pellegrini, S., Renzini, A., & D'Ercole, A. 1991, ApJ, 376, 380
13. Ciotti, L. & Ostriker, J.P. 1997, ApJ, 487, L105
14. Eracleous, M., Livio, M., Halpern, J.P., & Storchi-Bergmann, T. 1995, ApJ, 438, 610
15. Cappellari, M., Bertola, F., Burstein, D., Buson, L.M., Greggio, L., & Renzini, A. 2000, submitted to ApJ (Paper III)

Discovery of an X-Ray Outburst
in an Optically Non-Active Galaxy

Thomas H. Reiprich[1] and Jochen Greiner[2]

[1] Max-Planck-Institut für extraterrestrische Physik, Giessenbachstraße 1,
D-85740 Garching, Germany
[2] Astrophysical Institute Potsdam, An der Sternwarte 16, D-14482 Potsdam,
Germany

Abstract. The main X-ray and optical properties of a very luminous X-ray outburst source are summarized. The results are: large variability in X-ray emission
(count rate rise of a factor of 10 within 7 days), soft X-ray spectrum (photon index
= –3.8), giant peak X-ray luminosity (10^{44} erg/s), previously unknown non-active
optical counterpart (redshift = 0.05).

1 Introduction

Galaxies typically show constant X-ray emission. Variable X-ray emission is
common among AGN, typically about a factor of 2–3. Outbursts exceeding
a factor of ~ 10 are rare even among AGN. It was therefore quite surprising
when recently large UV/X-ray outbursts from optically 'normal' galaxies were
discovered (Renzini et al. 1995 [1]; Bade et al. 1996 [2]; Komossa & Greiner
1999 [3]; Grupe et al. 1999 [4]). Exciting possible outburst scenarios have
been discussed, including the tidal disruption of a star by a supermassive
black hole, based on suggestions of Rees (1988) [5]. However, it is always
difficult to safely identify an optical counterpart. Because of the intriguing
possible implications of these observations one would certainly rather rely
on a larger number of objects. Here we report the discovery of another soft
X-ray outburst in an optically non-active galaxy, thereby fortifying further
that this phenomenon deserves a detailed theoretical understanding.

2 Observations

X-ray observations: The outburst source was discovered during analysis of
5 archival ROSAT PSPC pointed observations of the galaxy cluster Abell
3560, where the source appeared only in 2 observations. Applying the source
detection algorithm incorporated in the EXSAS package (Zimmermann et
al. 1998 [6]) yields a source position R.A.: 13H31M57.6S, Dec.: –32D43M20S
(J2000). Within 7 days the source count rate rises about a factor of 10. Six
months before and after the burst no emission is detected. A power law fit
to the spectrum yields a photon index -3.8 ± 0.6. Using the *peak* count rate
(~ 0.1 cts/s) yields a giant luminosity $L_X(0.1 - 2.4\,\text{keV}) \sim 10^{44}\,h_{50}^{-2}$erg/s.

Optical observations: An optical image taken at the Danish 1.5m telescope, ESO/La Silla, 6 years after the burst shows only one bright object within the ROSAT PSPC error circle. The spectrum of this anonymous galaxy does not show any emission lines typical for AGN. Absorption lines yield a redshift $z = 0.051$, similar to the redshift of Abell 3560.

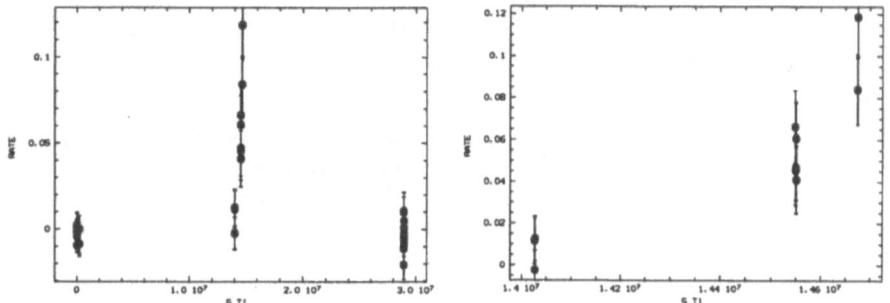

Fig. 1. Source light curve (cts/s over s) of the 5 PSPC observations and blow-up.

3   Conclusions

Our discovery brings the number of detected UV/soft X-ray outbursts from optically non-active galaxies to 5. Even though these objects differ in their properties, like X-ray spectrum, light curve and peak luminosity (for the latter, however, only lower limits can be derived from the data), they have in common that a large amount of energy is released in a relatively short time period from an otherwise boring looking galaxy. Several outburst scenarios have been discussed by Komossa & Bade (1999) [7] and Komossa & Greiner (1999) [3], who found that all successful models involve the presence of a supermassive black hole. Explicitly the tidal disruption of a star is favored, which may also explain the observations presented here.

Acknowledgements: We acknowledge enlightening discussions with Stefanie Komossa.
Images, spectra, light curves etc. of this source can be found at:
http://www.xray.mpe.mpg.de/~reiprich/act/publi.html

References

1. Renzini A., Greggio L. et al. (1995) Nature **378**, 39
2. Bade N., Komossa S., Dahlem M. (1996) A&A **309**, L35
3. Komossa S., Greiner J. (1999) A&A **349**, L45
4. Grupe D., Thomas H.-C., Leighly K. M. (1999) A&A **350**, L31
5. Rees M. J. (1988) Nature **333**, 523
6. Zimmermann U., Boese G. et al. (1998) EXSAS User's Guide, MPE Report July
7. Komossa S., Bade N. (1999) A&A **343**, 775

The Giant X-Ray Flare of NGC 5905 – a Tidal Disruption Event?

Stefanie Komossa

Max-Planck-Institut für extraterrestrische Physik, Giessenbachstr., D-85748 Garching; skomossa@xray.mpe.mpg.de

Abstract. We review our work on, and extent the discussion of, the giant X-ray flare from the HII-type galaxy NGC 5905 (Bade et al. 1996, Komossa & Bade 1999), which we consider to be presently the best candidate for a tidal disruption or tidal stripping event at the center of an otherwise *non-active* galaxy.

1 Observations

A giant X-ray flare from the center of the nearby spiral galaxy NGC 5905 was discovered during the *ROSAT* all-sky survey (Bade, Komossa & Dahlem 1996). Remarkably, the optical spectrum of NGC 5905 does *not* show any signs of Seyfert activity. The optical and X-ray observations can be summarized as follows:

- huge peak luminosity of $L_{x,0.1-2.4keV} \gtrsim 10^{42.5-43}$ erg/s during the outburst,
- giant amplitude of variability of a factor $\gtrsim 200$ in total,
- ultra-soft X-ray spectrum during outburst with $kT_{bb} \simeq 0.06$ keV when fit by a black body model,
- the spatial location of the X-ray emission is the nucleus of NGC 5905,
- no change in optical brightness simultaneous to the X-ray outburst, and and no long-timescale optical variability between ~1960 and 1996 (within the limits of the sensitivity of the photographic plates of *Sonneberg observatory*),
- high quality optical spectra of this galaxy prior to the X-ray flare (Ho et al. 1995), and several years after the outburst (Schombert 1998, Komossa & Bade 1999) are of HII-type, with *no signs of Seyfert activity.*

2 Outburst Scenario

Several outburst scenarios suggest themselves. However, most of them do not survive close scrutiny because they cannot account for the huge maximum luminosity (e.g., a supernova in dense medium), require extreme fine-tuning (e.g., a warm-absorbed hidden Seyfert nucleus), are inconsistent with the optical observations (gravitational lensing), or predict a different temporal behavior (X-ray afterglow of a Gamma-ray burst). A scenario consistent with the observations is the flare of a tidally disrupted star accreted by a supermassive black hole at the center of NGC 5905, or stripping of a star's atmosphere (see Komossa & Bade 1999 for details).

The idea of tidal disruption of stars by a supermassive black hole (SMBH hereafter) was originally studied as a possibility to fuel AGN, but dismissed. Rees (1988, 1990) proposed to use such events to detect SMBHs in nearby *non-active* galaxies. The debris of the disrupted star is accreted by the BH. This produces a flare of electromagnetic radiation, lasting on the order of months. The luminosity emitted if the BH is accreting at its Eddington luminosity can be estimated by $L_{edd} \simeq 1.3 \times 10^{38} M/M_\odot$ erg/s. In case of NGC 5905, a BH mass of at least $\sim 10^5$ $M_\odot$ would be required to produce the observed L_x, and a higher mass if L_x was not observed at its peak value. For comparison, BH masses of $M_{BH} \lesssim 10^{6-7} M_\odot$ have recently been reported by Salucci et al. (1999) for the centers of some late-type spiral galaxies.

NGC 5905 is the second *non-active galaxy* among the UV–X-ray outbursting ones (the first was the UV outburst of NGC 4552 observed with HST by Renzini et al. (1995, and this meeting); but note that there are several indications of low-level activity in this galaxy). A third example is RX J1242-1119 (Komossa & Greiner 1999, and these proceedings); see Grupe et al. 1999 and Reiprich & Greiner (this meeting) for two further candidates.

The X-ray outburst in the HII galaxy NGC 5905 then lends further support to the scenario that *all* galaxies passed through an active phase, with presently dormant SMBHs at their centers which are occasionally fueled by disrupted stars.

Acknowledgements: The *ROSAT* project has been supported by the German Bundesministerium für Bildung und Wissenschaft (BMBW/DLR) and the Max-Planck-Society. It is a pleasure to thank David L. Meier for discussions.
This and related papers can be retrieved from our webpage at
http://www.xray.mpe.mpg.de/~skomossa/

References

1. Bade N., Komossa S., Dahlem M., 1996, A&A 309, L35
2. Ho L.C., Filippenko A.V., Sargent W.L.W., 1995, ApJS 98, 477
3. Grupe D., et al., 1999, A&A 350, L31
4. Komossa S., Bade N., 1999, A&A 343, 775
5. Komossa S., Greiner J., 1999, A&A 349, L45
6. Rees M.J., 1988, Nature 333, 523
7. Rees M.J., 1990, Science 247, 817
8. Renzini A., et al., 1995, Nature 378, 39
9. Salucci P., et al., 1999, MNRAS, in press (astro-ph/9812485v2)
10. Schombert J., 1988, ApJ 116, 1650

A Giant, Ultra-Soft, and Luminous X-Ray Flare from the Optically Inactive Galaxy Pair RX J1242.6−1119

Stefanie Komossa[1] and Jochen Greiner[2]

[1] Max-Planck-Institut für extraterrestrische Physik, Giessenbachstr.,
  D-85748 Garching; skomossa@xray.mpe.mpg.de
[2] Astrophysikalisches Institut Potsdam, An der Sternwarte 16, D-14482 Potsdam

Abstract. We discuss the detection of a giant X-ray flare ($L_x \gtrsim 10^{44}$ erg/s) from the direction of the previously unknown, optically inactive galaxy pair RX J1242.6-1119 and investigate outburst scenarios. The most likely one is an accretion event onto a supermassive black hole in the center of one of the two galaxies, e.g., by a tidal disruption event.

1 X-Ray and Optical Observations

The X-ray source RXJ 1242.6-1119 showed strong variability between two *ROSAT* observations taken 2 yrs apart: Whereas the source is not detected during the *ROSAT* all-sky survey (countrate $CR < 0.015$ cts/s), it is present in a pointed PSPC observation with $CR = 0.3$ cts/s, revealing variability by at least a factor of 20. The high-state spectrum is ultra-soft ($kT_{bb} = 0.06$ keV when fit by a black body). We derive an intrinsic luminosity of $L_x > 9\,10^{43}$ erg/s in the ROSAT band (0.1–2.4 keV), using $H_0 = 50$ km/s/Mpc. Since we most likely have not caught the source exactly at maximum light, since the spectrum may extend into the EUV, and since we have conservatively assumed no X-ray absorption intrinsic to RXJ 1242-11, the real peak luminosity is likely to be much higher.

Optical imaging shows two galaxies in the X-ray error circle. Other objects are fainter then $\sim 22^m$. Our optical spectroscopy performed several years after the X-ray high-state observation reveals that both galaxies are *non*-active; no Seyfert-typical emission lines are detected (see Komossa & Greiner 1999 for details). We derive redshifts of $z=0.05$ for the two galaxies.

2 Outburst Scenarios

Both, the dramatic X-ray variability and the huge outburst luminosity of RXJ 1242-11 are very rare among *non-active* galaxies. The only previous cases are the UV outburst in NGC 4552 (Renzini et al. 1995) and the X-ray outburst of NGC 5905 (Bade, Komossa & Dahlem 1996, Komossa & Bade 1999). In the following, we discuss several potential outburst scenarios (see Komossa & Greiner 1999 for details):

- *Variable galactic foreground star:* This is unlikely, since further optical sources within the X-ray error circle are extremely weak. Given the high L_X/L_{opt} value, an ISM accreting neutron star might come to mind, but the strong X-ray variability would require an extreme ISM density gradient, thus leading us to reject this possibility.
- *Stellar sources within the galaxy pair* (like X-ray binaries, or a supernova in dense medium) fall several orders of magnitude short in explaining the huge outburst luminosity.
- *Accretion disk instability:* An accretion disk around a central SMBH in one of the two galaxies may exhibit burst-like oscillations due to, e.g., thermal instabilities. However, in case of repeated such outbursts in RX J1242-11, one would expect to see permanent AGN-typical NLR emission lines (like [OI] or [OIII]) in the optical spectrum of one of the two galaxies which are not detected.
- *Accretion event onto a SMBH:* The giant peak luminosity of at least ~ 10^{44} erg/s in the (0.1–2.4 keV) energy band suggests that we have seen the flare of a star that was tidally disrupted by a SMBH (Rees 1988, 1990) at the center of one of the two galaxies.

Such X-ray outbursts provide important information on the presence of SMBHs in non-active galaxies, the accretion history of the universe, and the link between active and normal galaxies. Future X-ray surveys (like the one that was planned with *ABRIXAS*, or the one that will be carried out with *MAXI*) will be valuable in finding further of these outstanding sources.

Acknowledgements: The *ROSAT* project has been supported by the German Bundesministerium für Bildung und Wissenschaft (BMBW/DLR) and the Max-Planck-Society. It is a pleasure to thank Joachim Trümper, Jules Halpern, Dirk Grupe, David L. Meier, and Weimin Yuan for fruitful discussions.
This and related papers can be retrieved from our webpage at
http://www.xray.mpe.mpg.de/~skomossa/

References

1. Bade N., Komossa S., Dahlem M., 1996, A&A 309, L35
2. Komossa S., Bade N., 1999, A&A 343, 775
3. Komossa S., Greiner J., 1999, A&A 349, L45
4. Rees M.J., 1988, Nature 333, 523
5. Rees M.J., 1990, Science 247, 817
6. Renzini A., et al., 1995, Nature 378, 39

New Evidence for Supermassive Binary Black Holes in the Blazar OJ287

Aimo Sillanpää

Tuorla Observatory, FIN-21500 Piikkiöö, Finland

Abstract. There have been many suggestions about the existence of supermassive binary black hole systems in the nuclei of Active Galactic Nuclei (AGN). One of the best candidates has been a blazar called OJ287 where we have found a possible periodicity of about 12 year in the optical major outbursts. However, there has been a confusion if the periodicity is exactly strict or not. In this new work we will show clear evidence that the periodicity is really amazingly strict and this gives strong support to the binary black-hole model developed for OJ287.

1 The Case of OJ287

Because of the suitable position on the sky quite close to the ecliptic, OJ287's historical light curve covers already more than 100 years [1]. Based on this light curve Sillanpää et al., 1988 [1] proposed a supermassive binary black hole (SMBBH) model for the optical outbursts being repeated every 12 years. In this model the outbursts are caused by tidally triggered mass inflow in the accretion disk around the primary black hole that is disturbed by the closest approach of the secondary black hole during its eccentric orbit around the primary. Based on the historical light curve Sillanpää et al., 1988 [1] predicted that the next outburst should occur at the end of the year 1994. This outburst also came just in time [2].

2 Strict Periodicity or not?

Because of the huge variability of OJ287 (by a factor of 150) it is not possible to use "normal" periodicity search methods [3]. The only way to check if the suggested periodicity is strict or not is to plot the best observed periods into one plot. This is shown in Fig.1. where we have used the time difference between the last two, best observed outbursts (11.86 year). This can be determined with an accuracy of about two weeks. The vertical line just in the middle of the graph shows the exact time when the outbursts should occur. It is very easy to see that the periodicity has been extremely strict, as strict as it can be with these time gaps. This result simply rules out all the models where the outburst period is changing more than two weeks during the last 60 years and on the other hand this gives very strong support to the SMBBH model behind the outbursts because there are only two systems *in astronomy* causing strict periodicities: single rotating body (cannot cause such outbursts) or a binary system.

OJ287 with an outburst periodicity of 11.86 year

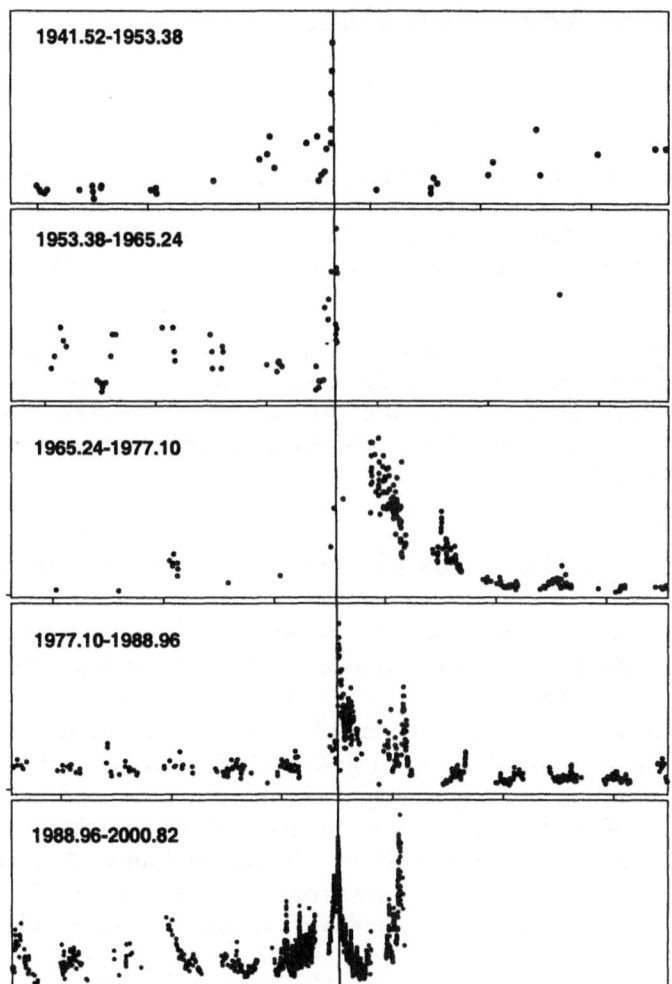

Fig. 1. The last five 11.86 year optical outburst periods. The vertical line just in the middle of the graph shows the proposed outburst times

References

1. A. Sillanpää et al. (1988), ApJ, **325**, 628
2. A. Sillanpää et al. (1996), A&AL, **305**, L17
3. B. Stothers and A. Sillanpää (1997), ApJL, **475**, L13

Gravitational Lens Diagnosis of Quasar Accretion Disk

Atsunori Yonehara[1][2], Shin Mineshige[1], and Edwin L. Turner[3]

[1] Department of Astronomy, Kyoto University, Sakyo-ku, Kyoto 606-8502, JAPAN
[2] Research Fellow of the Japan Society for the Promotion of Science
[3] Princeton University Observatory, Peyton Hall, Princeton NJ 08544, USA

Abstract. We present simulated microlensing light curves of a quasar accretion disk using two canonical accretion disk models. Furthermore, we investigate the method to reconstruct the accretion disk structure from observed microlensing light curves. Additionally, we also show a new method to limit the size of the origin of quasar variabilities from macrolens effects.

1 Microlens Diagnosis of Quasar Accretion Disk

Apparent angular size of a quasar accretion disk is comparable or smaller than that of the Einstein ring radius of quasar microlensing. Therefore, we can reveal the structure of quasar accretion disk by using microlensing [1].

There are two canonical accretion disk models, the so-called standard disk and advection-dominated accretion flow (ADAF). Owing to their different physics of accretion process, emergent emissivity profiles of these disks are significantly different, giving rise to distinct microlens light curves. If we observe microlensing events at multi-wavebands simultaneously, therefore, we will be able to discriminate these two accretion disk models (Fig. 1) [1] [2]. Furthermore, if we observe at X-ray waveband, we will also be able to obtain

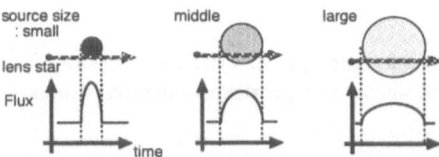

Fig. 1. Schematic view of the effect of effective source size difference to the expected light curve of microlensing.

informations regarding the X-ray emitting, innermost region of accretion disk (on $\sim$ AU scale). In the assumption that the X-ray emitting electron gas is trapped in a gravitational potential of a central black hole, we can limit the black hole mass [2].

In realistic situation, we should consider "caustic crossing" as quasar microlensing events. In such a situation, a part of an accretion disk in the vicinity of caustic is strongly amplified. Therefore, inversely, we can reconstruct the emissivity profile of the quasar accretion disk from observed microlensing light curves without any assumption about accretion disk model [3].

2 Macrolens Diagnosis of the Origin of Quasar Variabilities

Usually, in the case of quasar macrolensing, we treat a source as a point. Though a real source should have a finite size, we neglect the size for simplicity. If we include such an effect, in the case of multiply-imaged lensed quasar, time delay difference and flux ratio difference between images should appear. Moreover, if the source size is large enough (e.g., $\geq$ 100pc), differences will become significant (Fig. 2) [4]. To reproduce the observed, nicely correlated

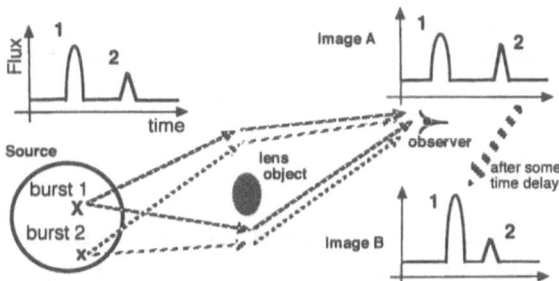

Fig. 2. Schematic view of the effect of time delay difference and flux ratio difference between lensed quasar images.

light curves of so-called "double quasar (Q0957+561A,B)", the source size of the origin of quasar variabilities should be smaller than 10pc [4] [5]. This fact indicate that the quasar variabilities originate from instabilities of an accretion disk in quasar.

References

1. Yonehara, A. et al. (1999) Microlens diagnostics of accretion disks in active galactic nuclei. A&A **343**, 41–50
2. Yonehara, A. et al. (1998) An X-ray microlensing test of AU-scale accretion disk structure in Q2237+0305. ApJ **501**, L41–L44 (erratum 511, L65–L66)
3. Mineshige, S., Yonehara, A. (1999) Gravitational microlens mapping of a quasar accretion disk. PASJ **51**, 497–504
4. Yonehara, A. (1999) Source size limitation from variabilities of a lensed quasar. ApJ **519**, L31–L34
5. Yonehara, A., Turner, E. L. (1999) in preparation.

X-Ray Evidence of an AGN
in the Starburst Galaxy M82

Takeshi G. Tsuru[1] and Hironori Matsumoto[2]

[1] Department of Physics, Faculty of Science, Kyoto University, Sakyo-ku, Kyoto 606-8502, Japan
[2] CSR, Massachusetts Institute of Technology, NE80-6045 77 Mass. Avenue, Cambridge, MA02139-4307, USA

Abstract. The X-ray spectrum of the famous starburst galaxy M82 consists of three components: soft, medium, and hard (Tsuru et al. 1997). We conducted a monitoring observation with ASCA in 1996. Although the X-ray flux of the soft and medium components remained constant, a significant time variability of the hard component was found between 3×10^{40} erg s^{-1} and 1×10^{41} erg s^{-1} at various time scales from 3 ks to one month. The spectrum of the variable source obtained by subtracting the spectrum of the lowest state from the highest state suggests strong absorption of $N_{\mathrm{H}} \sim 10^{22}$ cm^2, which means a variable source is embedded in the center of M82. These observations suggest existence of a LLAGN in M82.

1 Introduction

Tsuru et al. (1997) found that the ASCA spectrum of the archetypical starburst galaxy M82 observed in 1993 consists of three components: soft, medium, and hard. The soft and medium components are of thermal origin at temperatures of ~ 0.3 and ~ 1 keV and their images are extended. Thus, their origin would be the galactic wind. The hard component can be well described by either a power-law model ($\Gamma \sim 1.7$) or a thermal-plasma model ($kT \sim 14$ keV). This component is dominant in the X-ray spectra above the 2 keV band. The key to reveal the origin of the hard component is to clarify whether it shows a time variability or not. For these purposes, we made a monitoring observation of M82 with ASCA in 1996. The result shown here was already published in Matsumoto and Tsuru (1999).

2 Light Curve

We fitted SIS and GIS spectra for each observation simultaneously with the three-temperature thermal plasma model or with two-temperature thermal plasma + power-law model applied in Tsuru et al. (1997). The column densities, temperatures, and metal abundances of the soft and medium components are fixed to the best-fit values of Tsuru et al. (1997). Both of the two models could fit all the spectra quite well. While the luminosities of the soft and medium components are quite stable, we found that the hard component has a clear long time variability (figure). We also found a short time variability from the GIS and SIS light curves of #74049030 (1996 April 24) in the 3.0 – 10.0 keV band at a time scale less than a day, though we can see no clear time variability in the 0.7 – 1.5 keV band at that time. These results strongly suggest that the hard component of M82 detected by Tsuru et al. (1997) has short-term variability on a time scale of $\sim(1-2)\times10^4$ s.

3   Origin of the Hard Component

A collection of binaries, a young SNR, inverse Compton emission or hot interstellar medium cannot explain the time variability of M82. The highest luminosity of super Eddington sources is significantly lower than that of M82 (Okada et al. 1998). An LLAGN is the only possible origin that can explain the luminosity and time variability. However, the Ginga spectrum could be fitted by the thermal bremsstrahlung model but could not be fitted by the power-law model (Tsuru 1992). Therefore, it may also be possible that the origin of the hard component is a new type of accreting X-ray source.

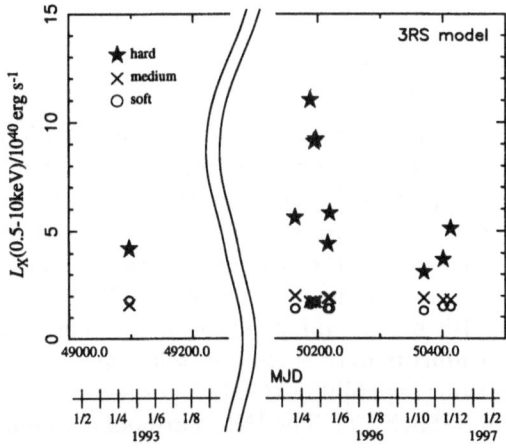

Fig. 1. Time variability of each component in the 3RS model. The stars, crosses, and circles show the hard, medium, and soft components, respectively.

References

1. Matsumoto H and Tsuru T.G. : 1999, PASJ 51, No.3
2. Okada K. et al. : 1998, PASJ 50, 25
3. Tsuru T. : 1992, PhD Thesis, The University of Tokyo
4. Tsuru T.G. et. al. : 1997, PASJ 49, 619

Gamma-Ray Observations of Black Holes: The Prospects of INTEGRAL

Norman R. Trams

INTEGRAL Science Operations, Astrophysics Division, Space Science Department of ESA, ESTEC, PO Box 299, 2200 AG Noordwijk, The Netherlands

Abstract. In this paper we describe the upcoming European gamma-ray observatory INTEGRAL (International Gamma-Ray Astrophysics Laboratory). This ESA satellite is scheduled to be launched in the Autumn of 2001 on board of a Russian Proton rocket. INTEGRAL hosts two main gamma-ray instruments: an imager (IBIS) and a spectrometer (SPI). These two gamma ray instruments are complemented with two monitoring instruments: an X-ray monitor (JEM-X) and an optical monitor (OMC). All four instruments will be observing simultaneously the same part of the sky The instruments will allow detailed imaging (angular resolution up to 12') and spectroscopic (energy resolution up to 2 keV at 1.3 MeV) observations in an energy band between 20 keV and 10 MeV. Possible targets for INTEGRAL include the galactic center, black hole binaries, and AGNs.

1 Introduction

Gamma-ray observations can provide important information about compact objects. Examples are the emission lines in supernovae and supernova remnants of ^{56}Co and ^{44}Ti, which give information on the layers just above the core collapse of the SN, and the emission from the central region of active galaxies, giving insights in the particle interactions in the inner regions of these objects. The observations of *Compton* GRO and GRANAT have already given a wealth of new information on the high energy sky in the gamma-ray range. Unfortunately these two telescopes were either sensitive, but with limited imaging capabilities (CGRO), or less sensitive, but with good imaging capabilities (GRANAT). The INternational Gamma-Ray Astrophysics Laboratory (INTEGRAL) observatory will combine imaging capabilities with sensitivity and spectral resolution.

2 The INTEGRAL Mission

INTEGRAL is a 20 keV – 10 MeV gamma-ray mission with concurrent source monitoring at X-rays (3 – 35 keV) and in the optical range (V, 500 – 600 nm). All instruments – co-aligned with large fields of view – cover simultaneously a very broad energy range of high energy sources (Table 1).

The scientific goals of INTEGRAL will be attained by fine spectroscopy with fine imaging and accurate positioning of celestial sources of gamma-ray

Table 1. INTEGRAL science and payload complementarity (overview)

Instrument	Energy range	Main purpose
Spectrometer SPI	20 keV - 8 MeV	Fine spectroscopy of narrow lines. Study diffuse emission on >degree scale
Imager IBIS	15 keV - 10 MeV	Accurate point source imaging Broad line and continuum spectroscopy
X-ray Monitor JEM-X	3 - 35 keV	Source identification X-ray monitoring of high energy sources
Optical Monitor OMC	500 - 600 nm (V-band)	Optical monitoring of high energy sources

emission. Fine spectroscopy over the entire energy range will permit spectral features to be uniquely identified and line profiles to be determined for physical studies of the source region. The fine imaging capability of INTEGRAL within a large field of view will permit the accurate location and hence identification of the gamma-ray emitting objects with counterparts at other wavelengths, and enable extended regions to be distinguished from point sources. On top of this it will provide considerable serendipitous science which is very important for an observatory-class mission. In summary the scientific topics will address:

Compact objects: white dwarfs, neutron stars, black-hole candidates, high-energy transients and GRBs.

Extragalactic astronomy: galaxies and clusters, AGN, Seyferts, blazars, cosmic diffuse background.

Stellar nucleosynthesis: hydrostatic nucleosynthesis (AGB and WR stars), explosive nucleosynthesis (supernovae and novae).

Galactic structure and the Galactic centre: cloud complex regions, mapping of continuum and line emission, ISM, cosmic-ray distribution.

Particle processes and acceleration: transrelativistic pair plasmas, beams, jets.

Identification of high energy sources: unidentified gamma-ray objects as a class.

Unexpected discoveries.

Each of the main gamma-ray instruments, the spectrometer (SPI) and the imager (IBIS), has both spectral and angular resolution, but they are differently optimised in order to complement each other and to achieve overall excellent performance. The two monitor instruments (JEM-X and OMC) will provide complementary observations of high-energy sources at X-ray and optical energy bands. A detailed description of INTEGRAL can be found in [5].

Table 2. Most important scientific characteristics of the INTEGRAL instruments.

	Field of View	Imaging resolution	Energy range	Spectral resolution
IBIS	$9^\circ \times 9^\circ$	12 arcmin	20 keV to 10 MeV	6% at 1 MeV
SPI	16° diameter	$2 - 3^\circ$	20 keV to 8 MeV	0.2% at 1 MeV
JEM-X	4.8° diameter	3 arcmin	3 to 35 keV	1.5% at 10 keV
OMC	$5^\circ \times 5^\circ$	17.6" × 17.6"	Johnson V filter	–

The imager IBIS (PI: P.Ubertini, IAS Rome, Italy) uses a coded mask in combination with two detector layers to achieve the high spatial resolution. The detector layers are a CdTe layer consisting of 16384 detector elements ($2600 \mathrm{~cm}^2$) and a CsI layer consisting of 4096 elements ($3100 \mathrm{~cm}^2$). The coded mask is located 3.1 meters above the detector plane. A passive shield is used to limit the field of view. A full description of the IBIS instrument can be found in [3].

The spectrometer SPI (PIs: V. Schoenfelder, MPE Garching, Germany and G. Vedrenne, CESR Toulouse, France) also uses a coded mask. However, in this case an array of 19 hexagonal Ge detectors is used. The detectors are cooled to an operating temperature of 85 K, to provide optimum sensitivity and resolution. To shield the detectors against background radiation and to limit the field of view a BGO anti-coincidence shield surrounds the Ge array. A pulse-shape discriminator system is used to reduce the background in the 200 keV to 1.5 MeV region. A full description of the SPI instrument can be found in [2].

The X-ray monitor JEM-X (PI: N. Lund, DSRI Copenhagen, Denmark) consists of two identical high-pressure gas chambers with microstrip detectors that view the sky through two aperture masks. The sensitive area of the detector is about $500 \mathrm{~cm}^2$. A full description of the X-ray monitor can be found in [4]. The X-ray monitor will observe the sources in the central region of the fields of view of the gamma-ray instruments.

The sensitivities for the three high-energy instruments are given in Table 3.

The optical monitor OMC (PI: A. Gimenez, INTA Madrid, Spain) consists of a 50 mm (diameter) telescope with a passively cooled CCD camera. The limiting magnitude for OMC is 19.7 (3 sigma on a low background). The photometric accuracy will be between 0.001 and 0.48 magnitudes, depending on source brightness and integration time. The OMC will perform monitoring of up to 100 sources in the field of view simultaneously. A full description of OMC can be found in [1].

Table 3. Continuum and narrow-line sensitivities of the INTEGRAL high-energy instruments. Sensitivities are 3σ in 10^6 seconds. The sensitivities are given in photons cm^2 s^{-1} keV^{-1} for continuum and in photons cm^2 s^{-1} for line sensitivities.

instrument	IBIS		SPI		JEM-X	
	Energy (keV)	Sensitivity	Energy (keV)	Sensitivity	Energy (keV)	Sensitivity
continuum	100	$3.8\ 10^{-7}$	100	$4\ 10^{-6}$	6	$1.4\ 10^{-5}$
	1000	$1.5\ 10^{-7}$	1000	$1.1\ 10^{-7}$	20	$3.3\ 10^{-6}$
line	100	$1.3\ 10^{-5}$	100	$2.0\ 10^{-5}$	6	$1.7\ 10^{-5}$
	1000	$4.0\ 10^{-5}$	1000	$6.8\ 10^{-6}$	30	$2.1\ 10^{-5}$

INTEGRAL (with a payload mass of 2019 kg and a total launch mass of 4000 kg) will be launched by a Russian PROTON launcher into a highly eccentric orbit with high perigee in order to provide long periods of uninterrupted observation with nearly constant background and away from trapped radiation. The parameters for the orbit are: period 72 hours, initial inclination 51.6^o, initial perigee height 10,000 km, and initial apogee height 153,000 km. Scientific observations will be carried out while the spacecraft is above an altitude of nominally 40,000 km thus utilising 90% of the orbital time for science. The particle background of the local spacecraft environment will be continuously measured by the on-board radiation monitor: this device allows the optimalization of the observing time before or after radiation belt passages and solar flare events, and provides essential information about the actual background.

INTEGRAL will be an observatory-type mission with a nominal lifetime of 2 years; an extension up to 5 years is technically possible. Most of the observing time (65% during year 1, 70% (year 2), 75% (year 3+)) will be awarded to the scientific community at large as the General Programme.

3   The Call for Proposals

The INTEGRAL Call for Observing Proposals (AO1) will be issued in 2001. Prior to the AO, a Call for Letters of Intent will be issued, to size the number of proposals that is to be expected for INTEGRAL. The full AO will consist of a number of documents, including detailed observer manuals for the spacecraft and the instruments. The AO will be issued electronically by ESA and proposals should be submitted to ESA. Observers who submit a Letter of Intent will be notified when the AO is available. Tools will be made available to enable observers to prepare proposals in the right format and to estimate observing times for the different INTEGRAL instruments. These will be made available through the internet at the time of the issue of the AO.

INTEGRAL also will have the possibility to observe Targets of Opportunity (TOO). TOO observations can be triggered by the INTEGRAL instruments or external triggers (e.g. ground based observations). The turn around time for TOOs is of the order of 10 hours (due to the necessary replan of the preplanned observing schedule). The Time Allocation Committee (TAC) will be advised to accept only a limited number of proposals for Targets of Opportunity.

The proposals will be reviewed by a single international peer review committee. This committee will consist of four panels, covering the four scientific topics addressed with INTEGRAL:

- Active Galactic Nuclei
- Compact objects
- Nucleosynthesis
- Miscellaneous, including Gamma-ray Bursts.

A technical feasibility check will be performed on the proposals. The TAC allocates observing time to proposals per observation. After the final TAC meeting the proposers are notified of the TAC decisions.

References

1. Gimenez, A., Mas-Hesse, J.M., Jamar, C., et al. (1997) in "2nd INTEGRAL Workshop: The transparent Universe", ESA SP-382, p.613.
2. Mandrou, P., Vedrenne, G., Jean, P., et al. (1997) in "2nd INTEGRAL Workshop: The transparent Universe", ESA SP-382, p.591.
3. Ubertini, P., di Cocco G., Lebrun, F. , et al. (1997) in "2nd INTEGRAL Workshop: The transparent Universe", ESA SP-382, p.599.
4. Westergaard, N.J., Budtz-Jorgensen, Schnopper, H.W., et al. (1997) in "2nd INTEGRAL Workshop: The transparent Universe", ESA SP-382, p.605.
5. Winkler, C. (1997) in "2nd INTEGRAL Workshop: The transparent Universe", ESA SP-382, p.573.

Jets from Black-Hole Binaries and Galactic Nuclei

I.F. Mirabel

Centre d'Etudes de Saclay / CEA/DSM/DAPNIA/SAP
91911 Gif/Yvette, France &
Instituto de Astronomía y Física del Espacio. Bs As, Argentina

Abstract. Relativistic outflows are a common phenomenon in accreting black holes. Despite the enormous differences in scale, stellar-mass black holes in binaries and supermassive black holes in galactic nuclei produce jets with analogous properties. In both two types of relativistic outflows are observed: 1) steady compact jets with flat-spectrum, and 2) sporadic extended jets with steep-spectrum and apparent superluminal motions. Besides, the most common class of gamma-ray bursts are afterglows from ultra-relativistic jets associated to the formation of black holes at cosmological distances.

1 Origin of the Idea of Stellar-Mass and Supermassive Black Holes

Black holes were first predicted by John Michell (1784) in the context of Newtonian physics and the corpuscular theory of light. He proposed that these objects could be detected in binary systems by the motion of nearby luminous objects. In the fourth year of the French revolution, Pierre-Simon Laplace (1795) went even further speculating on the possible existence of both, stellar-mass and supermassive black holes. In the fourth edition of "Exposition du Sytème du Monde" Laplace suggested: 1) that stellar-mass black holes could be as numerous as stars -"en aussi grand nombre que les étoiles"-, and 2) that the most massive objects of the universe could be black holes-"il est donc possible que les plus grands...corps de l'univers, soient par cela même, invisibles". In the nineteenth century the ondulatory conception of light became predominant and the idea of a black hole was forgotten for more than one hundred years, until it became a natural consequence of the general relativity conception of gravitation as a curvature of space-time.

2 The Quasar-Microquasar Analogy

The idea that supermassive black holes should exist at the centre of galaxies received strong support when the redshift of a quasi-stellar-radio-source (*quasar*) was measured for the first time by Maarten Schmidt (1963). The cosmological distance of quasars indicated that they must be powered by the

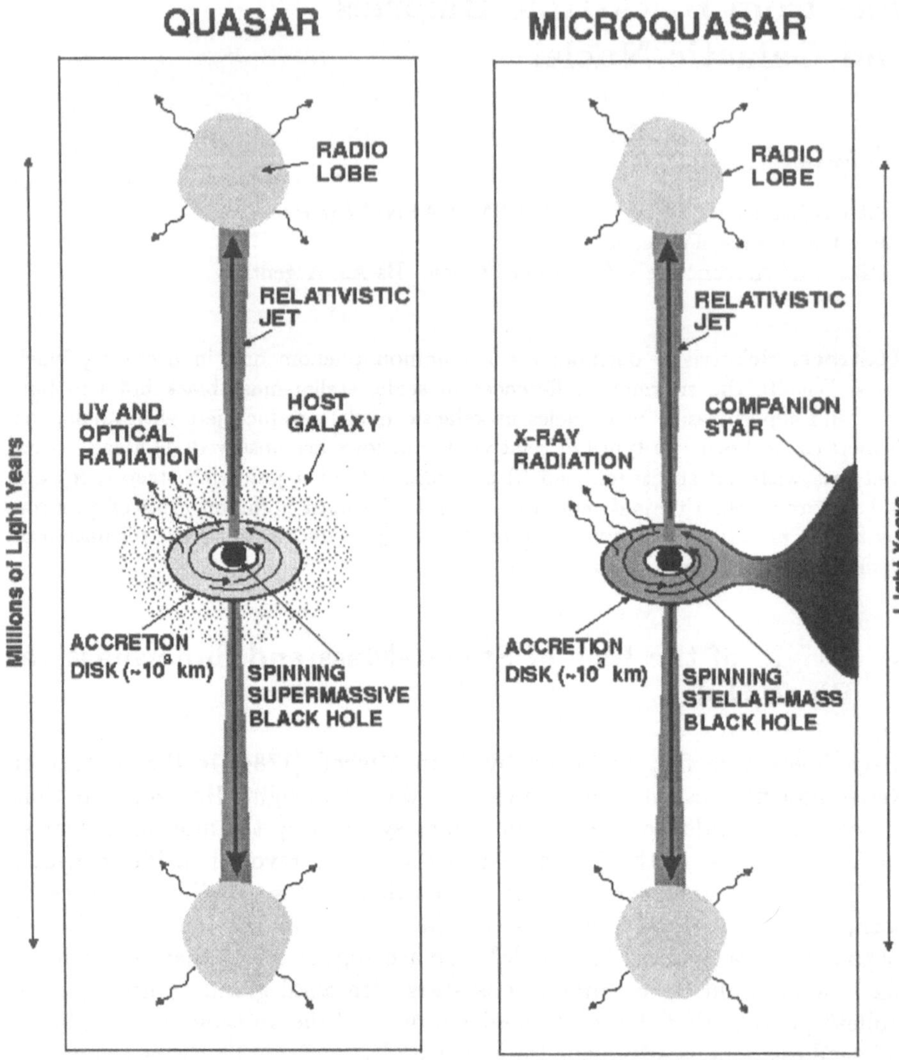

Fig. 1. Diagram illustrating the current ideas concerning quasars and microquasars (not to scale). As in quasars, in microquasars the following three basic ingredients are found: 1) a spinning black hole, 2) an accretion disk heated by viscous dissipation, and 3) collimated jets of relativistic particles. However, in microquasars the accreting black hole is only a few solar masses, while in quasars the central black-hole mass amounts several million solar masses. See text for more information.

release of gravitational energy around supermassive compact objects, rather than nuclear fusion in the interior of stars. On the other hand, the first observational evidence for the existence of stellar-mass black holes came from the discovery of X-ray sources beyond the solar system by Riccardo Giacconi et al. (1962). The accurate X-ray location of Cygnus X-1 lead to a precise radio position, which allowed the optical identification of the black-hole binary.

The recent finding in our own galaxy of *microquasars* (Margon 1994; Mirabel et al. 1992) with apparent superluminal motions (Mirabel & Rodríguez 1994) has opened new perspectives for the astrophysics of black holes (Mirabel & Rodríguez 1999 for a review). These scaled-down versions of quasars are believed to be powered by spinning black holes with masses of up to a few tens that of the Sun. The word *microquasar* was chosen to suggest that the analogy with quasars is more than morphological, and that there is an underlying unity in the physics of accreting black holes over an enormous range of scales, from stellar-mass black holes in binary stellar systems, to supermassive black holes at the centre of distant galaxies (Rees 1998).

Figure 1 illustrates the current ideas concerning quasars and microquasars (not to scale). As in quasars, in microquasars the following three basic ingredients are found: 1) a spinning black hole, 2) an accretion disk heated by viscous dissipation, and 3) collimated jets of relativistic particles. However, the black hole mass in microquasars is only a few solar masses, compared to several million solar masses in quasars; the accretion disk has a mean thermal temperature of several million degrees instead of several thousand degrees; and the particles ejected at relativistic speeds can travel up to distances of a few light-years only, instead of the several million light-years as in some giant radio galaxies. In quasars matter can be drawn into the accretion disk from disrupted stars or from the interstellar medium of the host galaxy, whereas in microquasars the material is accreted from the companion star in the binary system. In quasars the accretion disk has sizes of $\sim 10^9$ km and radiates mostly at ultraviolet and optical wavelengths, whereas in microquasars the accretion disk has sizes of $\sim 10^3$ km and the bulk of the radiation comes out as X-rays. It is believed that part of the spin energy of the black hole can be tapped to power the collimated ejection of magnetized plasma at relativistic speeds. This analogy between quasars and microquasars resides in the fact that in black holes the physics is essentially the same, irrespective of the mass, except that the linear- and time scales of the phenomena are proportional to the black hole mass. Because of the relative proximity and shorter time scales, in microquasars it is possible to firmly establish the relativistic motion of the sources of radiation, and to better study the physics of accretion flows and jet formation near the horizon of black holes.

At first glance it may seem paradoxical that relativistic jets were first discovered in the nuclei of galaxies and distant quasars and that for more than a decade SS433 was the only known object of its class in our Galaxy (Margon 1984). The reason for this is that disks around supermassive black holes emit strongly at optical and UV wavelengths. Indeed, the more massive the black hole, the cooler the surrounding accretion disk is. For a black hole accreting at the Eddington limit, the characteristic black-body temperature at the last stable orbit in the surrounding accretion disk will be given approximately by $T \sim 2 \times 10^7 \, M^{-1/4}$ (Rees 1984), with T in K and the mass of the black hole, M, in solar masses. Then, while accretion disks in AGNs have strong emission in the optical and ultraviolet with distinct broad emission lines, black-hole and neutron-star binaries usually are identified for the first time by their X-ray emission. Among these sources, SS 433 is unusual given its broad optical emission lines and its brightness in the visible. Therefore, it is understandable that there was an impasse in the discovery of new stellar sources of relativistic jets until the recent developments in X-ray astronomy. Strictly speaking, if it had not been for the historical circumstances described above, the acronym *quasar* would have suited better the stellar mass versions rather than their super-massive analogues at the centers of galaxies.

Since the characteristic time scales in the flow of matter onto a black hole are proportional to its mass, variations with intervals of minutes in a microquasar correspond to analogous phenomena with durations of thousands of years in a quasar of $10^9 \, M_\odot$, which is much longer than a human lifetime (Sams et al. 1996). Therefore, variations with minutes of duration in microquasars could be sampling phenomena that we have not been able to study in quasars. The repeated observation of two-sided moving jets in a microquasar (Rodríguez & Mirabel 1999) has led to a much greater acceptance of the idea that the emission from quasar jets is associated with moving material at speeds close to that of light. Furthermore, simultaneous multiwavelength observations of this microquasar are revealing the connection between the sudden disappearance of matter through the horizon of the black hole, with the ejection of expanding clouds of relativistic plasma (see Figure 2).

3   Compact Jets in Stellar-Mass and Supermassive Black Holes

The class of stellar-mass black holes that are persistent X-ray sources (e.g. Cygnus X-1, 1E 1740-2942, GRS 1758-258, etc.) and some supermassive black holes at the centre of galaxies (e.g. Sgr A* and many AGNs) do not exhibit luminous outbursts with large-scale sporadic ejections. However, despite the enormous differences in mass, steadily accreting black holes have analogous radio cores with steady, flat ($S_\nu \propto \nu^\alpha$; $\alpha \sim 0$) emission at radio wavelengths. The fluxes of the core component in AGNs are typically of a few Janskys (e.g. Sgr A* ~ 1 Jy) allowing VLBI high-resolution studies, but in stellar mass black holes the cores are much fainter, typically of less than a few mJy, which makes high-resolution observations of the core difficult.

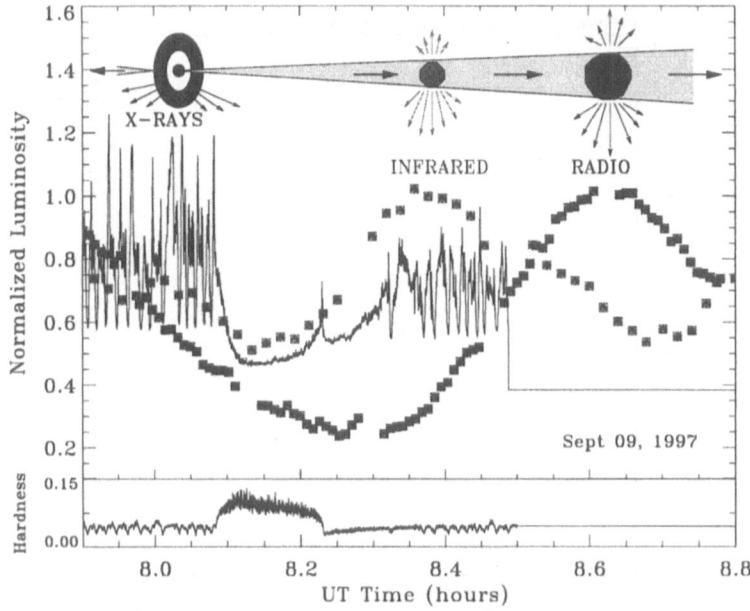

Fig. 2. Radio, infrared, and X-ray light curves for GRS 1915+105 at the time of quasi-periodic oscillations on 1997 September 9 (Mirabel et al. 1998). The infrared flare starts during the recovery from the X-ray dip, when a sharp, isolated X-ray spike is observed. These observations show the connection between the rapid disappearance and follow-up replenishment of the inner accretion disk seen in the X-rays (Belloni et al. 1997), and the ejection of relativistic plasma clouds observed as synchrotron emission at infrared wavelengths first and later at radio wavelengths. A scheme of the relative positions where the different emissions originate is shown in the top part of the figure. The hardness ratio (13-60 keV)/(2-13 keV) is shown at the bottom of the figure.

Although several multiwavelength studies have lead to speculations about the nature of the faint and steady compact radio emission in X-ray black hole binaries (e.g. Rodríguez et al. 1995; Fender et al. 1999, 2000), GRS 1915+105 is the black-hole binary where the core has been succesfully imaged at AU scale resolution (Dhawan, Mirabel & Rodríguez, 2000). GRS 1915+105 is the only X-ray binary where both a compact core with steady fluxes $\geq$20 mJy, and large-scale superluminal ejections are unambiguously observed. VLBA images during different states of the source always show compact jets with sizes $\sim$10λ_{cm} AU along the same position angle as the superluminal large-scale jets (see Figure 3). As in the radio cores of AGNs, the brightness temperature of the compact jet in GRS 1915+105 is $T_B \geq 10^9$ K. The VLBA images of GRS 1915+105 are consistent with the conventional model of a conical expanding jet with synchrotron emission (Hjellming & Johnston 1988; Falke & Biermann 1999) in an optically thick region of solar system size.

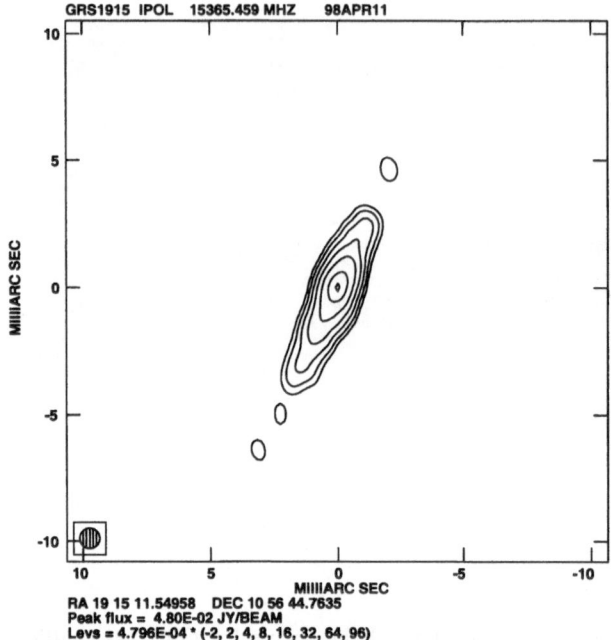

GRS1915 IPOL   15365.459 MHZ   98APR11

RA 19 15 11.54958   DEC 10 56 44.7635
Peak flux = 4.80E-02 JY/BEAM
Levs = 4.796E-04 * (-2, 2, 4, 8, 16, 32, 64, 96)

Fig. 3. Core of GRS 1915+105 resolved as a compact jet by 2-cm observations with the VLBA (Dhawan, Mirabel & Rodríguez 2000). At the 12 kpc distance of the source the angular resolution corresponds to 10 AU. Compact jets are observed during quiescent, as well as QPO states as the one shown in Figure 2. The length of the compact jet and the period of the oscillations are consistent with bulk motions $\geq 0.9c$, comparable with the velocities of the large-scale superluminal ejecta (Mirabel & Rodríguez 1994).

4   Microblazars and Gamma-Ray Bursts

In all three galactic microquasars where θ (the angle between the line of sight and the axis of ejection) has been determined, a large value is found (that is, the axis of ejection is close to the plane of the sky). This result is not inconsistent with the statistical expectation since the probability of finding a source with a given θ is proportional to $\sin \theta$. We then expect to find as many objects in the $60° \leq \theta \leq 90°$ range as in the $0° \leq \theta \leq 60°$ range. However, this argument suggests that we should eventually detect objects with a small θ. For objects with $\theta \leq 10°$ we expect the timescales to be shortened by $2\gamma^2$ and the flux densities to be boosted by $8\gamma^3$ with respect to the values in the rest frame of the condensation. For instance, for motions with $v = 0.98c$ ($\gamma = 5$), the timescale will shorten by a factor of ~ 50 and the flux densities will be boosted by a factor of $\sim 10^3$. Then, for a galactic source with relativistic jets and small θ we expect fast and intense variations in the observed flux.

These microblazars may be quite hard to detect in practice, both because of the low probability of small θ values and because of the fast decline in flux.

There is increasing evidence that the central engine of the most common form of gamma-ray burst (GRBs), those that last longer than a few seconds, are afterglows from ultra-relativistic jets produced during the formation of black holes (McFaden & Woosley 1999). Mirabel & Rodríguez (1999) propose that ultra-relativistic bulk motion and beaming are needed to explain: 1) the enormous energy requirements of $\geq 10^{54}$ erg if the emission were isotropic (e.g. Kulkarni et al. 1999; Castro-Tirado et al. 1999); 2) the statistical correlation between time variability and brightness (Ramirez-Ruiz & Fenimore 1999), and 3) the statistical anticorrelation between brightness and time-lag between hard and soft components (Norris et al. 1999). Beaming reduces the energy release by the beaming factor $f = \Delta\Omega/4\pi$, where $\Delta\Omega$ is the solid angle of the beamed emission. Additionally, the photon energies can be boosted to higher values. Extreme flows from collapsars with bulk Lorentz factors > 100 have been proposed as sources of γ-ray bursts (Mészáros & Rees 1997). High collimation (Dar 1998; Pugliese et al. 1999) can be tested observationally (Rhoads 1997), since the statistical properties of the bursts will depend on the viewing angle relative to the jet axis.

Recent multiwavelength studies of gamma-ray burst afterglows suggest that they are highly collimated jets. The brightness of the optical transient associated to GRB 990123 showed a break (Kulkarni et al. 1999), and a steepening from a power law in time t proportional to $t^{-1.2}$, ultimately approaching a slope $t^{-2.5}$ (Castro-Tirado et al. 1999). The achromatic steepening of the optical light curve and early radio flux decay of GRB 990510 are inconsistent with simple spherical expansion, and well fit by jet evolution. It is interesting that the power laws that describe the light curves of the ejecta in microquasars show similar breaks and steepening of the radio flux density (Rodríguez & Mirabel 1999). In microquasars, these breaks and steepenings have been interpreted (Hjellming & Johnston 1988) as a transition from slow intrinsic expansion followed by free expansion in two dimensions. Besides, linear polarizations of about 2% were recently measured in the optical afterglow of GRB 990510 (Covino et al. 1999), providing strong evidence that the afterglow radiation from gamma-ray bursters is, at least in part, produced by synchrotron processes. Linear polarizations in the range of 2-10% have been measured in microquasars at radio (e.g. Rodríguez et al. 1995), and optical (Scaltriti et al. 1997) wavelengths.

In this context, the jets in microquasars of our own Galaxy seem to be less extreme local analogs of the super-relativistic jets associated to the more distant gamma-ray bursters. However, there are caveats to this analogy and gamma-ray bursters are different from the microquasars found so far in our own Galaxy. The former do not repeat, seem to be related to catastrophic events, and have much larger super-Eddington luminosities. Therefore, the scaling laws in terms of the black-hole mass that are valid in the analogy between microquasars and quasars do not seem to apply to the case of gamma-ray bursters.

References

1. Belloni, T, Méndez, M, King, AR, van der Klis, M, van Paradijs, J. 1997, *Ap. J.* 479: L145-48
2. Castro-Tirado, AJ. et al. 1999, *Science* 283: 2069-73
3. Covino, S. et al. 1999, *IAU Circular 7172*
4. Dar, A. 1998, *Ap. J.* 500: L93-96
5. Dhawan, V, Mirabel, IF, Rodríguez, LF. 2000, To be submitted to *Ap. J.*
6. Falke, H. & Biermann, P.L. 1999, *Aastron. Astrophys.* 342, 49
7. Falke, H. et al. 1999, *astro-ph/9912436*
8. Fender, R.P. et al. 1999, *Ap. J.* 519, 165
9. Fender, R.P., Pooley, G.G., Durouchoux, P., Tilanus, R.P.J. & Brocksopp, C. 2000, *MNRAS* in press
10. Giacconi, R.H., Gursky, F., Paolini, R. & Rossi, B.B. 1962, *Phys. Rev. Lett* 9: 439
11. Hjellming, RM, Johnston, KJ. 1988, *Ap. J.* 328: 600-09
12. Kulkarni, S.R. et al. 1999, *Nature* 398: 389-94
13. Laplace, P.-S. 1795, *Exposition du Système du Monde, second edition* Volume II
14. Margon, BA. 1984, *Annu. Rev. Astr. Astrophys.* 22: 507-36
15. MacFayden, A.I. & Woosley, S.E. 1999, *Ap. J.* 524, 262
16. Mészáros, P, Rees, MJ. 1997, *Ap. J.* 482: L29-32
17. Michell, J. 1784, *Philosophical Transactions of the Royal Society* pages 35-57
18. Mirabel, IF, Dhawan, V, Chaty, S, Rodríguez, LF, Robinson, C, Swank, J, Geballe, T. 1998, *Astron. Astrophys.* 330: L9-12
19. Mirabel, IF, Rodríguez, LF., Cordier, B., Paul, J., Lebrun, F. 1992, *Nature* 358: 215-17
20. Mirabel, IF, Rodríguez, LF. 1994, *Nature* 371: 46-48
21. Mirabel, IF, Rodríguez, LF. 1999, *Annu. Rev. Astr. Astrophys.* 37: 409
22. Norris, J.P. Marani, G.F. & Bonell, J.T. 1999, *submitted to Ap. J.* astro-ph/9903233
23. Pugliese, G, Falcke, H, Biermann, PL. 1999, *Astron. Astrophys.* 344: L37-40
24. Ramirez-Ruiz, E., Fenimore, E.E. 1999, in Proc. Vth Compton workshop on GRBs
25. Rees, MJ. 1966, *Nature* 211: 468-70
26. Rees, MJ. 1984, *Annu. Rev. Astr. Astrophys.* 22, 471-506
27. Rees, MJ. 1998, in *Black Holes and Relativistic Stars*, ed. Wald, RM, University of Chicago, 79-101
28. Rhoads, JE. 1997, *Ap. J.* 487: L1-4
29. Rodríguez, LF, Gerard, E., Mirabel, IF, Gómez, Y., & velázquez, A. 1995, *Ap. J. Supp.* 101: 173-79
30. Rodríguez, LF, Mirabel, IF. 1999, *Ap. J.* 511: 398-404
31. Sams, BJ, Eckart, A, Sunyaev, R. 1996 *Nature* 382: 47-49
32. Scaltriti, F, Bodo, G, Ghisellini, G, Gliozzi, M, Trussoni, E. 1997, *Astron. Astrophys.* 327: L29-31
33. Schmidt, M. 1963, *Nature* 197, 1040

Black-Hole States and Radio-Jet Formation

R.P. Fender

Astronomical Institute 'Anton Pannekoek' and Center for High Energy
Astrophysics, University of Amsterdam, Kruislaan 403,
1098 SJ Amsterdam, The Netherlands

Abstract. An empirical relation between 'canonical' X-ray states in black hole
X-ray binaries and radio emission is presented. In the Low/Hard state a quasi-
continuous radio-emitting jet is produced. In the High/Soft state the jet is not
observed. In the Very High state and at major state transitions, which may correspond physically to rapid and/or significant changes in the inner accretion-disc
radius, discrete ejection events, which manifest themselves as radio flares, are observed. These relations appear to hold for both transient and persistent systems.
Furthermore, it is argued that the extremely strong coupling between hard X-rays
and radio emission from black-hole X-ray binaries implies that the comptonising
corona is simply the base of the jet.

1 Introduction: 'Canonical' Black-Hole States

Black-hole-candidate (BHC) X-ray binaries (Tanaka & Lewin 1995) are believed to consist of an accreting black hole with a mass in the range $3M_\odot \leq$
$M_{\mathrm{BH}} \leq 20M_\odot$ (see e.g. Charles 1999 for a review of the dynamical evidence
for black holes in these systems). Four BHC systems in our Galaxy (Cyg X-
1, GX 339-4, 1E1740.7-2942 and GRS 1758-258) and two in the LMC (LMC
X-1 and LMC X-3) are persistent X-ray sources, in the sense that their X-
ray emission is more or less constant over time (which means, in effect, over
several decades since their discoveries). Several more systems, the soft X-ray
transients (SXTs), are known only to have undergone a handful of X-ray outbursts during the history of X-ray astronomy, but it is these systems which
offer the most convincing dynamical evidence for black holes. Charles (1999)
lists several such systems, which include A 0620-00, GRO J0422+32, GS
2000+25 and GRS 1124-68. Generally speaking, these X-ray outbursts of the
SXTs are accompanied by radio outbursts (Hjellming & Han 1995; Kuulkers
et al. 1999).

It was initially realised that black holes displayed at least two spectral
states, 'high' (in this paper High/Soft) and 'low' (Low/Hard), based upon
the strength of their soft ($\leq$ few keV) X-ray emission (Tanaka & Lewin 1995).
An 'Off' state was first reported for GX 339-4 by Markert et al. (1973). Additionally, the 'Very High' (Miyamoto et al. 1991) and 'Intermediate' (Méndez
& van der Klis 1997) states have been identified. Méndez, Belloni & van der
Klis (1998) have shown that even unusual systems like the 'superluminal'
transient GRO J1655-50 display these 'canonical' states.

The physical interpretation of the different states is based upon an origin for the soft X-ray component in an accretion disc and the hard X-ray component via comptonisation of softer photons in a corona of high-energy electrons. The most popular current models invoke a truncated accretion disc in the Low/Hard X-ray state, inside of which may be an advection-dominated flow (e.g. Esin et al. 1998), and the presence of a strong comptonising corona. In the High/Soft state the disc component dominates and the corona is smaller and cooler. In the Very High state both the disc and corona are strong; the Intermediate state may be similar but at a lower overall luminosity.

2   Radio Emission and Black-Hole State

2.1   State Transitions

How does radio emission relate to black hole state transitions? For the transient sources the radio emission is generally (although not always) associated with one or more discrete ejections, the first (and generally foremost) of which is associated with the rapid rise (corresponding to something like a Off→High/Soft state transition in the space of a day or so) of X-ray emission at the time of the outburst. The evidence for discrete ejections comes from both direct imaging and the radio light curves (discrete peaks) and radio spectra (rapidly evolving to optically thin and subsequently decaying) observed (e.g. Kuulkers et al. 1999). As well as the transients, the persistent sources Cyg X-1 (Zhang et al. 1998) and maybe also GX 339-4 (Corbel et al. in prep) show flaring at times of state transitions.

2.2   Off and Low/Hard States

How does radio emission relate to canonical state outside of state transitions? We know from the X-ray transients that radio emission is pretty weak (often undetectable) during X-ray quiescence ≡ Off state. This is supported by weak radio detections of GX 339-4 when at very low X-ray flux levels (Corbel et al. in prep; Fig 1).

Cyg X-1 and GX 339-4 are generally observed in the Low/Hard X-ray state, and reveal very similar characteristics, namely weak (few mJy at cm wavelengths) but steady radio emission with a flat/inverted radio spectrum (i.e. spectral index $\alpha = \Delta \log S_\nu / \Delta \log \nu \geq 0$) which is correlated with both soft and hard X-ray emission over approximately a factor of two in intensity (Brocksopp et al. 1999 and references therein; Hannikainen et al. 1998; Corbel et al. in prep). The radio emission from Cyg X-1 is additionally modulated at the 5.6-day orbital period (Pooley, Fender & Brocksopp 1999; Brocksopp et al. 1999), probably due to free-free absorption in the dense stellar wind of the massive companion star. Models for flat/inverted radio emission generally invoke a partially self-absorbed outflow of the kind originally envisaged for

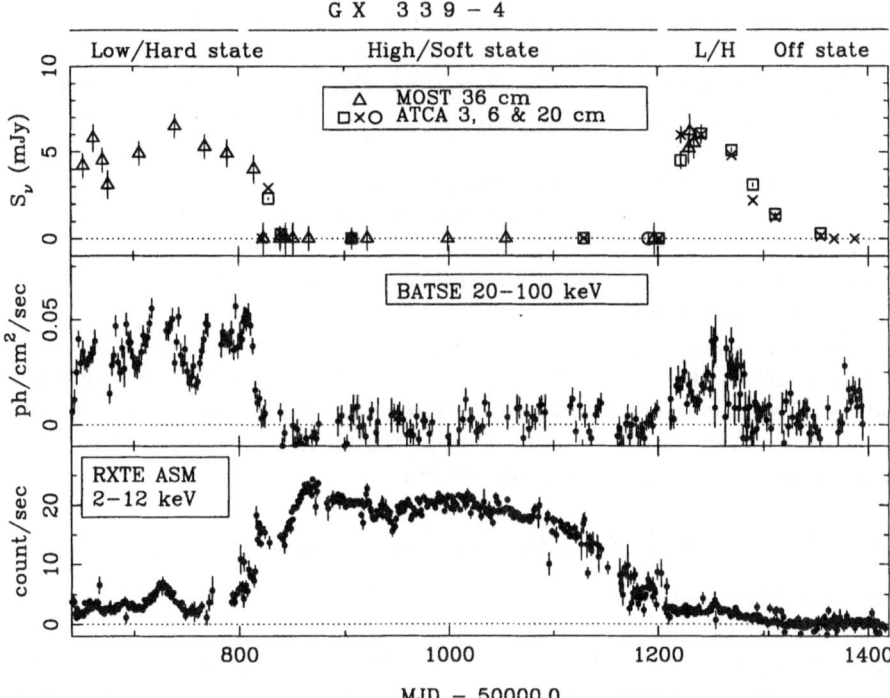

Fig. 1. Radio, soft- and hard-X-ray observations of the Low/Hard, High/Soft and Off states in GX 339-4. From Fender et al. (1999) and Corbel et al. (in prep).

AGN cores by Blandford & Königl (1979). This is supported by the large size scale required (a sphere with radius greater than the binary separation of Cyg X-1 is required to generate the observed cm-wavelength emission). Final confirmation of the jet hypothesis has come from VLBA observations of an extension on mas-scales from Cyg X-1 (Stirling et al. 1997; de la Force et al. in prep).

As well as the persistent sources Cyg X-1 and GX 339-4, some X-ray transients also spend extended periods in the Low/Hard state (e.g. GRO J0422+32, GS 2023+338/V404 Cyg). These systems also display a flat radio spectrum with comparable luminosity to Cyg X-1 and GX 339-4, when in the Low/Hard state (Fender, in prep).

2.3   The High/Soft State

It was already suspected that the radio flux density in Cyg X-1 dropped significantly when the system was in the High/Soft X-ray state. Dramatic confirmation that this was a phenomenon shared by other BHCs came from long-term monitoring of GX 339-4 at radio, soft- and hard-X-ray energies, in which it was found that the radio emission from the system was reduced by a factor ≥ 25 during a year-long period in the High/Soft state (Fender et al. 1999). Figure 1 demonstrates this result, encompassing Low/Hard, High/Soft, Low/Hard and finally Off states during the period of our monitoring program (Corbel et al. in prep), and dramatically reveals the extremely strong correlation between hard X-ray and radio emission. Is the outflow physically suppressed during this state or cooled (via inverse compton losses) so rapidly by the more luminous disc that the electrons no longer produce synchrotron emission ? The answer to this is unclear, but at least one (naive) argument, the evidence for radio emission during the (even more luminous) Very High State (see below), suggests physical suppression of the jet.

Note that while many X-ray transients appear, at face value, to be radio-bright whilst in the High/Soft state, more careful scrutiny reveals that the radio emission is usually just the decaying tail of emission from the flare event at the start of the outburst (i.e. at the state transition), which is by then physically decoupled from the system.

2.4   The Very High (and Intermediate?) State(s)

The Very High State is much rarer than the Low/Hard or High/Soft states, and few clear examples exist of radio observations of this state. GX 339-4, while having previously entered this state (Miyamoto et al. 1991), has not done so since radio monitoring began. GS 1124-683 also spent an extended period in the Very High State (Miyamoto et al. 1993), during which time there *was* radio coverage (for at least 30 days – Kuulkers et al. 1999 and references therein), revealing fairly bright and variable emission. However, the radio coverage does not appear to have been good enough to distinguish between rapid radio variability (as in GRS 1915+105, see below), or simply two large, discrete ejections.

Belloni (1998) has suggested that GRS 1915+105, when exhibiting its highly variable 'dipping' behaviour, may be in the Very High State. This pattern of behaviour is associated with bright and repeated flare events with a flat spectrum from the radio – mm – infrared, which are most likely associated with discrete ejection events $\equiv$ transient jet formation (Fender & Pooley 1998 and references therein). The intermediate state may be similar to the Very High State, but at lower luminosity levels, but again there has been no good radio coverage of this state.

3   Discussion

We are now in a position to be able to characterise the relation of radio emission from BHC X-ray binaries to their 'state' as defined observationally by their X-ray spectral and timing properties, except for the Intermediate state, for which there is no clear case of simultaneous radio observations. This relation is summarised in Fig. 2 and Table 1. Several implications are quite clear:

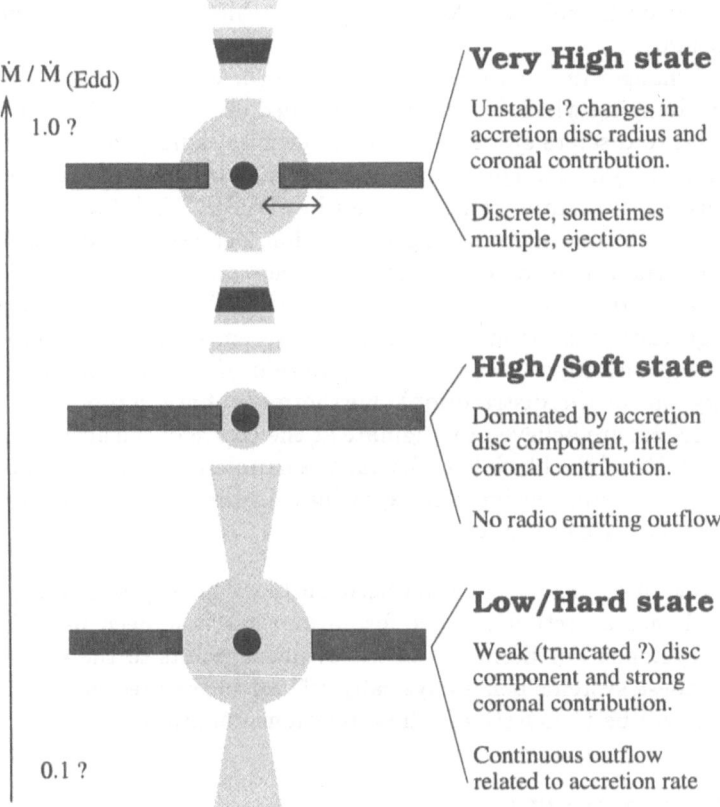

$\dot{M} / \dot{M}_{(Edd)}$

1.0 ?

0.1 ?

Very High state

Unstable ? changes in accretion disc radius and coronal contribution.

Discrete, sometimes multiple, ejections

High/Soft state

Dominated by accretion disc component, little coronal contribution.

No radio emitting outflow

Low/Hard state

Weak (truncated ?) disc component and strong coronal contribution.

Continuous outflow related to accretion rate

Fig. 2. Relation between models for BHC X-ray states and observed radio emission

Table 1. The relation of radio emission to black hole X-ray state.

State	Radio properties
Very High	Bright ejections, spectral evolution from absorbed $\rightarrow$ optically thin
High/Soft	Radio suppressed by factor ≥ 25
Intermediate	?
Low/Hard	Low level, steady, flat spectrum extending to at least sub-mm
Off	Weak; similar to Low/Hard but reduced $\propto \dot{m}$

- The Low/Hard state supports a quasi-continuous outflow whose contribution to the overall energetics of the system is likely to be significant. In this state there is a three-way correlation between radio, soft- and hard-X-rays, which may reflect small changes in the accretion rate. The Off state may simply be the Low/Hard state 'turned down' to lower accretion rates.
- In the High/Soft state the radio emission drops below detectable levels, probably corresponding to the physical disappearance of the jet. It is this

state, in both radio and X-ray emission, which is most different to the other states.

- Rapid changes in the accretion disc radius (or whatever it is that is physically changing which is currently modelled as an inner disc radius!) correspond to discrete ejections, which appear as radio (and sometimes mm and infrared) flares. This mechanism may be operating in analogous ways in both general state transitions and in the Very High State.
- There is an extremely strong correlation between radio and hard X-ray emission. The radio emission has been *directly observed* to arise in outflows and to originate in synchrotron emission from a population of high-energy electrons. The hard X-ray emission is *inferred* to arise via comptonisation by a similar population of electrons (albeit the low-energy tail of the distribution). Furthermore, both components, the jet and corona, are believed to originate at the centre of the accretion disc, in the vicinity of the black hole. By far the simplest interpretation therefore is that the comptonising corona, at least in the case of BHC systems, is simply the base of the jet.

A simple observational relation between black hole state and radio emission, and hence accretion and jet formation, has now been established for the BHC systems. Furthermore, all the evidence points to the comptonising corona in these systems being physically related to the presence of a jet. The next stage will be to investigate these relations quantitatively.

Acknowledgements

RPF would like to thank Mariano Méndez, Eric Ford, Jeroen Homan and Michiel van der Klis for stimulating discussions.

References

1. Belloni T., 1998, New Astronomy Reviews, 42, 585
2. Blandford R., Königl A., 1979, ApJ, 232, 34
3. Brocksopp C., Fender R.P., Larionov V., Lyuty V.M., Tarasov A.E., Pooley G.G., Paciesas W.S., Roche P., 1999, MNRAS, 309, 1063
4. Charles P.A., 1999, *Black Holes in our Galaxy : observations*, In : Abramowicz, M. A., Björnsson, G., Pringle, J. E. (Eds), Theory of Black Hole Accretion Discs, Cambridge Contemporary Astrophysics, CUP, 1998, p.1
5. Esin A.E., Narayan R., Cui W., Grove J.E., Zhang S.N., 1998, ApJ, 505, 854
6. Fender R.P., Pooley G.G., 1998, MNRAS, 300, 573
7. Fender, R., Corbel, S., Tzioumis, T., McIntyre, V., Campbell-Wilson, D., Nowak, M., Sood, R., Hunstead, R., Harmon, A., Durouchoux, P., Heindl, W., ApJ, 1999, 519, L165
8. Hannikainen D.C., Hunstead R.W., Campbell-Wilson D., Sood R.K., 1998, A&A, 337, 460

9. Hjellming, R. M., Han, X., 1995, *Radio properties of X-ray binaries*, In : Lewin, W. H. G., van Paradijs, J., van der Heuvel, E. P. J. (Eds.), X-ray binaries, Cambridge University Press, Cambridge, 308–330

10. Kuulkers, E., Fender, R. P., Spencer, R. E., Davis, R. J., Morison, I., 1999, MNRAS, 306, 919

11. Markert T.H., Canizares C.R., Clark G.W., Lewin W.H.G., Schnopper H.W., Sprott G.F., 1973, ApJ, 184, L67

12. Méndez M., van der Klis M., 1997, ApJ, 479, 926

13. Méndez M., Belloni T., van der Klis M., 1998, ApJ, 499, L187

14. Miyamoto S., Kimura K., Kitamoto S., Dotani T., Ebisawa K. 1991, ApJ, 383, 784

15. Miyamoto S., Iga S., Kitamoto S., Kamado Y., 1993, ApJ, 403, L39

16. Pooley G.G., Fender R.P., Brocksopp C., 1999, MNRAS, 302, L1

17. Stirling, A., Spencer, R., Garrett, M., 1998, New Astronomy Reviews, 42, 657

18. Tanaka Y., Lewin W.H.G., 1995, *Black-hole binaries*, In : Lewin, W. H. G., van Paradijs, J., van der Heuvel, E. P. J. (Eds.), X-ray binaries, Cambridge University Press, Cambridge, 308–330

19. Zhang S.N., Mirabel I.F., Harmon B.A., Kroeger R.A., Rodríguez L.F., Hjellming R.M., Rupen M.P., 1997, *Galactic Black Hole Binaries : Multifrequency Connections*, In : Dermer C.D., Strickman M.S., Kurfess J.D. (Eds.), Proc. 4th Compton symposium, AIP, p.141

Black Hole X-Ray Binaries:
A New View on Soft-Hard Spectral Transitions

Friedrich Meyer[1], Bifang Liu[1,2], and Emmi Meyer-Hofmeister[1]

[1] Max-Planck-Institut für Astrophysik,
   Karl-Schwarzschild-Str. 1, D-85740 Garching, Germany
[2] Yunnan Observatory, Chinese Academy of Sciences, P.O.Box 110,
   Kunming 650011, China

Abstract. The theory of coronal evaporation predicts the formation of an inner hole in the cool thin accretion disk for mass accretion rates below a certain threshold and the sudden disappearance of this hole when the mass accretion rate rises above that threshold. This appears to quantitatively account for the observed transitions between hard and soft spectral states at critical luminosities.

1   Spectral Transitions, Observation and Interpretation

Cyg X-1 undergoes occasional transitions between a high state (soft thermal spectrum) and a low state (hard power law spectrum). Studies of X-ray novae during decline from outburst show a change from a soft state at high luminosities to a hard state at low luminosities. These transitions have been observed both for neutron star and black hole systems [2]. The phenomenon always occurs at a luminosity of about 10^{37}erg/s, which corresponds to a mass accretion rate of about 10^{17}g/s.

The two spectral states are thought to be related to different states of accretion. Model spectra based on an ADAF fit the observations well ([1] and references therein). We suggest that features in the physics of equilibrium between cool disk and corona explain the triggering of the transition between the two states.

2   The Physics of the Corona Above a Thin Disk

Frictional heat released in the corona flows down into cooler and denser transition layers. There it is radiated away if densities are sufficiently high. If densities are too low cool matter is heated up and evaporated into the corona until equilibrium density is established. In the corona as in an accretion disk differential rotation removes angular momentum outwards and mass flows inwards. The mass drained from the corona is replaced by mass evaporating from the disk as the system establishes equilibrium. When the mass evaporation rate exceeds the mass flow rate in the cool disk the cool disk terminates, inside only a hot coronal flow exists.

The corona above the innermost disk region dominates this exchange. In a one-zone-model it is described by a set of ordinary differential equations for mass, momentum and energy that uniquely determines mass accretion rate, wind loss, and temperature in the corona as a function of radius.

3   Results

The new result is the relation between the mass flow rate in the disk $\dot{M}$ and the location of the inner edge of the thin disk r_{min} for higher $\dot{M}$. The radius of the disk hole decreases with increasing $\dot{M}$, but for a certain $\dot{M}_{crit}$ the thin disk suddenly reaches inwards to the last stable orbit (see Fig. 1).

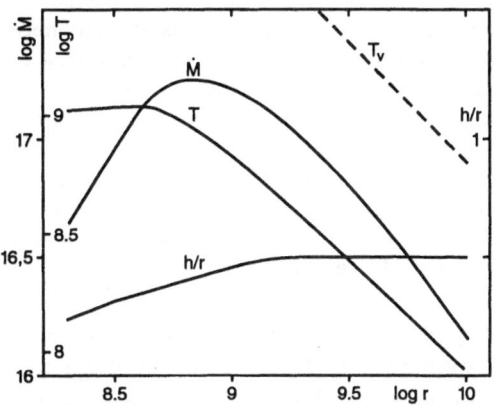

Fig. 1. Rate of inward mass flow $\dot{M}$ (in $M_\odot/yr$) in the corona at the inner edge r_{min} of the cool disk for a $6M\odot$ black hole, T_v virial temperature.

The predictions from the evaporation model are: When the mass accretion rate is high during the outburst the thin disk extends inward to the last stable orbit. A corona exists above the thin disk, but the mass flow in the thin disk is so high that no hole appears. In decline from outburst the mass accretion rate decreases. When $\dot{M}_{crit}$ is reached a hole form: the transition soft/hard occurs. *Our value for the critical mass accretion rate agrees with the observations.* If the mass accretion rate varies up and down we expect transitions hard/soft and soft/hard, as observed in Cyg X-1.

References

1. Esin, A.A., Narayan, R., Cui, W. et al. (1998) Spectral transitions in Cygnus X-1 and other black hole X-ray binaries. ApJ **505**, 854–868
2. Tanaka Y., Shibazaki N., (1996) X-ray novae. ARA&A **34**, 607–644

Orbital, Precessional and Flaring Variability in Cygnus X-1

Catherine Brocksopp[1], Rob Fender, Valeri Larianov, Viktor Lyuty, Anatoli Tarasov, Guy Pooley, William Paciesas, and Paul Roche

[1]Department of Physics & Astronomy, Open University, Walton Hall, Milton Keynes MK7 6AA

Abstract. We present the results of a 2.5-year multiwavelength monitoring programme of Cygnus X-1, making use of hard and soft X-ray data, optical spectroscopy, *UBVJHK* photometry and radio data. In particular we confirm that the 5.6-day orbital period is apparent in all wavebands and note the existence of a wavelength-dependence to the modulation, in the sense that higher energies reach minimum first. We also find a strong modulation at a period of 142 ± 7 days, which we suggest is due to precession and/or radiative warping of the accretion disc. We present the basic components required for more detailed future modelling of the system – including a partially optically thick jet, quasi-continuous in the low state. In addition, we choose two periods of flaring to study in further detail and find that the hard and soft X-rays are well-correlated in the first and that the soft X-rays and radio are correlated in the second.

We have monitored Cyg X-1 for 2.5 years; for the first three months the system was in its high/soft X-ray state, although we do not have radio monitoring through this period. Since then Cyg X-1 has been in the low/hard X-ray state, characterised by fairly constant X-ray, radio and optical fluxes superimposed by orbital (5.6 days) and precessional (142 days) modulations. There are also flaring periods with various types of correlated behaviour between the wavelength bands. The results of this monitoring programme have enabled us to explain various characteristics of Cyg X-1 that have not been accounted for in previous models of the system. These are outlined below.

- Orbital modulation across the spectrum in the hard state
  - Optical/IR – ellipsoidal modulation, due to shape of star/wind
  - Radio/X-rays – ν-dependent $\Rightarrow$ line-of-sight absorption by stellar wind (radio requires continuous jets)
- Long period modulation across the spectrum in the hard state
  - Precession and/or radiative warping of disc/jet system
- Lack of optical and soft X-ray orbital modulation during the soft state
  - X-ray – flaring is of a magnitude $\gg$ modulation. Therefore we just don't see it
  - Optical – becomes single-peaked. Either there is extra emission at phase 0.5 due to significant heating/ionization of the supergiant, or we are seeing the accretion stream

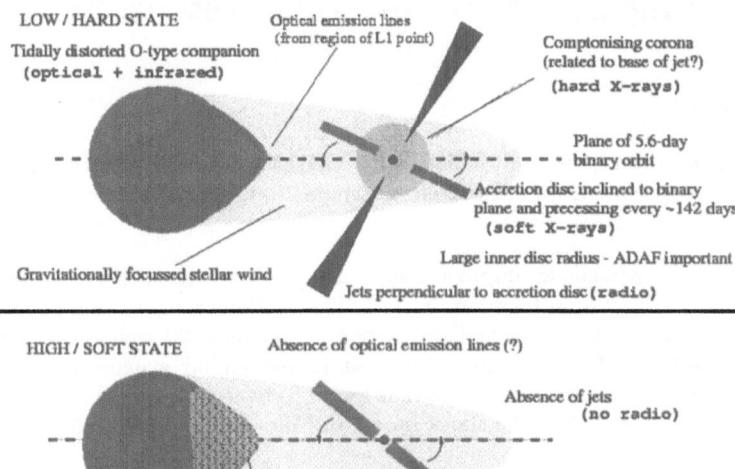

Fig. 1. Multiwavelength model for Cyg X-1

- Diminishing (disappearance?) of optical emission lines, hard X-rays and radio during the soft state
  - Optical – additional wind-ionization or stronger free-free component
  - X-ray/radio – jet and corona disrupted by inner edge of disc or by highly ionized wind. Material advected onto black hole
- Correlated behaviour between X-ray and radio
  - Requires disc/jet coupling – small flares due to slight increase in $\dot{M}$, but not enough to produce significant additional ionization
- Occasional lack of correlated behaviour between the hard and soft X-rays
  - Possible hybrid between small flare and full state change. Accompanied by spectral softening. May be a 'failed' state change. Requires disc/jet model
- Relation between the base of the jet and the corona
  - Both arise near the inner disc, both are quenched during the soft state. The corona is inferred to consist of high energy electrons – the observed synchrotron radio emission proves the existence of high energy electrons at the base of the jet. Must be some jet/corona coupling
- Lack of flaring in the optical and infrared
  - Supergiant is bright enough for the overall luminosity to appear unaffected by additional ionization due to the disc

Further details are given in [1] and references within.

References

1. Brocksopp C. et al., 1999, MNRAS, 309, 1063

Modelling Synchrotron Outbursts in 3C 273

Marc Türler

Geneva Observatory, ch. des Maillettes 51, CH-1290 Sauverny, Switzerland
INTEGRAL Science Data Centre, ch. d'Écogia 16, CH-1290 Versoix, Switzerland

Abstract. We present an improved approach to derive the observed evolution of a typical synchrotron outburst in 3C 273. The method consists in a simultaneous decomposition of 13 long-term light curves – spanning 20 years of observations from 2.7 GHz to 0.35 mm – into a series of 15 self-similar synchrotron outbursts. The spectral evolution of the outbursts is found to be in remarkable agreement with the evolution expected by the shock model of Marscher & Gear (1985).

1 Introduction

Submillimetre-to-radio light curves of blazars show evidence of prominent structures apparently propagating from high to low frequencies. Marscher & Gear [1] proposed that such features are synchrotron outbursts emitted behind a shock front propagating down a relativistic jet. According to their shock model, the turnover (or maximum) of the self-absorbed synchrotron spectrum should evolve with time along a typical three-stage track. The dominant cooling process of the electrons changes from one stage to the other from inverse-Compton scattering to synchrotron radiation and finally to adiabatic expansion energy losses. While the turnover frequency is expected to decrease steadily, the turnover flux density is first rising during the Compton stage, then peaking at nearly constant flux during the synchrotron stage, before declining during the adiabatic stage.

To test and constrain this physical model, it is important to extract from the observations the properties of a typical outburst. To be able to still follow the outburst's evolution after the onset of a new one, we defined in [2] a new approach consisting in a complete decomposition of several light curves into a series of self-similar outbursts. After the promising results of the two first approaches, we present here the results of an hybrid approach which aims at including only the positive aspects of the two earlier parameterizations.

2 The Hybrid Approach

The approach presented here models strictly the shape of the synchrotron spectrum with both a high- and a low-frequency break as described in [3]. But it leaves the evolution of the spectral turnover in the $(\log S, \log \nu, \log t)$-space (thick solid line in Fig. 1) as free as possible by using 15 free parameters to define two cubic splines parameterized at 5 unfixed times. The

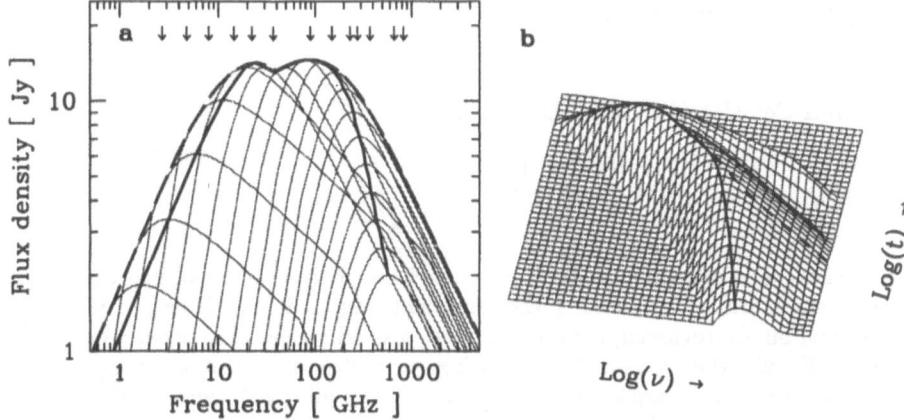

Fig. 1. Spectral evolution of a typical outburst in 3C 273 as obtained by the hybrid approach. Spectra at different times (**a**) and the three-dimensional evolution in the $(\log S, \log \nu, \log t)$-space (**b**). The arrows in (**a**) mark the frequency distribution of the 13 light curves used to constrain the overall evolution of the outbursts

optically thin spectral index is allowed to change logarithmically with time from $\alpha_{\text{thin}}(t \leq t_1)$ to $\alpha_{\text{thin}}(t \geq t_2)$ between t_1 and t_2, while the optically thick spectral index α_{thick} is assumed to be constant. Three other parameters are used to define the evolution of the two spectral breaks in a similar way as in [3]. The observational material and all other features of the parameterization are exactly as in [3] with the emission from the underlying jet assumed to be negligible.

The obtained fit of the 13 light curves is very similar to the fit shown in [3]. The spectral evolution is shown in Fig. 1. The spectral turnover rises approximately as $S_{\text{m}} \propto \nu_{\text{m}}^{-1.7}$ and declines as $S_{\text{m}} \propto \nu_{\text{m}}^{+0.9}$ at the end of the outburst's evolution. The optically thin spectral index is found to flatten from $\alpha_{\text{thin}}(t \leq t_1) = -0.83$ to $\alpha_{\text{thin}}(t \geq t_2) = -0.37$ ($S_\nu \propto \nu^{+\alpha}$) between $t_1 = 0.7$ and $t_2 = 1.4$ year, i.e. just around the transition from the peaking to the declining phase. The flattening of the optically thin spectral index by $\Delta\alpha_{\text{thin}} \approx +0.5$ at this particular point of the evolution and the fact that the path followed by the spectral turnover has a nearly flat peaking phase give strong support to the physical shock model of [1].

References

1. Marscher A.P., Gear W.K. (1985) Models for high-frequency radio outbursts in extragalactic sources. ApJ **298**, 114–127
2. Türler M., Courvoisier T.J.-L., Paltani S. (1999) Modelling the submillimetre-to-radio flaring behaviour of 3C 273. A&A **349**, 45–54
3. Türler M., Courvoisier T.J.-L., Paltani S. (2000) Modelling 20 years of synchrotron flaring in the jet of 3C 273. A&A, submitted

On the Formation of Jets

Annalisa Celotti[1] and Roger D. Blandford[2]

[1] SISSA, via Beirut 2-4, I-34014 Trieste, Italy
[2] Caltech, Pasadena, CA 91125, USA

Abstract. The phenomenology of jets associated with a variety of black hole systems is summarized, emphasizing the constraints imposed on their origin. Models of jet formation are reviewed, focusing in particular on recent ideas concerning MHD models. Finally, the potential for advancing our understanding of jets both through future observations – especially forthcoming X–ray missions – and for elucidating some crucial theoretical questions is highlighted.

1   What We (Probably) See

Jets – in the broad qualitative meaning of collimated/elongated structures of outflowing plasma – are observed in a variety of systems. Other contributions to these proceedings focus on Galactic sources and thus what follows is biased toward extragalactic jets. For exhaustive reviews see e.g. [1], [2], [3].

1.1   Active Galactic Nuclei

Extended double radio emitting structures (lobes) were observed and associated with galaxies and quasars more than 40 years ago, and eventually recognized to be physically connected, through jets supplying them with energy and momentum, to the activity taking place in the nucleus in (about 10% of) active galaxies [4]. Analysis of the extended structures provided us with an estimate of the energy supplied by the nuclear engine.

A step toward the understanding of the (magneto)–hydrodynamics of jets on large (arcsec) scales and their interaction with the environment was the recognition that differences in radio morphologies correspond to differences in power. Edge darkened, low power or type I sources, usually show two jets plausibly at most transonic, while edge brightened, powerful type II sources are mostly one–sided and appear to be supersonic and mildly relativistic.

The rather poor information on large scale jets in other spectral bands has been significantly filled in by HST, which has so far detected more than a dozen jets, and X–ray images from Chandra are just starting to be obtained (PKS 0637-752 and Cen A so far). These observations, which show rather similar radio and optical morphologies, reveal: a) that particles are accelerated to energies $\sim 10^8 m_e c^2$ [5] and b) a constancy of the radio–optical spectra along jets, suggesting that the acceleration of synchrotron emitting *particles occurs* all along the jet, although it is not clear whether this is due to many (relativistic) shocks or wave modes.

However it has been the study of the predicted motion of features in the inner jets at relativistic speed – relative to the flat spectrum compact source – which has provided major clues and constraints on the jet formation mechanism, in terms of the (bulk) acceleration and collimation required. The strongest pieces of evidence include the detection of components apparently moving at superluminal speeds $\sim$ few–10 $c\,h^{-1}$, brightness temperatures exceeding 10^{12} K, and one-sided jets. Often extragalactic jets are aligned over many decades in scale, are collimated within a few degrees and reach bulk Lorentz factors ~ 10.

When the emission from the highly relativistic (non–thermal) plasma is beamed in the direction of the observer, it dominates the isotropic line and continuum emission associated with the accreting gas and the stars. This is observed in blazars. The implied anisotropy in the emission has led to identify the misaligned counterparts of blazars with radio galaxies and quasars, providing some degree of unification among jetted sources of both types I and II. Another source of anisotropy, due to a putative obscuring torus, is invoked to account for the lack of broad emission lines in the spectra of powerful radio galaxies (and type 2 Seyferts).

A strong stimulus to this field has come from the observations of γ–ray emission in a significant number of blazars. Blazar jets are commonly observed at GeV energies and, when located within an intergalactic absorption length, also at TeV energies. (Note that TeV emission has been observed to vary on 15 min timescales [6], suggesting that these high energy γ–rays originate within $\sim 10^{2-3}$ m.) These observations allow us to observe most of the energy that is radiatively dissipated (of which the radio emission contributes a negligible amount). Furthermore, constraints derived from the implied opacity for γ–rays to pair production, locate the emitting region at $\sim 10^{2-4}$ m and thus limit the possible radiative processes involved.

Jet structures have been also increasingly found in radio–quiet AGN. Although these are not radio silent, they appear to be a separate class from the radio–loud AGN. These are less powerful and less collimated, with plasma moving at sub or at most mildly relativistic velocities (up to $\sim 0.1c$). They are observed as radio outflows and indirectly through Broad Absorption Lines (BAL) in the optical–UV spectra in about 10% of the powerful objects, possibly those observed at low latitudes through equatorial outflows.

1.2 X-Ray Binaries

In the last decade, evidence has also accumulated for jets being commonly associated with Galactic X–ray binary systems, possibly as many as 20 % of them [7], and among these, a few transient objects intriguingly showing apparent superluminal motion. A peculiar case – although other neutron star binaries might present a similar behavior – is represented by the precessing jet associated with SS433, with the best – although still unexplained – measured velocity of $0.26\,c$. See the contributions [7], [8], [9] on jets in Galactic sources.

1.3   And More...

The zoology of jets should also include less powerful systems, in particular bipolar outflows associated with protostars (YSO), comprising different components, with gas in different ionizing states, and reaching velocities of a few hundred km s^{-1}. At the other extreme, jets or relativistic outflows seem to account best for the gamma–ray burst phenomenon and relativistic afterglow, with Lorentz factors believed to reach a few hundreds.

Although here we focus on jets associated with candidate black hole systems, the richness and diversity of conditions and environments in which jets are observed on one side reveals that jets are common features, relatively easy structures to form, but on the other cautions against the temptation of following the (morphological) similarity to understand their origin as rather different detailed mechanisms are likely to be at work in different systems.

2   What We Dream

What can we infer from the wealth of observations on why jets form, how they are energized, accelerated, collimated and confined, and what is the gas flow around the black hole ?

The main piece of evidence which emerges from observing jets in different objects is the invariable association with an accretion disc, although possibly reflecting different accretion regimes. While not all disk systems appear to produce powerful quasi–stationary jets, weaker jets/outflows might always be present at some level, as a necessary condition for accretion to take place, by extracting angular momentum from the inflow (indeed a large scale outflow rather than a powerful well collimated jet on small scales might dominate this process). If so, this would also indicate that the powerful flows are an extra ingredient. The other indication is that jets have speeds comparable with the escape velocity from their sources. (Speaking loosely, this is ultrarelativistic in the case of black holes.)

There are three proposed general mechanisms for jet formation.

i) Hydrodynamic acceleration: An adiabatic fluid propagating in an external medium with decreasing pressure, provides a relatively simple and direct way of achieving hydrodynamical self–collimation and acceleration, due to the requirement for the fluid to pass through the sonic point [10]. However, the gas that would be required to confine the most powerful extragalactic jets would radiate an X–ray flux far larger than has been observed. This mechanism could be appropriate for low power jets.

ii) Radiative acceleration: An alternative possibility is to consider the intense radiation field as responsible for the acceleration. Two of the difficulties associated with this hypothesis are: a) many sources with powerful jets have luminosities well below the conventional Eddington limit – and consequently insufficient for acceleration, even when more efficient absorption processes are considered (e.g.[11]); b) the drag caused by the radiation severely limits the attainable velocity. An independent confining mechanism would, in any case, be required if the jet is accelerated by a thin disk. Alternatively, while a funnel, which might form in the central part of a flow accreting onto a black hole when the fluid has enough (radiation or ion) pressure, can provide the initial collimation, this structure is possibly subject to instabilities which can mass–load and decelerate the outflow, and would also imply a more isotropic – thus less efficient – accelerating field.

iii) Hydromagnetic acceleration: Therefore, hydromagnetic models appear as more promising at least to account for the production of the most powerful jets. Magnetic fields provide a natural mechanical link between disks and jets and can account for the launching, confinement and collimation of jets. The power can be extracted from (and symmetry provided by) the rotation of either/both an accretion disk, giving rise to an MHD wind over a large range in radii [12], [13], or/and – being limited to the inner radii – a spinning black hole threaded by a large scale magnetic field [14]. Much of the current debate involves hydromagnetic models.

Indeed the efficiency of disk vs. black hole energy extraction has been discussed, as the former mechanism may produce more power depending upon the assumptions made [14], [15]. The simplest picture involves the existence of a field component frozen in and threading the disk at large enough angle that matter is centrifugally launched along the field lines. The differential rotation of the disk and inertia of the gas lead to the wrapping up of the field lines, whose hoop stress due to the toroidal component thus generated could then provide the collimation, while the pressure gradient would help the acceleration. Solutions for the structure of the field and resulting MHD flow have been found even for the relativistic case. Although dependent on the inner and outer boundary conditions, they seem to confirm the efficiency of this process in generating collimated flows, asymptotically converting a large fraction of Poynting flux into bulk kinetic power (e.g. [16]), although doubts have been cast on their survival against pinch and helical instabilities [17]. Angular momentum, energy and mass can thus be removed from the accreting flow with an efficiency that depends upon the ratio of the mean value of the open magnetic field to the surface mass density in the disk.

However a good reason to consider still the extraction of the spin energy of the hole is that the high latitude outflow from near a black hole is unlikely to be loaded with baryons, unlike that from an accretion disk corona, and can plausibly attain an ultrarelativistic asymptotic speed. Furthermore as the energy flux is likely to be dominated by the Poynting component close to the hole, radiative drag can be avoided. A particularly attractive picture is that the ultrarelativistic cores of the jets, observed at high radio frequency and γ–ray energy, are powered by the spin of the hole and collimated by a mildly relativistic hydromagnetic outflow launched by the inner disk, which is, in turn, successively collimated by slower winds from larger radii. Note that it is not necessary for the hole power to dominate the disk radiative or hydromagnetic power to account for ultrarelativistic jets, though our current understanding of black hole/disk electrodynamics does not preclude this.

Fundamental questions remain unanswered. They mainly reflect the difficulty in determining the field origin and configuration. The simplest picture involves large scale unipolar fields. Observational evidence for ordered components exists on arcsec and milli arcsec scales (consistent with the effect of shearing), but no indication of the field structure can be inferred for the inner region. It has been suggested that a large field component dragged in from the outer disk might not reach the inner parts as the inflow timescale may exceed the diffusion and reconnection timescales. However, as numerical simulations show [18], following magneto–rotational instabilities in the disk, loops of toroidal and radial field might be generated on a rotational timescale, with scale height comparable to the pressure one. These are then likely to emerge from the disk through buoyancy and reconnect. Interestingly it has been suggested that small scales/unordered field might still provide the requested conditions to launch a jet [19], [20]. The stability of these structures present a further unclear issue. However an interesting possibility is that the formation of a suitable configuration and consequent ejection of matter is non–stationary (e.g. [12], [21]), on the line of what might be hinted from the behavior observed in the micro–quasar GRS 1915+105.

3   What We Hope to Learn

Let us now consider both theoretical and observational issues which currently constitute the most promising steps forward to shed light on the jet formation and the accretion–ejection connection.

• An appealing possibility recently proposed and much debated, is that the formation of jets/outflows might be a natural and necessary condition for accretion to occur. In particular, it has been pointed out that whenever the flow is adiabatic, i.e. radiative dissipation of energy in the flow is inefficient – because either the density is too low or radiation is produced but trapped within the flow due to the long diffusion timescale ([22], [23], [24]) – then energy has to be extracted from the flow mechanically and/or electromagnetically (ADIOS [25]). Hydrodynamic simulations show that the net mass accretion rate increases roughly linearly with radius, though, in the absence of a rapid source of dissipation, the surplus mass escapes as a subsonic breeze rather than a supersonic wind [26]. It will be interesting to see if hydromagnetic simulations exhibit centrifugally–driven super-Alfvénic outflows.

• It is appropriate at this meeting, to remark that the three X–ray observatories, Chandra, XMM and Astro–E, with their complementary capabilities, should revolutionize our understanding of the high energy properties of extragalactic disks and jets. As discussed by [27], X–ray reflection features provide the strongest current evidence for the presence of optically thick disks in AGN, thus setting constraints on the geometry and regime of accretion. Iron line profiles are starting to provide a diagnostic of the spacetime around black holes, i.e. a direct measure of their spin, and the system geometry. XMM should allow much improvement and reverberation mapping should become possible with Constellation X or XEUS.

The characteristics of iron lines, strength and profile, are currently of much lower quality in radio–loud objects, for which not even widths can be robustly determined, e.g. [28], [29]. The reflection features appear so far to be comparably weaker than in Seyfert galaxies, and thus still compatible with reprocessing occurring in an optically thin medium (e.g torus, wind). Observations with high sensitivity and spectral resolution broad band X–ray detectors are thus of primary importance.

An independent measurement of the mass and spin of the hole should eventually be provided by QPO in binary X–ray sources, although we do not have a good understanding of how the normal mode frequencies reflect the spacetime geometry, nor of which modes are likely to be excited. Perhaps numerical simulations will be very helpful here. QPO might be starting to be detected in AGN too, on timescales of the order of ~ 10 minutes [30].

• Precession of jets has been clearly established only for SS433 and possibly for the blazar OJ 287, e.g. [31], [32]. Other evidence, however, is accumulating from radio imaging of jets which can be interpreted as helical and inverted symmetric structures, possibly originating from precession.

Theoretically the coupling between a disk and a spinning black hole, and in particular the interaction between the two whenever their spin axes are not aligned (Bardeen–Petterson mechanism) is still an open issue. In particular, it has been suggested [33] that the hole might be rapidly spun down by the interaction, thus inhibiting the efficiency of extracting the spin energy of the hole. (Note here the different role that accretion might have in changing the hole mass and spin for galactic and extragalactic objects [34], [35]).

HST imaging indicates that jets and disks are aligned on large scales; higher resolution radio imaging will be necessary to determine what happens closer to the hole and whether the jet is aligned with the spin axis of the disk within the distance at which the hole–disk coupling is plausibly effective.

• As mentioned above, the inner jet speed, as well as changes in velocity field, are crucially linked to the possible baryon loading. Relativistic jet speeds might be even more extreme than commonly assumed.

Recent observations of the jet in M87 have revealed apparent motion on arcsec scales implying $\Gamma \sim 6$, significantly larger than observed before [36].

More extreme bulk velocities have also been invoked to account for extreme values of brightness temperatures inferred from observation of intraday variability at GHz frequencies, if the emission is due to incoherent synchrotron radiation [37]. $\Gamma > 100$ would be needed even in the most conservative case (in terms of total jet power implied [38]) that interstellar refractive scintillation affects the brightness temperature values. Coherent processes would also require ad hoc conditions. However new results on variable polarization might imply that more exotic processes are at work [39].

Radio outflows at mildly relativistic velocities are more commonly observed and might be present in as many as 30% of the radio–quiet sources, although with powers typically three orders of magnitude smaller than in the radio–loud ones. Even more surprising and relevant would be the detection of superluminal motion associated with parsec-scale jets in radio–quiet AGN as recently suggested [40].

• It is clearly important to determine whether jets on different scales can be confined by the external gas or if magnetic fields are needed. Cases of overpressured (powerful) jets have been found, which involve large sections of the flow and thus seem unlikely to be just transient regions. Alternatively, the estimated internal pressure could be lower for a small filling factor of the emitting plasma. Much is expected from the improved (factor $\sim$ ten) spatial resolution of Chandra with regards to the detection of X–ray jets and especially the estimate of the external pressure on small (arcsec) scales, thus possibly determining the scales on which magnetic confinement is indeed compulsory. Note that if magnetic hoop stresses (e.g. in a MHD wind) confine the jet, the required external gas pressure might be reduced by orders of magnitude, corresponding to the radial extension of the wind, i.e. the scale over which the jet is formed and collimated.

Another diagnostic for field confinement would be the determination of its toroidal structure through Faraday rotation mapping, though foreground effects make this quite difficult to carry out in practice.

High resolution radio imaging constitutes also the most direct evidence for the collimation scale. In M87 it has been possible to trace the jet down to scales $\sim 10^{-2}$ pc [41], and recent evidence obtained with high resolution 7mm VLBI observations indicates that indeed collimation occurs on these scales, $\sim 60 - 100$ m [36], thus involving a significant part of the disc.

The degree of collimation could be in principle also determined from statistical arguments within the frame of unification scenarios. However, the likely existence of velocity gradients (polar and possibly radial) limits the robustness of any conclusion.

• A crucial quantity to be determined is obviously the average power and mass flux in jets. Model–dependent estimates can be inferred from the radiative dissipation in the inner jet of blazars and lead to kinetic powers in some cases comparable to and often exceeding the observed radiation from the accreting flow (most conspicuously in low power sources). A further relevant piece of information would be the ratio between the jet power and the Eddington limit: direct mass measures of nearby radio–loud objects might allow this. The emission models adopted assume stationary flows with filling factors of order unity and the results also strongly depending on the low energy end of the particle distribution. Tight constraints on the latter can be derived from the soft X–ray spectrum of high power blazars; here XMM and Astro–

E should provide the best limits, as well as assess the claimed presence of absorption features by relatively dense and cold gas in and/or around jets.

Closely connected to the power is the jet composition, which is still undetermined. The main initial energy carrier is, possibly, electromagnetic (although this might limit the formation of strong shocks), as this has to be in any case invoked to accelerate and collimate the flow. Radiative constraints from the lack of soft X-ray features exclude a large pair contribution. (Note however that none of the radiative and dynamical constraints on the paucity of pairs seem to apply to low power radio sources). On larger scales, where the bulk of the dissipation occurs and further out on VLBI scales, a significant fraction of the electromagnetic energy is likely to have been converted into kinetic power of an ordinary plasma [42]. Alternatively – as observations of circular polarization might imply [43] – the plasma could be energetically dominated by electron–positron pairs loaded in the jet, although spectral constraints imply that the loading might not be easily achieved [44], [45].

• Galactic superluminal sources provide a promising site to hunt for clues on the disk–jet connection because the variations are so rapid compared with those associated with AGN and it is possible to perform statistical studies with relatively short stretches of data. Furthermore, in these systems a better estimate of the mass inflow might presumably be inferred. Of interest for the galactic vs. extragalactic analogy is the recent determination of highly relativistic bulk velocities in GRS 1915+105 [7], [9].

Indeed GRS 1915+105 provides the strongest case for a tight inner disk–jet connection ([7], [46]), thanks to the detection of episodic accretion–ejection events. These findings strongly call for time dependent models. Nevertheless, jets are also observed in binaries during normal states and transitions between them. Information is still too scarce to infer a clear connection between mass inflow and outflow. In the AGN case the nature of ejection (quasi–stationary or impulsive) is unclear. Flaring events are observed – although with poorly constrained duty cycles – but there is some evidence for a quasi stationary underlying emission component. The corresponding timescales suggest that these events involve only local jet instabilities and no apparent connection with the disk emission has been found. Clues on longer scale trends might be inferred from statistics and the study of young radio sources, e.g. [47].

• Some of the issues concerning the relation between the inflows and outflows will be probably clarified through the interpretation of the results of numerical simulations, which are certainly becoming the necessary support to the understanding of extremely complex physical problems requiring 3D treatment with high dynamical resolution and including MHD, special and general relativity effects. Here great progress has been already achieved on several issues, such as simulation of the behavior of magnetic fields in disks [18], 2D hydrodynamical accreting flows [26], the inner black hole magnetosphere [48] and the propagation of relativistic jets [49].

• Finally, let us mention the recent findings [50] concerning the host galaxies of radio–quiet quasars which, contrary to previous belief, appear to be elliptical, as are the hosts of jetted sources. This evidence restricts the parameter space for the origin of the radio–loud/radio–quiet dichotomy, which has now to be ascribed to nuclear or evolutionary properties (e.g. geometry and angular momentum of the inflow, magnetic flux, spin, accretion rate, black hole mass). It is even possible that it is the black hole activity which determines the structure of the host galaxy [51].

One possible speculation is that a rapidly spinning hole is a necessary but not sufficient condition for the formation of a powerful jet, and that a second parameter would be involved, namely the accretion rate over mass ratio [51]. Highly super-Eddington flows could give rise to strong winds (e.g. BAL systems) and strongly ionized disks – accounting for the paucity of X–ray reflection features in highly luminous objects [52] – and a strong enough radiation field to inhibit the formation of jets. The latter would instead occur in $\sim$ Eddington limit systems whenever the black hole spins rapidly enough. Very sub-Eddington flows would finally allow for the formation of outflows and jets despite of being radiatively inefficient – as there is growing evidence in low power radio–loud sources.

Acknowledgments
AC thanks the organizers for setting up this interesting meeting. They are acknowledged together with the MURST for financial support. RB thanks the Institute of Astronomy for hospitality and support through the Beverly and Raymond Sackler Foundation and NASA under grant 5-2837 for support.

References

1. Begelman M. C., Blandford R. D., Rees M. J. (1984) Rev. Mod. Phys. **56**, 255
2. Burgarella D., Livio M., O'Dea C. P. (Eds.) (1993) Astrophysical Jets. Cambridge University Press, Cambridge
3. Ostrowski M., Sikora M. et al. (Eds.) (1997) Relativistic Jets in AGN. Jagellonian University, Cracow
4. Rees M. J. (1971) Nature **229**, 312
5. Harris D. E., Biretta J. A., Junor W. (1997) MNRAS **284**, L21
6. Gaidos J. A., Akerlof C. W. et al. (1996) Nature **383**, 319
7. Fender R. P. (1999) these proceedings
8. King A. R. (1999) these proceedings
9. Mirabel I. F. (1999) these proceedings
10. Blandford R. D., Rees M. J. (1974) MNRAS **169**, 395
11. Ghisellini G., Bodo G. et al. (1990) ApJ **362**, L1
12. Blandford R. D., Payne D. G. (1982) MNRAS **199**, 883
13. Königl A. (1989) ApJ **342**, 208
14. Blandford R. D., Znajek R. L. (1977) MNRAS **179**, 433
15. Livio M., Ogilvie G. I., Pringle J. E. (1999) ApJ **512**, 100
16. Chiueh T., Li Z., Begelman M. C. (1991) ApJ **377**, 462

17. Begelman M. C. (1998) ApJ **493**, 291
18. Balbus S. A., Hawley J. F. (1998) Accretion Processes in Astrophysical Systems: Some Like it Hot. Holt S.S., Kallman T.R. (Eds.), AIP Conf Proc. 431, 79
19. Tout C. A., Pringle J. E. (1996) MNRAS **281**, 219
20. Heinz S., Begelman M. C. (1999) 19th Texas Symp. on Relativistic Astrophysics and Cosmology. Paul J., Montmerle T., Aubourg E. (Eds.), in press
21. Begelman M. C. (1993) Astrophysical Jets. Burgarella D., Livio M., O'Dea C. P. (Eds.) Cambridge University Press, Cambridge, 305
22. Rees M. J., Begelman M. C., et al. (1982) Nature **295**, 17
23. Begelman M. C., Meier D. L. (1982) ApJ **253**, 873
24. Narayan R. (1999) this workshop
25. Blandford R. D., Begelman M. C. (1999) MNRAS **303**, L1
26. Stone J. M., Pringle J. E., Begelman M. C. (1999) MNRAS, in press
27. Fabian A. C. (1999) this workshop
28. Wozniak P. R., Zdziarski A. A. et al. (1998) MNRAS **299**, 449
29. Sambruna R. M., Eracleous M., Mushotzky R. (1999) ApJ, in press
30. Iwasawa K., Fabian A. C. et al. (1998) MNRAS **295**, L20
31. Vermeulen R. (1993) Astrophysical Jets. Burgarella D., Livio M., O'Dea C. P. (Eds.), Cambridge University Press, Cambridge, 241
32. Hughes P. A., Aller H. D., Aller M. F. (1999) BL Lac Phenomenon. Takalo L. O., Sillanpää A. (Eds.), ASP Conf. Ser. 159, San Francisco, 273
33. Natarajan P., Pringle J. E. (1998) ApJ **506**, L97
34. King A. R., Kolb U. (1999) MNRAS **305**, 654
35. Wilson A., Colbert E. J. M. (1995) ApJ **438**, 62
36. Junor W., Biretta J. A. (1999) AAS 194.50.17
37. Kedziora–Chudczer L., Jaumcey D. L. et al. (1997) ApJ **490**, L9
38. Begelman M. C., Rees M. J., Sikora M. (1994) ApJ **429**, L57
39. Macquart J.–P., Kedziora–Chudczer L. et al. (1999) Nature, submitted
40. Blundell K. M., Beasley A. J. (1999) AAS 193.110.04
41. Junor W., Biretta J. A. (1995) AJ **109**, 500
42. Celotti A. (1997) Relativistic jets in AGN. Ostrowski M., Sikora M. et al. (Eds.), Jagellonian University, Cracow, 270
43. Wardle J. F. C., Homan D. C. et al. (1998) Nature **395**, 457
44. Blandford R. D., Levinson A. (1995) ApJ **441**, 79
45. Sikora M., Madejski G. (1999) ApJ, submitted
46. Belloni T., Mendez M., et al. (1997) ApJ **488**, L109
47. Reynolds C. S., Begelman M. C. (1997) ApJ **487**, L135
48. Koide S., Meier D. L. et al. (1999) ApJ, submitted
49. Aloy M. A., Ibanez J. M. A. et al. (1999) ApJ **523**, L125
50. McLure R. J., Kukula M. J. et al. (1999) MNRAS **308**, 377
51. Blandford R. D. (1999) Royal Society, London
52. Fabian A. C. (1998) Accretion Processes in Astrophysical Systems: Some Like it Hot. Holt S.S., Kallman T.R. (Eds.) AIP Conf Proc. 431, 246

The Importance of Rapid Black Hole Spin in Relativistic Jet Formation

David L. Meier[1] and Shinji Koide[2]

[1] Jet Propulsion Laboratory, Caltech, Pasadena, CA 91109, USA
[2] Toyama University, Toyama, Japan

Abstract. We present a summary of several recent papers on the simulation of relativistic jet formation in the vicinity of rotating and non-rotating black holes. We find that the strongest and fastest jets ($\Gamma \sim 3$) occur when the hole is rotating *and* the disk plunges rapidly into the ergosphere.

We have performed several investigations ([1], [2], [3], [4]) of jet production by both rotating (Kerr) and non-rotating (Schwarzschild) black holes and rotating (Keplerian) and non-rotating (Keplerian) accretion disks. Our MHD code uses the simplified total variation diminishing method, axisymmetry, ideal magneto-hydrodynamics (MHD), relativistic flow, and a background Kerr metric.

Our simulations begin with a magnetized disk of material encircling a black hole with inner edge at $4.5\,r_g$, where $r_g \equiv GM/c^2$ is one gravitational radius for a black hole of mass M. The initial magnetic field is given by a Wald [5] (uniform) solution with a relatively small Alfén velocity of $0.01c$ inside the disk. Filling the region above the disk is a freely-falling corona, also permeated with the magnetic field. All Kerr holes below have a normalized angular momentum parameter of $j \equiv J/(GM^2/c) = 0.95$.

1   Results for Different Cases

Schwarzschild hole, non-rotating disk. [3] Even in this simplest of tests, some outflow is still produced. The disk is drawn inward toward the black hole and tidally compressed. While most of the material accretes into the hole, the compression produces an overpressured region at $\sim 5\,r_g$ which temporarily halts the inflow and expels a small amount in an uncollimated, hydrodynamical (HD) wind or "splash" above the disk at a velocity $\sim 0.45c$.

Schwarzschild hole, Keplerian disk. [1] In addition to the HD splash, when the disk is imparted with an initial Keplerian rotation, a classic Blandford-Payne [6] mechanism is set up near the last stable orbit at $\sim 6\,r_g$. The rotating magnetic field lines embedded in the disk coil around the rotation axis, producing an upward magnetic pressure and a collimating pinch. Confined by this magnetic bottle, the HD splash reaches a somewhat higher speed of $0.88c$ ($\Gamma \equiv (1 - v^2/c^2)^{-1/2} \sim 2.1$), with the outer MHD jet at a speed of $0.4c$.

Kerr hole, non-rotating disk. [2] The influence of black hole rotation on jet production is most clearly seen by studying the case when the hole is rotating but the disk is not. As in the Schwarzschild case, an HD splash is produced near $5\,r_g$ with outflow speed of $0.45c$. However, as the magnetized material is drawn toward the static limit at $2\,r_g$, the dragging of inertial frames imparts a rotation to the (initially non-rotating) disk with respect to the external frame. At first, a frame-dragged Blandford-Payne condition is set up, with an MHD jet ejected by the disk as it is rotated by the hole. Then, when the magnetic field lines are pressed onto the hole horizon, the Blandford-Znajek [7] process also occurs, coupling the acceleration directly to the rotating hole, in addition to indirectly through the frame-dragged disk. The MHD jet in this case emanates from the region $R < 3\,r_g$ — *inside* the splash rather than outside — and attains a terminal velocity of $0.93c$ ($\Gamma = 2.7$). A detailed analysis shows that this process extracts angular momentum (reducible mass) but not energy (irreducible mass) from the black hole.

Kerr hole, counter-rotating disk. [4] Adding Keplerian rotation in a sense opposite to the hole's spin shows all three of the above outflows: BP MHD jet from $R \sim 6\,r_g$; HD splash confined by the BP jet; and a strong, collimated BZ jet from $R < 3\,r_g$. The counter-rotating case behaves similarly to the non-rotating case because the former is unstable to collapse inside $\sim 9\,r_g$.

Kerr hole, co-rotating disk. [4] Because the last stable orbit for a co-rotating disk in Kerr geometry is near the horizon, the inner edge of this disk is stable at the beginning of the simulation. Evolution, therefore, occurs on a secular, not dynamic, time scale, and appears very similar to the Schwarzschild Keplerian case for the times we have calculated so far. Eventually we expect the disk to accrete inside the static limit and produce a BZ jet. However, it is still unclear whether it will be as fast or as powerful as that produced by a disk that violently collapses into the ergosphere.

2   Conclusions

We see three basic types of outflow in black hole systems with magnetized accretion disks: a BZ MHD jet from $R < 3r_g$ when the hole is rotating, an HD splash from $R \sim 4\,r_g$, and a BP MHD jet from $R \sim 6\,r_g$ when the disk is rotating. The latter can collimate the HD splash into a fairly fast ($0.88c$) outflow, but highest velocities are produced by the BZ jet, which can achieve $\Gamma \geq 3$ and power densities ≥ 10 times that in the outer $v \sim 0.4c$ BP jet.

References

1. Koide, S., Shibata, K., & Kudoh, T. (1998) ApJ Letters, 495, L63-L66.
2. Koide, S., Meier, D., Shibata, K., & Kudoh, T. (1999a) 19th Texas Symp.
3. Koide, S., Shibata, K., & Kudoh, T. (1999b) ApJ, 522, 727-752.
4. Koide, S., Meier, D., Shibata, K., & Kudoh, T. (1999c) ApJ Letters, in press.
5. Wald, R. M. (1974) Phys. Rev. D, 10, 1680-1685. 522, 727.
6. Blandford, R. D. & Payne, D. G. (1982) MNRAS, 199, 883-903 (BP).
7. Blandford, R. D. & Znajek, R. (1977) MNRAS, 179, 433-456 (BZ).

Radio Cores in Low-Luminosity AGN: ADAFs or Jets?

Heino Falcke[1], Neil M. Nagar[2], Andrew S. Wilson[2], Luis C. Ho[3], and Jim S. Ulvestad[4]

[1] Max-Planck-Institut für Radioastronomie, Auf dem Hügel 69, D-53121 Bonn, Germany (hfalcke@mpifr-bonn.mpg.de)
[2] Dept. of Astronomy, University of Maryland, College Park, MD 20742-2421, USA (wilson,neil@astro.umd.edu)
[3] Carnegie Observatories, 813 Santa Barbara Street, Pasadena, CA 91101, USA (lho@ociw.edu)
[4] NRAO, P.O. Box O, 1003 Lopezville Road, Socorro, NM 87801 (julvesta@aoc.nrao.edu)

Abstract. We have surveyed two large samples of nearby low-luminosity AGN with the VLA to search for flat-spectrum radio cores, similar to Sgr A* in the Galactic Center. Roughly one third of all galaxies are detected (roughly one half if HII transition objects are excluded from the sample), many of which have compact radio cores. Follow-up observations with the VLBA have confirmed that these cores are non-thermal in origin, with brightness temperatures of $\geq 10^8$ K. The brightest of these are resolved into linear structures. The radio spectral indices of the cores are quite flat ($\alpha \sim 0$), with no evidence for the highly inverted radio cores predicted in the ADAF model. Spectrum and morphology of the compact radio emission is typical for radio jets seen also in more luminous AGN. The emission-line luminosity seems to be correlated with the radio core flux. Together with the VLBI observations this suggests that optical and radio emission in at least half the low-luminosity Seyferts and LINERs are black-hole powered. We find only a weak correlation between bulge luminosity and radio flux and an apparently different efficiency between elliptical and spiral galaxies for producing radio emission at a given optical luminosity.

1 Introduction

What powers the nuclei of nearby galaxies? Many of them show evidence for emission lines similar to those seen in active galactic nuclei (AGN) but on a much lower level (Ho et al. 1997a) — therefore they are called low-luminosity AGN (LLAGN). In some cases broad lines are seen and hence one infers the presence of a central black hole (Ho et al. 1997b). In most cases, however, even a moderate starburst might be able to explain the observed optical spectra (Alonso-Herrero et al. 1999), especially those residing in LINER galaxies (Heckman et al. 1983).

Another method to identify the nature of the activity is to search for compact, flat-spectrum radio cores with high brightness temperatures, since this is a typical feature of many AGN and cannot be explained by star formation. For LLAGN the nature of these radio cores is largely unclear. It has been proposed that the compact radio emission could be produced either by emission from an Advection Dominated Accretion Flow (ADAF; e.g. Narayan et al. 1998) or from scaled-down AGN jets (Falcke & Biermann 1996; 1999).

We have therefore performed a VLA survey of two samples of nearby galaxies with optical emission lines to identify such compact radio cores. Follow-up observations with the VLBA of these cores have been made that shed further light on their nature.

2   Samples and Observations

The first sample we observed consisted of 48 galaxies with mainly LINER-like emission spectra that were part of ongoing studies at other wavelengths. In a second project we expanded this sample to a distance-limited sample of galaxies with emission lines within 19 Mpc.

Both samples were observed with the VLA in its largest configuration at 15 GHz. In the final data reduction we reached a 10σ detection limit of ~ 1.1 mJy. The resolution was about 0.15″ which corresponds to a linear scale of 14 pc for a galaxy at a distance of 19 Mpc. All sources which were detected with compact emission above 3 mJy in either sample were then observed with the VLBA at 5 GHz with a resolution of 2.5 mas (~ 0.2 pc at 19 Mpc distance) and a detection limit around 2 mJy.

3   Results

We are going to restrict the following discussion to the detection of compact core emission. The detection rate in our first LLAGN sample was 35% (17 of 48), higher than similar deep surveys of normal galaxies (Wrobel & Heeschen 1991). Only two sources had steep spectra and only one out of eighteen sources with optical classification as transition sources (Ho et al. 1997a) was detected. The other detections are all in LLAGN with either Seyfert or LINER spectra. This is confirmed by the results of our distance-limited survey: 44% of LLAGN with Seyfert or LINER spectra have compact cores, but only 12% of transition objects do.

These results suggest that galaxies with Seyfert and LINER spectra are black hole powered, while transition objects are dominated by star formation. The evidence for black hole powered engines is further strengthened by our VLBA results. Even though our detection limit was close to our selection threshold, 19 out of 20 galaxies[1] showed compact radio emission with brightness temperatures of the order $T_B \geq 10^8$ K. The one non-detected

---

[1] This includes M81 and M87 which are part of the sample but have well-known radio cores and were not observed by us.

source had a steep spectrum and hence is the exception which confirms the rule that also in LLAGN flat-spectrum radio cores are a sign of high-T_B AGNs. We find that the six brightest sources in our VLBI sample all show typical core-jet structures. The fainter cores probably have too low dynamic range and signal-to-noise to show any significant extended structure. Figure 1 (left panel) shows the distribution of spectral indices between our total 6 cm (VLBA) and 15 GHz (VLA) flux densities ($S_\nu \propto \nu^\alpha$). Even though we are comparing VLBA with VLA fluxes and that our selection at 15 GHz is biased towards highly inverted spectra, none of the spectral indices has $\alpha > 0.25$, in conflict with the prediction of the ADAF model (e.g. Yi & Boughn 1998) but quite consistent with the predictions of jet models (Falcke & Biermann 1999). The average is $\langle \alpha \rangle = 0.0$.

For the VLBI sample, i.e. the well-detected cores above 3 mJy, for which we have basically established that the radio emission is AGN-related, we also looked at correlations between radio, emission-line, and bulge luminosities. Figure 1 (right panel) shows that there is a trend for galaxies with stronger Hα emission to have more luminous radio cores. Interestingly, elliptical and spiral host galaxies are offset from each other. Does this reflect a radio-loud/radio-quiet dichotomy for LLAGN?

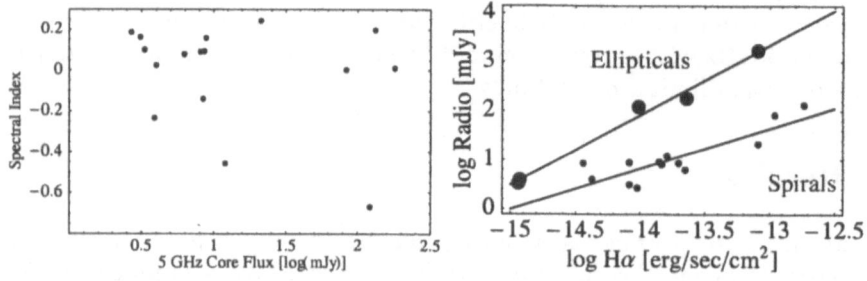

Fig. 1. Left: Spectral indices of LLAGN in our sample with $S_{15GHz} > 3$ mJy between 5 GHz (VLBI) and 15 GHz (VLA) as a function of radio core flux at 5 GHz. Right: S_{15GHz} plotted versus Hα flux for the same sample; ellipticals and spirals are distinguished by big and small dots, respectively.

However, there is another important factor: the galaxy bulge luminosity. We do see a weak trend for the radio luminosity to be related to bulge luminosity; also the ratio between radio and Hα luminosity tends to increase with increasing bulge luminosity. Hence, galaxies apparently become more efficient in producing radio emission relative to Hα in bigger bulges. This also holds if we look at the entire VLA detected sample (Fig. 2). Whether this is due to increasing obscuration, effects intrinsic to the AGN, or a selection effect is unclear. Since ellipticals and spirals in our sample are nicely separated between the top and bottom end of the bulge luminosity distribution, an apparent dichotomy in Fig. 1 is a natural consequence of this trend.

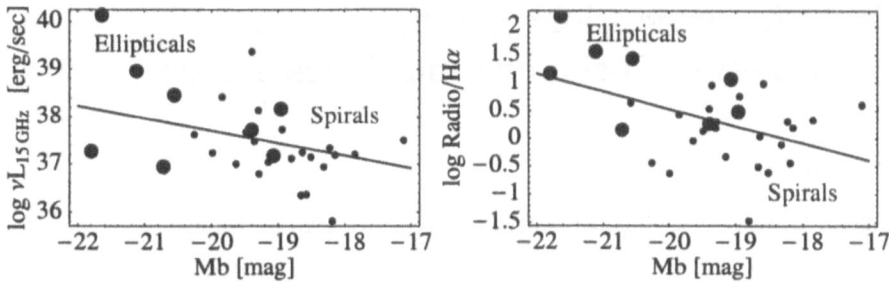

Fig. 2. Left: Radio luminosity (νL_ν) at 15 GHz of LLAGN in our sample with $S_{15GHz} > 1.5$ mJy as a function of blue bulge magnitude. Right: Ratio between 15 GHz radio core and $H\alpha$ flux as a function of blue bulge magnitude in the same sample. Ellipticals and spirals are distinguished by big and small dots respectively.

4   Discussion & Summary

We find that at least 40% of the optically selected LLAGN with Seyfert and LINER spectra have compact radio cores. VLBI observations show that these cores are similar to radio jets in more luminous AGN with high brightness temperatures, jet-like structures, and flat radio spectra (e.g. Falcke & Biermann 1996; 1999). The radio emission seems to be related to the luminosity of the emission-line gas and hence both are probably powered by genuine AGN operating at low power. We found no evidence for high frequency components with highly inverted spectra as predicted in ADAF models. Hence, for these models one should probably not include radio fluxes in broad-band spectral fits. We also find only a weak correlation between radio and bulge luminosity. Together with the radio-$H\alpha$ correlation this makes it very unlikely that the black-hole mass could be reliably determined from the radio data—in contrast to what is occasionally suggested.

References

1. Alonso-Herrero, A., Rieke, M.J., Rieke, G.H., Shields, J.C. 1999, ApJ, in press [astro-ph/9909316]
2. Falcke, H., & Biermann, P.L. 1996, A&A 308, 321
3. Falcke, H., & Biermann, P.L. 1999, A&A 342, 49
4. Heckman, T. M., Van Breugel, W., Miley, G. K., Butcher, H. R. 1983, AJ, 88, 1077
5. Ho, L. C., Filippenko, A. V., & Sargent, W. L. W. 1997a, ApJ, 487, 568
6. Ho, L. C., Filippenko, A. V., & Sargent, W. L. W. 1997b, ApJS, 112, 391
7. Narayan, R., Mahadevan, R., Grindlay, J. E., Popham, R.G., & Gammie, C. 1998, ApJ 492, 554
8. Wrobel, J. M., & Heeschen, D. S. 1991, AJ, 101, 148
9. Yi, I., & Boughn, S. P. 1998, ApJ, 499, 198

The Magnetic Field Configuration of Accretion Disks around Black Holes

Nazar R. Ikhsanov

Pulkovo Observatory, Pulkovo 65/1, St. Petersburg 196140, Russia

Abstract. A hot disk corona forms due to the dynamo amplification of the magnetic field in the differentially rotating, turbulent disk and buoyancy instability of the magnetic field tubes. The resulting configuration of the magnetic field in the disk corona is explained in terms of Z-pinch. The energy release in Z-pinch leads to the formation of relativistic particle beams and collimated plasma outflows.

1 Magnetic Field of the Keplerian α-Disk

Differential character of Keplerian rotation of accretion disk matter and the turbulence result in the rapid amplification of initial magnetic field by the dynamo-mechanism up to the threshold value $B_{\rm cr} \simeq \sqrt{8\pi nkT}$, at which the disk becomes unstable with respect to Parker instability. Under the certain conditions this may lead to a disintegration of the magnetized disk to a system of blobs bounded by the magnetic lines of force through the hot magnetic corona of the accretion disk (Ikhsanov & Pustil'nik 1999).

2 The Magnetized Blobby Disk

Once the disc has disintegrated into the system of blobs, the exchange of rotational momentum between them can take place only via the tension of field lines, which connect the blobs through the magnetic corona. The blob's rotation being differential, the connecting magnetic fluxes get entangled with current layers forming at the surface of the blobs. However, owing to high coronal plasma conductivity, the dissipation of magnetic field via reconnection does not occur at this stage. On the contrary, the process of field amplification is going on, with generation of a stronger azimuthal component, B_{φ}, and corresponding amplification of the poloidal component, caused by extension of field lines and the self-compression of the forming structure by $B_{\varphi}^2/8\pi$.

As a result, on the axis of accretion disc, in the region above and under the gravitating centre, a configuration of Z-pinch type is formed, parallel to the disc rotation axis (Fig. 1), in which the tension of the toroidal component is balanced by the counterpressure of the disc poloidal field. The continuing Keplerian rotation of the blobs leads to further winding of magnetic lines of force around the axial Z-pinch. The value of the magnetic field strength in

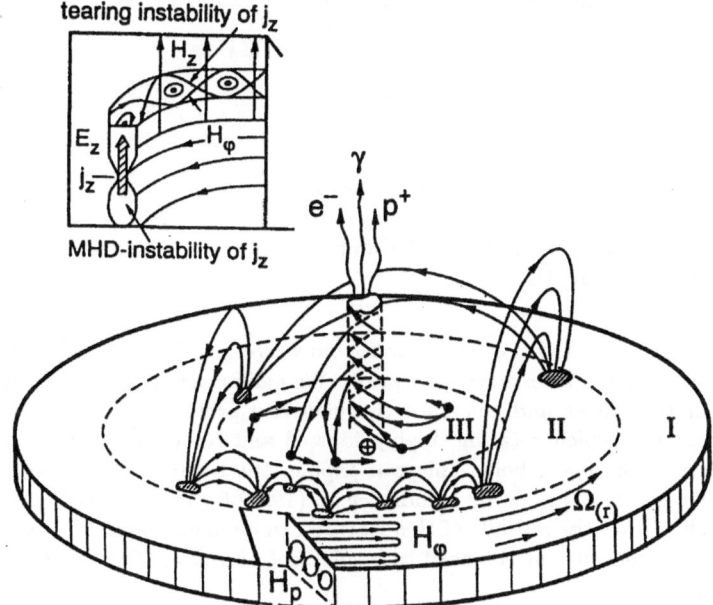

Fig. 1. Magnetized accretion disk: I, region of α-disk; II, intermediate region of Parker instability; III, central region of the polar Z-pinch.

the Z-pinch has the upper limit $B_G = \sqrt{8\pi\rho GM/R}$, at which the tension of the field lines becomes equal to the gravitational force and the blobs begin to move nearly radially. At this value of the field plasma turbulence in the Z-pinch essentially reduces the effective conductivity σ_{eff} and, therefore, turns on rapid processes of non-thermal dissipation of the magnetic energy concentrated in it. Under the action of the chain of MHD- and resistive instabilities (sausage and screw modes $\Longrightarrow$ tearing modes $\Longrightarrow$ plasma turbulization with current discontinuity and formation of double layers) powerful electric fields accelerating the particles are generated. Hence, in the framework of this scheme, the basic energy release occurs not in the accretion disc, but in the axial Z-pinch (see also Ikhsanov & Pustil'nik 1994).

I acknowledge the support of the Followup program of the Alexander von Humboldt Foundation. The work was partly supported by RFBR under the grant 99-02-16336, the Federal program "INTEGRATION" under the grant KO 232 and by INTAS under the grant YSF 99-4004.

References

1. Ikhsanov N.R., Pustil'nik L.A., 1994, ApJS 90, 959
2. Ikhsanov, N.R., Pustil'nik, L.A., 1999, in "BL Lac Phenomenon",
   eds. L.O. Takalo & A. Sillanpää, ASP Conf. Ser. 159, 321

The Influence of Resonant Absorption on the Fe Emission Line Profiles of Accreting Black Holes

Mateusz Ruszkowski and Andrew C. Fabian

Institute of Astronomy, University of Cambridge, Madingley Road,
Cambridge CB3 OHA, UK

Abstract. Diffuse plasma existing above an accretion disc can affect the X-ray spectrum by iron Kα resonance absorption. We embark on a fully relativistic computation of this effect and calculate the iron line profile in the framework of a specific model in which rotating, highly ionized and resonantly-absorbing plasma occurs close to the black hole. This can explain the features seen in the iron Kα line profile recently obtained by Nandra et al. (1999) for the Seyfert 1 galaxy NGC 3516. We show that the redshift of this feature can be mainly gravitational in origin and accounted for without the need to invoke fast accretion of the matter onto the black hole.

We assume that the accretion disc is optically thick and geometrically thin and the primary source of X-ray radiation is an optically thin and hot corona located just above the accretion disc. The primary flux is incident on the disc and produces fluorescent photons. The line and continuum photons can then interact with the cooler ($\sim 10^7$K) Thomson thin plasma in which the disc, corona and the black hole are embedded. Under such conditions a significant fraction of iron ions are in the hydrogen and helium-like states and can resonantly absorb. Due to the large relativistic energy shifts (as seen by the observer comoving with the plasma ($u^r \equiv 0$)), the resonant absorption occurs locally and we may use the Sobolev approximation. In our model the absorbing medium is optically thin for resonant scattering. We use the ray back-tracing method and calculate the optical depth as a function of energy. We separately include a contribution to the iron emission line from the deexciting ions in the ionized cloud.

Examples of the computed iron line profiles for the parameters similar to that obtained by [1] for NGC 3516 are shown in Fig. 1. The thick black line on the left panel denotes the total iron line profile which may provide an acceptable fit to the temporal profile obtained by [1](compare with the inset in their Fig. 1). The observed redshift of the absorption feature can be mostly accounted for by the strong gravitational redshift alone. This has to be contrasted with the main interpretation originally given by [1] who suggest that the redshift of the resonant absorption feature is due to the scattering of radiation by matter rapidly infalling onto the black hole. Note that such matter would have to fall in along the black hole rotation axis with low angular momentum in order to efficiently absorb radiation from the central parts

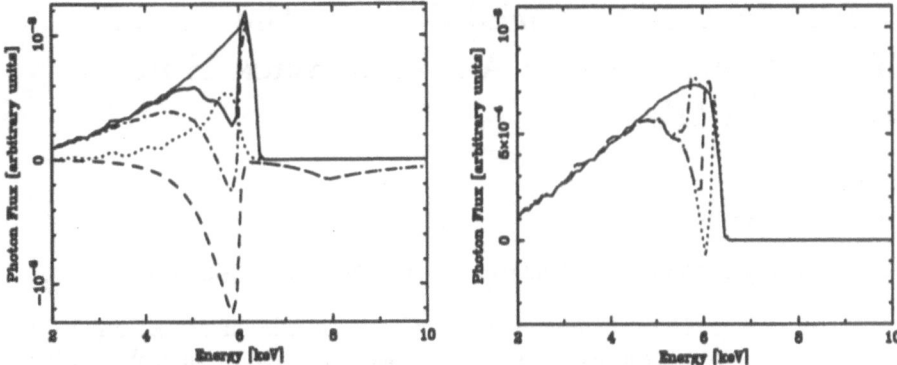

Fig. 1. The iron line profiles (see text for explanations)

of the disc. Our results suggest that the current observational data do not
necessarily give *direct* evidence for fast accretion of matter onto the black hole
in NCG 3516. The other profiles seen on the left panel correspond to: disc
fluorescent emission (thin black line), resonant absorption of the continuum
(dashed line), deexcitation correction (dotted line) and the line absorbed by
resonant absorption with the absorption edge smeared out by the strong relativistic effects (dash-dotted line). Note that there is no intrinsic resonant
absorption of the fluorescent disc emission line and only the continuum radiation is resonantly absorbed due to the low inclination of the disc ($\sim 10°$).
The characteristic skewed shape of the profile of resonant absorption of the
continuum is the result of the systematic effect of the gravitational redshift
and is also due to the variations of the Sobolev length across the disc as seen
by the observer. The right panel shows the effect of the size of the cloud
on the line profile. Large cloud sizes lead to deeper absorption because the
absorption coherence length increases with distance from the black hole. Of
course the smaller the cloud the greater the redshift of the continuum absorption feature, however the magnitude of absorption decreases as the size
of the cloud gets smaller. It is plausible that the absorbing plasma exists in
the form of small magnetically-confined clumps. An ensemble of very small
clouds or filaments distributed at a range of distances from the central black
hole could in principle produce a number of absorption features superimposed on the emission line, an effect similar to the 'Lyα forest' observed in
the spectra of distant quasars. In the present case such features are likely to
be time variable.

References

1. Nandra K., George I.M., Mushotzky R.F., Turner T.J., Yaqoob T., 1999, ApJL,
   in press
2. Ruszkowski M., Fabian A.C., 1999, submitted to MNRAS

X-Ray Iron Line Variability for the Model of an Orbiting Flare Above a Black Hole Accretion Disc

Mateusz Ruszkowski

IoA, University of Cambridge, Madingley Road, Cambridge CB3 0HA, UK

Abstract. We calculate the temporal response of the iron fluorescent line to the illuminating flux from orbiting flares above a black hole accretion disc. We take into account all relevant general relativistic effects. We suggest that the complex temporal behaviour of Fe line profiles seen in future data from XMM or Constellation-X may not necessarily imply a complicated model but could be explained for example in the framework of the orbiting flare model.

The X-ray Fe fluorescent line may be due to the illumination of the accretion disc by active flares located above it. For an observer at infinity any variations in the primary X-ray source would be 'echoed' by different locations on the disc leading to time evolution in the line profile. The Fe line variability was recently calculated by [1] assuming an arbitrary black hole spin; they searched for observational signatures of the spin but mainly focused on the case of a static, instantaneous, on-axis flare. *The duration of bright flares can be of order $100m_7$ in geometric time units ($GM/c^3 \approx m_7 49s$ for the black hole mass $M = 10^7 m_7 M_\odot$) which is comparable to the orbital period close to the black hole.* It is very likely that any variability associated with blobs above accretion discs comes from moving sources. Therefore, in the present work, we relax these assumptions and consider reverberation effects from corotating, off-axis, non-instantaneous flares. The figure shows the iron line variability corresponding to one full revolution of the flare above the accretion disc. The iron line extends to lower frequencies in the Kerr case as a result of the cold iron line being produced very close to the black hole. For a Schwarzschild black hole the disc may be highly ionized below the radius of marginal stability ($r_{ms} = 6m$) and the line may not be produced. The two 'bumps' (see lower right panel) are due to the blueshifted 'hot' iron lines from the ionized regions and *could be detected* by future X-ray satellites. The most prominent feature seen in all diagrams is the drifting maximum of the flux. This sinusoid-shaped feature is due to the fluorescence from the most strongly illuminated part of the disc located just below the flare. Its shape and brightness distribution is the result of special and general relativistic effects. The *high energy maximum* however (e.g. at 50-100 GM/c^3 on lower right panel), is a consequence of the bending of light which is focused on the opposite side of the black hole relative to the actual position of the flare and leads to an enhanced illumination of the disc in this region. This radiation is then further amplified by Doppler boosting. This effect could be readily

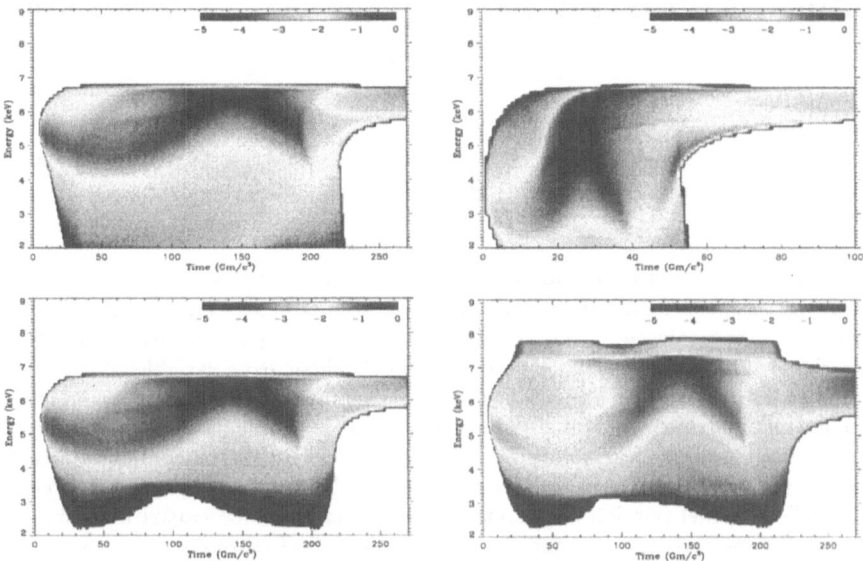

Fig. 1. Fe line time sequences $(\log(Flux/Flux_{\max}))$ for: a=0.998 (upper panels), a=0.0 (lower panels), $\theta = 70°$ (the inclination of the flare relative to the symmetry axis), $i = 30°$ (except for the lower right panel for which $i = 50°$), r=10(except for the upper right panel for which r=3)

observable by future high throughput spectrometers. An interesting feature can be seen when the flare is receding. The main redshifted flux maximum *broadens* or *splits* into two maxima (compare lower panels at 50-100 GM/c^3). The middle maximum is a result of the strong bending of light which is focused from the back side of the disc relative to the observer. As in previous cases, this effect *could be observed* by future X-ray spectrometers. The magnitude of the energy shift of the main maximum is related to the inclination i of the accretion disc (compare lower panels) and also to the distance of the flare from the centre. The bright, short-duration Λ-shaped feature extends to lower energies for smaller distance r of the flare from the centre (see upper right panel). For a very close flare, the middle and strong lower maxima on the right side of the Λ-shaped feature are due to strong gravitational focusing. The lack of the uppermost and middle maxima on the left hand side of this feature is related to the very large overall gravitational and transverse Doppler redshift. It is also caused by the faster motion of the flare combined with the relatively slow propagation of signals slowed by the Shapiro delay.

References

1. Reynolds C.S., Young A.J., Begelman M.C., Fabian A.C, 1999, ApJ, 514, 164
2. Ruszkowski M., 1999, MNRAS, in press

Causal Model for Relativistic Accretion Flow

Jochen Peitz[1][2]

[1] Harvard-Smithsonian Center for Astrophysics, Cambridge, MA 02138, USA
[2] Landessternwarte Königsstuhl, D-69117 Heidelberg, Germany

Abstract. Accretion onto compact objects is a prominent astrophysical problem involving dissipative fluid flow in strong gravitational fields. The generic approach for modeling is by the relativistic Navier-Stokes-Fourier equations, which are of non-hyperbolic type. Consequently, fluctuations in the dissipative variables (shear stress and heat flux) propagate at causality violating infinite speeds, and thermodynamic equilibrium states are unstable. This description is problematic in particular for systems that undergo variability on timescales shorter than or comparable to the dissipative relaxation times. To overcome these difficulties we proposed (Peitz & Appl [3]) to model relativistic accretion flows using extended causal fluid theories. Here we review such a theory and the corresponding 3+1 representation appropriate for numerical implementation.

1 Field Equations

Relativistic kinetic theory can motivate a description for the dynamics of a non-ideal fluid that consists of the three equations

$$0 = \nabla_\alpha N^\alpha \,, \tag{1}$$

$$0 = \nabla_\alpha T^{\alpha\beta} \,, \tag{2}$$

$$I^{\beta\gamma} = \nabla_\alpha F^{\alpha\beta\gamma} \,, \tag{3}$$

where N^α, $T^{\alpha\beta}$, and $F^{\alpha\beta\gamma}$ are the first three moments of the distribution function, i.e., the baryon flux vector, the stress-energy tensor, and the tensor of fluxes, respectively, with $I^{\alpha\beta}$ the corresponding collisional production density. Kinetic theory suggests also that $T^{\alpha\beta}$ and $F^{\alpha\beta\gamma}$ (and thus $I^{\alpha\beta}$) are symmetric and that $F^{\alpha\beta}{}_\alpha = N^\beta$ (i.e., $I^\alpha{}_\alpha = 0$). Fluid theories of the form (1–3) are referred to as *divergence-type* theories (cf. Geroch & Lindblom [1]). Within the *particle-frame* description, N^α and $T^{\alpha\beta}$ are represented by [1]

$$N^\alpha = nu^\alpha \,, \tag{4}$$

$$T^{\alpha\beta} = wu^\alpha u^\beta + \tilde{p}h^{\alpha\beta} + 2q^{(\alpha}u^{\beta)} + \tau^{\alpha\beta} \,, \tag{5}$$

where u^α is the velocity ($u^\alpha u_\alpha = -1$) and $h^{\alpha\beta} = g^{\alpha\beta} + u^\alpha u^\beta$ the projection into the *local rest frame*. The local thermodynamic state is characterized by

---

[1] We use geometrized units $G = c = 1$ and metric signature $(-, +, +, +)$. Indices in round brackets represent (normalized) symmetrization.

the baryon number density n or the rest mass density $\rho = nm$ (with m the baryon rest mass), the energy density $w = \rho(1 + \epsilon)$ (with ϵ the specific internal energy), the equilibrium pressure p, together with the non-equilibrium contributions by the non-ideal bulk pressure $\Pi = \tilde{p} - p$, the heat flux vector q^α, and the trace-free symmetric stress tensor $\tau^{\alpha\beta}$. Equilibrium is defined as a reversible thermodynamic process, and a process with vanishing production, $I^{\alpha\beta} = 0$, is indeed reversible. This imposes nine conditions which can be satisfied setting $\Pi = q^\alpha = \tau^{\alpha\beta} = 0$ and specifying a relation among the equilibrium state functions p, n, ϵ (the equilibrium EOS). Closure of the system (1–3) requires representations for $F^{\alpha\beta\gamma}$ and $I^{\alpha\beta}$, and for the entropy current vector S^α, in agreement with the second law $\nabla_\alpha S^\alpha \geq 0$.

Assuming that $F^{\alpha\beta\gamma}$, $I^{\alpha\beta}$, and S^α are constitutive variables, Liu, Müller, & Ruggeri [2] consider the case where $F^{\alpha\beta\gamma}$ and $I^{\alpha\beta}$ are linear functionals of the *dissipative fluxes* Π, q^α, and $\tau^{\alpha\beta}$, which is consistent with S^α being a quadratic functional. This representation introduces a set of 14 thermodynamic coefficients, which are a priori unknown functions of the independent state variables (e.g., n and ϵ), constrained by the entropy principle, the principle of relativity, and the requirement for symmetric hyperbolicity. For the case of a relativistic Boltzmann gas some of these functions can be derived from the distribution function ([2]).

2 Application to Accretion Flow

A relativistic system is preferentially evolved w.r.t. a globally timelike parameter, i.e., w.r.t. the proper time of a congruence of appropriately specified time-like observers (*fiducial congruence*). The corresponding 3+1 decomposition for the equations of dissipative fluid dynamics (Peitz & Appl [3], [4]) contain only three-dimensional tensor fields plus geometrical source terms compensating for the fiducial velocity field.

In astrophysical flow turbulent transport dominates over molecular transport. The corresponding turbulent transport coefficients are generally unknown. We propose to model turbulent astrophysical flow by the above divergence-type fluid theory assuming an appropriate α-parametrization for the turbulent transport coefficients, and to gauge these parametrizations on the corresponding parametrization frequently used for the Fourier-Navier-Stokes turbulent transport coefficients (e.g., the α-viscosity). This can guarantee subluminal signal speeds for fluctuations in the turbulent dissipative fluxes, thereby resolving the causality problem in relativistic accretion discs.

References

1. Geroch R., & Lindblom L., 1991, *Ann. Phys.* **207**, 394
2. Liu I., Müller I., & Ruggeri T., 1986, *Ann. Phys.* **169**, 191
3. Peitz J., & Appl S., 1998, MNRAS **296**, 231
4. Peitz J., & Appl S., 1999, *Class. Quantum Grav.* **16**, 979

Black Holes or Supermassive Compact Objects Without Event Horizon?

Leonid Verozub

Kharkov State University, Kharkov, 310077, Ukraine

Abstract. It was previously shown that our gravitation equations [1] lead to the existence of supermassive stable compact configurations of degenerated electronic gas [2],[3] without an event horizon. In the present paper the simplest model of such objects in a gas environment has been considered. It is shown that for spherically symmetric accretion onto the object the luminosity is about $10^{37} erg/s$ for a mass accretion rate of the order of $\dot{M} = 10^{-6} M_\odot/year$. The wavelength of the radiation maximum is $400 - -500$ Å. There is an ionization zone around the central objects.

The gravitation equations whose spherically symmetric solution have no physical singularity in flat space-time from the viewpoint of a remote observer are given in [1]. According to the equations the event horizon is absent for the spherically symmetric solution. The radial component F of the gravity force affecting a test particle with mass m in a spherical coordinate system in flat space-time is given by

$$F = -m \left[c^2 B_{00}^1 + (2B_{01}^0 - B_{11}^1)v^2 \right]. \tag{1}$$

Here B_{00}^1, B_{01}^0 and B_{11}^1 are the components of the strength tensor $B_{\alpha\beta}^\gamma$ of the gravitational field in flat space-time :

$$B_{00}^1 = \frac{1}{2} \frac{\alpha f' f^4 (1 - \alpha/f)}{f^2 r^4}, \qquad B_{11}^1 = \frac{1}{2} \frac{\alpha f'}{f^2 (1 - \alpha/f)}, \tag{2}$$

$B_{01}^0 = 2/r$, $f = (\alpha^3 + r^3)$, $\alpha = 2GM/c^2$ is the Schwarzschild radius and v is the radial component of the particle velocity.

We consider an object without an event horizon with a mass of $2.5 \times 10^6 M_\odot$ in the center of a spherically symmetric gas medium.

The velocity of the gas falling from infinity to the center reaches a sound velocity a at a distance r_s which is defined by the equations

$$2\frac{a_s^2}{r_s} = F(r_s, a_s), \qquad r_s^2 a^{\frac{\gamma+1}{\gamma-1}} = \frac{\dot{M}}{4\pi Q}, \tag{3}$$

where a_s is the value of a at $r = r_s$, $Q = \rho_\infty / a_\infty^{2/(\gamma-1)}$, v_∞ and ρ_∞ are the velocity and density at infinity. In contrast to the Newtonian formalism, equations (3) have two solutions. At $v_\infty = 10^7$ cm/s, at the density of the particles number $n = 10^2$ cm^{-3} and at $\dot{M} = 10^{-6}$ M$_\odot$/yr, the numerical solution of eqs. (3) yields: $r_{1s} = 0.83 \times 10^{18}$ cm, $r_{2s} = 1.4 \times 10^{11}$ cm and $a_{1s} = 1.40 \times 10^{7}$ cm, $a_{2s} = 1.1 \times 10^{9}$ cm.

The reason for the second solution is that the gas velocity of a particle falling from infinity increases up to the distances of the order of the Schwarzschild radius and after that decreases according to the peculiarity of the gravitational force.

Fig. 1 shows the velocity v as a function of r from $r = r_{2s}$ to the surface of the central object (in cgs units). The radius R of the central object is found by numerical solution of the equation of hydrodynamical equilibrium and is equal to 0.04α, or 10^{11} cm.

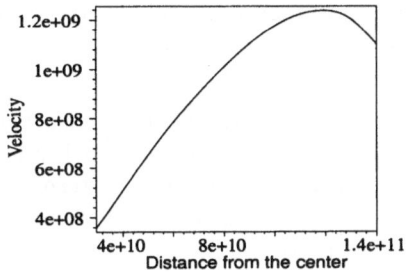

Fig. 1. The dependence of the gas velocity v on r.

Fig. 2. The dependence of the degree of ionization on $\lg(r)$.

The luminosity of the central object in the absence of magnetic fields is of the order of $L = v^2(R)\dot{M}$. For the object under consideration $v(R) = 2.3 \times 10^8$ cm/s and at $\dot{M} = 10^{-6}$ $M_\odot$/yr which yields $L = 0.3 \times 10^{37}$ erg/s. Thus, in spite of a sufficiently large accretion rate the object has a low luminosity.

There must be an ionization zone around a central object without an event horizon which depends on the temperature of the central object and the physical conditions in the gas environment. Fig. 2 shows the degree of ionization X as a function of the distance $z = \lg(r)$ from the center. The maximum of the radiation corresponds to a wavelength of about $\lambda = 500$ Å.

References

1. Verozub L.V. (1991), Phys.Lett. A **156**, 404
2. Verozub L.V. (1966), Astr. Nach. **317**, 107
3. Verozub L.V., submitted to ApJ

Outflowing Coronae Above Accretion Discs in AGN and GBH

Agnieszka Janiuk, Piotr T. Życki, and Bożena Czerny

Copernicus Astronomical Center, Bartycka 18, 00-716 Warsaw, Poland

Abstract. We discuss the role of the coronal outflow perpendicular to the disc surface and study the problem of its influence on spectral shape and on the amplitude of reflected component in hard X-ray spectrum. We compare our results with the predictions of model of X-ray emission from the central spherical source, surrounded by the cold disc which inner part is disrupted.

The X-ray spectra of galactic black hole candidates (GBH) and active galactic nuclei (AGN) consist of two characteristic components: black body emission, and a power law tail. The latter can be explained in terms of Compton upscattering of soft photons on thermal electrons in hot, optically thin medium. There are two basic geometries of the hot and cold matter distributions: hot corona above a cold accretion disc and hot, spherical cloud surrounded by disrupted cold disc. In the case of 'disc plus corona' model the hard X-ray spectra require the outflow of the hot material.

Outflow Velocity and Reflection Amplitude

Part of the hard X-ray flux that is scattered back toward disc surface produces so called 'reflection hump' seen in hard X-ray spectra over 10 keV. Both in Seyfert galaxies and X-ray binaries in the hard state we observe a correlation between spectral index and amplitude of reflection (Zdziarski et al. 1999): the harder the spectrum the lower the amplitude. This can be modelled by the outflow velocity in the corona or by the transition radius between hot and cold phase in 'disc plus sphere' model.

In the hot plasma net electron velocity is a sum of thermal chaotic motion and systematic (bulk) outflow. The reflection amplitude, averaged over the whole range of electron's velocity directions, may be expressed as a function of electron temperature kT_e, bulk velocity β and observer's viewing angle i. In case of static corona ($\beta{=}0$) in the single scattering approximation the reflection amplitude is enhanced due to anisotropy of radiation field. On the other hand, large bulk velocity causes relativistic beaming of radiation in the direction of electron's motion and the reflection amplitude drops rapidly. For large plasma temperature and mildly relativistic outflow ($v_{term}/c \approx \beta_{bulk}$) the first reflection is significantly enhanced.

In X-ray spectra the roughly power-law shape of the primary component is likely due to multiple Compton scatterings in the hot plasma of a certain optical depth. In that case the anisotropy effect is very weak and in order to explain large values of reflection amplitude the bulk velocity directed towards the disc is required.

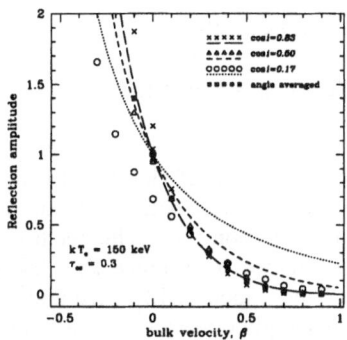

Fig. 1. Amplitude of reflection as a function of the bulk outflow velocity. Points show results of our Monte Carlo simulations, while curves show result of Beloborodov (1999)

Radiation Spectra

Since the first scattering dominates soft X-ray band for galactic sources, the backscattered radiation in this band is enhanced, as predicted by analytical approximation. However, hard X-ray part is dominated by multiple scattered photons, anisotropy effect is smeared off and the backscattered component is not enhanced. Therefore, the continuum formed in the corona and emitted towards an observer is slightly curved, particularly in the soft X-ray band. On the other hand, the 'disc plus sphere' model spectrum is concave.

The two considered scenarios: 'disc+sphere' and outflowing corona predict subtly different shapes of hard X-ray continuum. Clearly the two spectra are not perfect power laws and they curve in opposite directions. The ratios can be as large as 10 per cent, well above the size of statistical and systematic errors in current quality data.

References

1. Beloborodov, A.M., 1999, ApJ, 496, L105
2. Done, C., Mulchaey, J.S., Muschotzky, R.F., Arnaud K.A., 1992, ApJ, 395, 275
3. Ghisellini, G., George, I.M., Fabian, A.C., Done, C., 1991, MNRAS, 248, 14
4. Zdziarski, A.A., Lubiński, P., Smith, D.A., 1999, MNRAS, 303, L11
5. Życki, P.T., Done, C., Smith, D.A., 1998, ApJ, 496, L25

Formation of Broad Line Clouds from Turbulent Shocks in the Accretion Flows of Active Galactic Nuclei

Michael J. Fromerth and Fulvio Melia

The University of Arizona, Tucson, AZ 85721, USA

Abstract. We find that the formation of clouds capable of producing the broad emission lines seen in the spectra of most active galactic nuclei (AGNs) is possible within an unsteady, turbulent accretion flow. The resulting clouds display the correct range in densities and velocities to produce the observed line shape. We also find that the modeled broad line region (BLR) size and differential line response, with the blue and red wings responding fastest to continuum changes, are consistent with the results of reverberation studies.

1 Turbulent Shock Model

Recent 3D hydrodynamic simulations [1] of the gas dynamics in the Galactic Center suggest that the highly variable winds accreting onto a central black hole are subject to several shock-producing effects, including turbulence and stellar wind-wind collisions. In the context of a different model, Perry & Dyson [2] showed that, under some circumstances, shocked gas is initially compressed and heated out of thermal equilibrium with the ambient radiation field. A cooling instability sets in as the gas loses energy via inverse-Compton and bremsstrahlung processes. If the cooling time, t_{cool}, is less than the dynamical flow time, t_{dyn}, through the shock region, the gas may clump to form clouds.

We assume that the accretion flow is roughly spherical at large radii, but with angular momentum parameter $\lambda \equiv l/cr_g \neq 0$, the gas circularizes and settles onto an accretion disk at a radius $r_{circ} \simeq 2\lambda^2 r_g$. Using the prescription of Perry & Dyson [2] in our new picture, we find that the shocks in the accretion flow are viable sites for the production of BLR clouds. The density of clouds produced is a function of the density and temperature of the shocked gas from which they cool, which in turn change with the flow characteristics as a function of radius.

The most important diagnostic for any dynamical model of the BLR is the emission line shape. To model the line profile, the radial cloud distribution function is assumed to be a power law, $n_c(r) \propto r^\alpha$. After specifying the free parameters λ and α, a cloud population is constructed from $n_c(r)$ using Monte Carlo techniques. The photoionization code CLOUDY [3] is then used to calculate the CIV $\lambda1549$ line emission intensity from each cloud. This line intensity must be corrected for anisotropic emission from optically thick clouds. Each cloud's emission line profile is then calculated by including the effects of Doppler and Lorentz line broadening and gravitational redshifting. Finally, the total line profile is calculated by adding the contributions from each cloud. A good fit to the observed data, using $\alpha = -2$ and $\lambda = 10$, is shown in Figure 1.

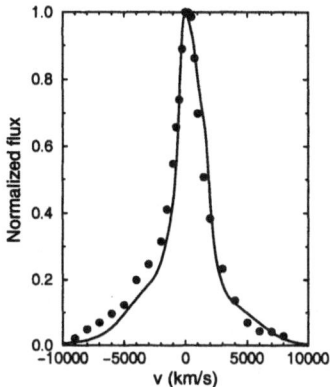

Fig. 1. A comparison of the modeled CIV λ1549 line profile using $\alpha = -2$ and $\lambda = 10$ (solid line) with the observed Fairall 9 (black dots) line profile [4].

The inner extent of our BLR is determined by the circularization radius, which for Fairall 9 is predicted to be $r_{circ} \approx 5\ (\lambda/10)^2$ light-days, consistent with the observed $\sim$ 4 light-days inferred on the basis of the HeII λ1640 response [5]. Our model is also consistent with the differential line response seen in most sources [4], with the blue and red wings (originating in rapidly moving clouds at small radii) responding fastest to continuum changes before the central peak. Considering that our approach here has been semi-analytical, our results hold promise for a more serious attempt at modeling the BLR with a full multi-dimensional calculation in the future.

References

1. Coker, R. F., Melia, F. (1997) Spherical Accretion onto Sgr A*. ApJL **488**, L149–L152
2. Perry, J. J., Dyson, J. E. (1985) Shock formation of broad emission-line regions. MNRAS **213**, 665–710
3. Ferland, G. J. (1996) Hazy, a Brief Introduction to CLOUDY. (Univ. of Kentucky, Lexington)
4. Türler, M., Courvoisier, T.J.-L. (1998) Principal Component Analysis of two UV emission-lines. A&A **329**, 863–872
5. Rodríguez-Pascual, P. M., et al. (1997) Size of Broad-Line Region in AGNs. IX. ApJS **110**, 9–20

line intensity must be corrected for interstellar reddening (i.e. optical) thin clouds. Each cloud's emission line profile is then enhanced by including the effect of Doppler and thermal line broadening and gravitational redshifting. finally, the total line profile is obtained by adding the contributions from each cloud, weighted by its mass.

Fig. 1. A comparison of the modeled C IV λ1550 line profile over the ±2 kms boxcar and with the observed emergent 6 kms data with redshift H.

The inner events of our cILH is determined by the de-polarization factor, which for cILH is predicted to be $z = z_{max} = (\Delta/10^4)$, in broad agreement with the observed z. A little data is figured on the basis of the cILH λ1640 response [2]. Our model is also consistent with the de-localization line structure seen in most sources [3], with the little additional stage 1 similarity in particular. Very little redshift of small redshift (emanating factor) to emit rays, results below the compact peak. One notes that one approach here has been semi-analytic, one result could provide the a most natural attempt at modeling the cILH with a full multi-dimensional calculation in the future.

References

1. Capes, W. F. et al. "Theory and model of Accretion onto Hot Gas" ApJL 458, 5140, 1997.

2. Perry, J. J., Tysen, D. J. (1988) State formation of interstellar emission line regions. MNRAS 215, 685–670.

3. Beaver, G. E. (1996) story a new interpretation to CILV models. J. Univ. of Ker. Today, Cambridge.

4. Lurley, K., Oppenheimer, T. L. (1994) "Spectral component Analysis of Far spectral transmission ApJL 442, 845–872.

5. Baldmann-Jackson, P. M., et al. (2000) Size of Broad Line Region in AGN ApJ ApJL 644, L1–L4.

Part 4

BLACK-HOLE DEMOGRAPHY

Wolf-Rayet Stars and Black Holes, an Overview

Karel A. van der Hucht

Space Research Organization Netherlands,
Sorbonnelaan 2, NL-3584 CA Utrecht, the Netherlands

Abstract. We review the literature of the past two decades on the possible evolution from the massive star Wolf-Rayet phase to the black hole phase, and on the masses of WR stars observed in binary systems. Of the 52 WR binaries in the VIIth Catalogue of Galactic Wolf-Rayet stars, the 19 WR binaries with measured radial-velocity (RV) orbits have masses $M_{WN} = 2$–$55\,M_\odot$ and $M_{WC} = 9$–$16\,M_\odot$. We discuss four candidate WR+BH systems: Cyg X-3, HD 197406, R 140a2, and Mk 34.

1   Introduction

Although the notion that massive stars may evolve into black holes is over three decades old, it was not until the late 70-ies that it was realized that Wolf-Rayet (WR) stars represent a normal phase in the evolution of massive stars, i.e., the phase just before the supernova explosion. We review here the literature on the relation between WR stars and black holes. Of the WR binaries 19 have RV-solutions and mass determinations, against which evolutionary models of massive stars can be tested.

2   Wolf-Rayet Single Stars

Maeder (1981) presented evolutionary models with mass loss for initially massive ($M_i > 30\,M_\odot$) stars, showing that WR stars are likely to be SN precursors. With WR stars having lost their hydrogen envelope, such explosions will have a spectrum showing no hydrogen lines and thus correspond to Type I SNe. Maeder & Lequeux (1982) determined the fraction of supernovae originating from WR progenitors in three different ways: (i) based on an estimated number of 1200 WR stars in the Galaxy and a mean WR lifetime of $\sim 5 \times 10^5$ yr; (ii) based on SN remnants; and (iii) based on the mass spectrum and mass limits, with $M_i = 8\,M_\odot$ as the lower limit initial mass of a star evolving toward a SN explosion and $M_i = 23\,M_\odot$ as the lower mass limit to form a SN from a WR progenitor. All three estimates indicate that one out of 3–7 SN events may result from a WR progenitor. This agrees with the fact that of the five Galactic SNe of the past millennium, at least one, Cas A (Langer & El Eid 1986; Fesen et al. 1987), may have originated from a WR star. El Eid & Langer (1986) proposed for Cas A a SN explosion by $e^\pm$-pair-creation instability in a $45\,M_\odot$ WN star (with $M_i = 100\,M_\odot$), its final fate being complete disruption by explosive oxygen burning. Maeder (1983) and Maeder & Meynet (1987) presented three scenarios to form SNe:

$40\,M_\odot < M_i$ O - Of - BSG - LBV - WNL - WNE - WC - (WO) - SN
$35\,M_\odot < M_i < 40\,M_\odot$ O - BSG - YSG - RSG - (BSG) - WNE - WC - SN
$15\,M_\odot < M_i < 35\,M_\odot$ O - (BSG) - RSG - YSG/Cepheid - RSG - SN

Maeder & Meynet (1989) concluded from further evolutionary models for massive stars with mass loss by stellar wind, that $M_i(WR) = 20$–$40\,M_\odot$ is the minimum initial mass allowing the evolution to the WR phase. The minimum initial mass allowing evolution to the black hole phase is: (i) $M_i(BH) = 25\,M_\odot$ from theoretical models (e.g., Woosley 1986); (ii) $M_i(BH) = 40$–$80\,M_\odot$ from two observed X-ray binaries (van den Heuvel & Habets 1984); and (iii) $M_i(BH) = 50\,M_\odot$ from pulsars in young associations (Schild & Maeder 1985). Thus, they concluded that the direct progenitors of black holes should be WR stars, notably WNL and WCE stars. Maeder (1991) specified massive-star evolution as a function of both initial mass and metallicity.

Langer (1989) calculated models with mass-dependent mass-loss rates and concluded that WNE and WC stars have masses in the range 5–20 $M_\odot$ and that the mean final $M_{WR} = 5$–$10\,M_\odot$. This makes them candidates for SN Ib progenitors, since Ensman & Woosley (1988) derived that light-curves of SN Ib explosions require 4–7 $M_\odot$ progenitors with $M_i = 15$–$20\,M_\odot$. Langer (1991) concluded that: (i) stars with $M_i > 100\,M_\odot$ are supposed to explode during their WNL phase as peculiar SNe II due to the pair-creation mechanism, accounting for $\sim$ 1% of the observed SNe; and (ii) the bulk of the stars with $35\,M_\odot < M_i < 100\,M_\odot$ is supposed to end their evolution as WC stars with $M_{WC} < 10\,M_\odot$. Some of them explode due to the core-collapse mechanism, to become visible as possible SNe Ib. Woosley et al. (1993) calculated a synthetic light-curve of an exploding 4.25 $M_\odot$ WC star, evolved from a $M_i = 60\,M_\odot$ star with mass-dependent mass loss rate, revealing similarities to observed SN Ib light-curves. Woosley et al. (1995) narrowed down the final mass range of helium (WR) stars, with $M_i = 4$–$20\,M_\odot$ and mass-dependent mass-loss rates, to 2.3–3.6 $M_\odot$, and identified these as progenitors of SNe Ib (no hydrogen lines) and, perhaps, SNe Ic (not even helium lines). Up to 25% of all SNe are of Type Ib or Ic (typical light-curves are shown by Vanbeveren et al. 1998, p. 45), i.e., powered by the iron-core collapse in a hydrogen-free star.

An important parameter in the evolutionary models mentioned above is the mass-loss rate during the WR phase. WR mass loss rate determinations have come down in recent years when accounting for clumping effects, by factors of 3–4 with respect to homogeneous atmosphere models (e.g., Hillier 1996; Crowther 1999). Therefore, perhaps, SNe Ic originate only from mass-transfer binaries, rather than single massive stars.

3   Wolf-Rayet Binaries

A significant fraction of the WR stars is contained in binaries, mostly SB2 WR+OB objects, in a few cases possibly as SB1 WR+c objects (c: neutron star or black hole). Van der Hucht et al. (1988) determined a WR binary frequency of 37 % within 2.5 kpc from the Sun, or 34 % within 3 kpc. Moffat (1995) counted 42 %. Incompleteness of both WR star detection and binary classification in these volumes renders these numbers only lower limits.

The occurrence of WR+c systems was proposed from evolutionary considerations for massive early-type binaries (van den Heuvel & Heise 1972). However, van den Heuvel (1994) noted that the evolution beyond the HMXB phase is uncertain due to lack of observed WR+c objects, apart from Cyg X-3 (see below). The most dependable test to check whether a given close binary does comprises a black hole is a large mass combined with strong X-ray emission. Indirect evidence for black holes may be obtained for WR stars with a massive unseen companion (SB1), even if the object is not a strong X-ray source, but has a ring nebula or lies high above the Galactic plane (Tutokov & Cherepashchuk 1985). Stevens & Willis (1988) calculated the expected X-ray luminosity of the proposed WR+c system WR6 (HD 50896, WN4b) and found that, while the dense WR wind will absorb a large fraction of the X-rays originating from accretion on to the neutron star, the observed X-ray luminosity should be considerably higher than is actually observed, suggesting that an accreting neutron star is not present in WR6. However, Vanbeveren et al. (1998, p. 238–245) argued that lack of X-rays can be explained if the compact companion is a rapidly spinning neutron star.

Mosqueda & Koenigsberger (1990) derived constraints that restrict the number of close massive binary systems in which black-hole formation is allowed. They showed that in co-rotating massive close binaries, provided that no more than 40 % of the specific angular momentum of the exploding star is lost, black holes can form only from binary systems with pre-explosion orbital periods larger than 15 d. De Donder et al. (1997) predicted that (i) about 75 % of the observed WR+OB binaries will disrupt due to the SN explosion of the WR component. The majority of the OB components of these disrupted binaries will evolve into WR stars, i.e., single WRs with a binary history of possible accretion; and (ii) less than 3 % of the observed WR+OB binaries will form WR+c binaries, with periods in the range 0.05–3.6 d.

MacFadyen & Woosley (1999), discussing the association of GRB 980425 with the SN 1998bw, argued that a black hole formed by incomplete explosion of a rapidly rotating massive star — a collapsar — produces a Type Ib/c supernova like SN 1998bw, but that 99% of SNe Ib/c are made in 3–4 $M_\odot$ helium cores (WR stars) that leave neutron stars and make no GRBs. Langer & Heger (1999) concluded that the majority of SNe Ib/c occur in post-mass-transfer close binaries. Yet the observed Galactic binary frequency is < 50%.

Thus, although SN statistics show that there must be many more WR+OB binaries than single WR stars, we have not yet detected all those binaries.

The VIIth Catalogue of Galactic Wolf-Rayet stars (van der Hucht 2000, replacing the VIth Catalogue of van der Hucht et al. 1981), lists 227 WR stars. Among those are at least 52 WR binaries, 25 more than in the WR binary census of Smith & Maeder (1989). New WR binary discoveries have been made by optical detection of absorption lines (e.g., Williams & van der Hucht 2000), by long-term IR photometric monitoring (Williams 1999), and by high-resolution IR imaging (Tuthill et al. 1999; Monnier et al. 1999). Among the WR binaries all WR subtypes are represented, apart from WN9-11 types. Among the 52 WR binaries are 19 spectroscopic binaries with RV solutions and derived masses. The 13 WN stars have masses in the range 2.3–55 $M_\odot$ with $\overline{M_{\mathrm{WN}}} = 22 \pm 17\,M_\odot$. Three WN5–7 stars (WR 22, WR 47 and WR 141) have masses $> 40\,M_\odot$. The 6 WC stars have masses in the range 9–16 $M_\odot$ with $\overline{M_{\mathrm{WC}}} = 12 \pm 3\,M_\odot$. In agreement with the evolutionary scenarios referred to above, $\overline{M_{\mathrm{WC}}} < \overline{M_{\mathrm{WN}}}$. For these WR binaries the mass ratio $q = M_{\mathrm{WR}}/M_{\mathrm{O}}$ ranges as $q_{\mathrm{WN}} = 0.3$–4.8, $q_{\mathrm{WC}} = 0.3$–0.6 and $q_{\mathrm{WO}} = 0.15$. Four (WN) stars have $q > 1$. The orbital periods among the 52 WR binaries range from 0.2 d (Cyg X-3) to 4820 d (WR 137) and larger (lower limits of 5500 d and 6200 d for, respectively, WR 125 and WR 48a). Especially the census of very-short-period WR binaries ($P < 1$ d) and long-period WR binaries ($P > 100$ d) is suffering from observational bias.

4   Four Wolf-Rayet + Black-Hole Candidates

WR 145a (Cygnus X-3, WN4-7+c, P = 0.1997 d)
At a distance of 10 kpc, Cyg X-3 is one of the brightest X-ray sources in the Galaxy: $\log(L_X/L_\odot) \simeq 3$–5 (van der Klis 1993). The companion of Cyg X-3 was predicted (van den Heuvel & de Loore 1973) and shown (van Kerkwijk et al. 1992; Fender et al. 1999) to be a WR star. Schmutz et al. (1996) found $f(M) = 2.3\,M_\odot$. Assuming $M_{\mathrm{WR}} \simeq 13\,M_\odot$ they derived $M_c \simeq 17\,M_\odot$, suggesting the companion to be a black hole. However, Mitra (1996, 1998) objected against the WR classification for the visible component of Cyg X-3, stating that a WR wind should be opaque to low-energy X-rays (see also Vanbeveren et al. 1998, p. 245–251; Vanbeveren 1999). Verbunt & van den Heuvel (1995) argued that Cyg X-3 is an ideal progenitor for a close double neutron star, and that after spiral-in the binary will consist of a helium star and a black hole. Upon exploding, the WR star may leave a neutron star or a black hole. In the first case, an eccentric close binary will result, consisting of a young pulsar and a black hole. Ergma & Yungelson (1998) argued that for a model with two massive helium stars as immediate progenitors of the Cyg X-3 system, the requirement of having two WR stars in the post-common-envelope orbit, each with strong mass loss, would prevent formation of BH+WR sys-

tems with orbital periods less than several days.

WR 148 (HD 197406, WN8h+B3IV/c, $P = 4.32$ d)
Drissen et al. (1986) derived from $f(m) = 0.28 \, M_\odot$, $i = 66°.7$ and an assumed $M_{WN} = 60 \, M_\odot$, that the mass of the unseen companion is $12.4 \, M_\odot$. The system is a runaway star with a net peculiar velocity of $52 \, \mathrm{km \, s}^{-1}$ (Moffat & Seggewiss 1980) and located 870 pc above the Galactic plane (van der Hucht et al. 1988). If the unseen companion of WR 148 is a black hole, then the recoil from a preceding supernova explosion may be the cause of its present runaway status. We would expect WR+BH systems to be strong X-ray sources. However, $L_X/L_\odot = 0.16$ (Pollock et al. 1995), i.e., only three times the average WR X-ray luminosity. Yet, Marchenko et al. (1996) argued in favor of a black hole scenario, based on timing arguments of the orbiting ionized cavity (; see also Vanbeveren 1998, pp. 245, 286; Vanbeveren 1999).

R 140a2 (WN6+O) and Mk 34 (WN5h) in 30 Doradus (LMC)
Wang (1995) found with *ROSAT*-HRI two X-ray sources in 30 Doradus in the LMC, each with $L_X \simeq 260 \, L_\odot$. One is associated with R 140a2 (WN6+O, $P = 2.76$ d, $f(M) = 0.10$), with a mass of the unseen companion in the range 2.4–$15 \, M_\odot$. The other is associated with Mk 34 (WN5h) and of similar nature. Thus, these two X-ray sources may represent WR+BH binaries.

References

1. Crowther, P.A. (1999): in: K.A. van der Hucht, G. Koenigsberger & P.R.J. Eenens (eds.), *Wolf-Rayet Phenomena in Massive Stars and Starburst Galaxies*, Proc. IAU Symp. No. 193 (San Francisco: ASP), p. 116
2. De Donder, E., Vanbeveren, D., Van Bever, J. (1997): A&A **318**, 812
3. Drissen, L., Lamontagne, R., Moffat, A.F.J., Bastien, P., Seguin, M. (1986): ApJ **304**, 188
4. El Eid, M.F., Langer, N. (1986): A&A **167**, 275
5. Ensman, L.M., Woosley, S.E. (1988): ApJ **333**, 754
6. Ergma, E., Yungelson, L.R. (1998): A&A **333**, 151
7. Fender, R.P., Hanson, M.M., Pooley, G.G. (1999): MNRAS **308**, 473.
8. Fesen, R.A., Becker, R.H., Blair, W.P. (1987): ApJ **313**, 388
9. van den Heuvel, E.P.J., Heise, J. (1972): Nature **239**, 67
10. van den Heuvel, E.P.J., de Loore, B. (1973): A&A **25**, 387
11. van den Heuvel, E.P.J., Habets, G.M.H.J. (1984): Nature **309**, 598
12. van den Heuvel, E.P.J. (1994): in: D. Vanbeveren, W. van Rensbergen & C. de Loore (eds.), *Evolution of Massive Stars: A Confrontation between Theory and Observation*, Proc. Brussels Workshop, SSR **66**, 309
13. Hillier, D.J. (1996): in: J.-M. Vreux, A. Detal, D. Fraipont-Caro, E. Gosset & G. Rauw (eds.), *Wolf-Rayet Stars in the Framework of Stellar Evolution*, Proc. 33rd Liège Int. Astroph. Coll. (Liège: Univ. of Liège), p. 509
14. van der Hucht, K.A., Conti, P.S., Lundström, I., Stenholm, B. (1981): SSR **28**, 227

15. van der Hucht, K.A., Hidayat, B., Admiranto, A.G., Supelli, K.R., Doom, C. (1988): A&A **199**, 217
16. van der Hucht, K.A. (2000): New Astronomy Reviews (submitted)
17. van Kerkwijk, M.H., Charles, P.A., Geballe, T.R., King, D.L., Miley, G.K., Molnar, L.A., van den Heuvel, E.P.J., van der Klis, M., van Paradijs, J. (1992): Nature **355**, 703
18. van der Klis, M. (1993): SSR **62**, 173
19. Langer, N., El Eid, M.F. (1986): A&A **167**, 265
20. Langer, N. (1989): A&A **220**, 135
21. Langer, N. (1991): in: S. Woosley (ed.), *Supernovae*, Proc. 10th Santa Cruz Workshop (New York: Springer), p. 549
22. Langer, N., Heger, A. (1999): in: K.A. van der Hucht, G. Koenigsberger & P.R.J. Eenens (eds.), *Wolf-Rayet Phenomena in Massive Stars and Starburst Galaxies*, Proc. IAU Symp. No. 193 (San Francisco: ASP), p. 187
23. MacFadyen, A.I., Woosley, S.E. (1999): ApJ **524**, 262
24. Maeder, A. (1981): A&A **99**, 97
25. Maeder, A., Lequeux, J. (1982): A&A **114**, 409
26. Maeder, A. (1983): A&A **120**, 113
27. Maeder, A., Meynet, G. (1987): A&A **182**, 243
28. Maeder, A., Meynet, G. (1989): A&A **210**, 155
29. Maeder, A. (1991): A&A **242**, 93
30. Mitra, A. (1996): MNRAS **280**, 953
31. Mitra, A. (1998): ApJ **499**, 385
32. Moffat, A.F.J., Seggewiss, W. (1980): A&A **86**, 87
33. Moffat, A.F.J. (1995): in: K.A. van der Hucht & P.M. Williams (eds.), *Wolf-Rayet Stars: Binaries, Colliding Winds, Evolution*, Proc. IAU Symp. No. 163 (Dordrecht: Kluwer), p. 213
34. Monnier, J.D., Tuthill, P.G., Danchi, W.C. (1999), ApJ (Letters) **525**, L97
35. Mosqueda, A., Koenigsberger, G. (1990): RMAA **20**, 79
36. Pollock, A.M.T., Haberl, F., Corcoran, M.F. (1995): in: K.A. van der Hucht & P.M. Williams (eds.), *Wolf-Rayet Stars: Binaries, Colliding Winds, Evolution*, Proc. IAU Symp. No. 163 (Dordrecht: Kluwer), p. 512
37. Schild, H., Maeder, A. (1985): A&A (Letters) **143**, L7
38. Schmutz, W., Geballe, T.R., Schild, H. (1996): A&A (Letters) **311**, L25
39. Smith, L.F., Maeder, A. (1989): A&A **211**, 71
40. Stevens, I.R., Willis, A.J. (1988): MNRAS **234**, 783
41. Tuthill, P.G., Monnier, J.D., Danchi, W.C. (1999): Nature **398**, 487
42. Tutokov, A.V., Cherepashchuk, A.M. (1985): Astron. Zh. **62**, 1124 (= Sov. Astron. **29**, 654)
43. Vanbeveren, D., van Rensbergen, W., de Loore, C. (1998): *The Brightest Binaries*, Astrophysics & Space Science Library Vol. 232 (Dordrecht: Kluwer)
44. Vanbeveren, D. (1999): in: K.A. van der Hucht, G. Koenigsberger & P.R.J. Eenens (eds.), *Wolf-Rayet Phenomena in Massive Stars and Starburst Galaxies*, Proc. IAU Symp. No. 193 (San Francisco: ASP), p. 196
45. Verbunt, F., van den Heuvel, E.P.J. (1995): in: W.H.G. Lewin, J. van Paradijs & E.P.J. van den Heuvel (eds.), *X-Ray Binaries* (Cambridge: CUP), p. 457
46. Wang, Q.D. (1995): ApJ **453**, 783
47. Williams, P.M. (1999): in: K.A. van der Hucht, G. Koenigsberger & P.R.J. Eenens (eds.), *Wolf-Rayet Phenomena in Massive Stars and Starburst Galaxies*, Proc. IAU Symp. No. 193 (San Francisco: ASP), p. 267

48. Williams, P.M., van der Hucht, K.A. (2000): MNRAS (in press)
49. Woosley, S.E. (1986): in: B. Hauck et al. (eds.), *Nucleosynthesis and Chemical Evolution*, 16th Saas Fee Conference, p. 1
50. Woosley, S.E., Langer, N., Weaver, T.A. (1993): ApJ **411**, 823
51. Woosley, S.E., Langer, N., Weaver, T.A. (1995): ApJ **448**, 315

Structure of the Globular Cluster M15 and Constraints on a Massive Central Black Hole

Roeland P. van der Marel

Space Telescope Science Institute, 3700 San Martin Drive,
Baltimore, MD 21218, USA

Abstract. Globular clusters could harbor massive central black holes (BHs), just as galaxies do. So far, no unambiguous detection of a massive BH has been reported for any globular cluster. However, the dense core-collapsed cluster M15 seems to be a good candidate. I review the available photometric and kinematic data for this cluster. Both are consistent with a BH of mass $M_\bullet \approx 2000\,M_\odot$, although such a BH is not unambiguously required by the data. I discuss some ongoing studies with Keck and HST which should shed more light on this issue in the coming years.

1 Massive Central Black Holes in Globular Clusters

Massive BHs have been convincingly detected in the centers of some nearby galaxies (e.g., Miyoshi et al. 1995), including our own (Genzel et al. 1997). In certain galaxies the BH directly reveals itself through its associated accretion and activity. Such activity is never observed in globular clusters, but nonetheless, it may well be that (some) globular clusters also contain BHs. There are many ways in which globular cluster evolution at high densities (Meylan & Heggie 1997) can lead to the formation of a massive BH in the center (Rees 1984). For example, core collapse induced by two-body relaxation may lead to sufficiently high densities for individual stars or stellar-mass black holes to interact or collide, with a single massive BH as the likely end product (Quinlan & Shapiro 1987; Lee 1993, 1995).

The black hole mass $M_\bullet$ in galaxies correlates with galaxy (bulge) mass M such that $M_\bullet/M \approx 10^{-3\pm1}$ (Kormendy & Richstone 1995; Magorrian et al. 1998; van der Marel 1999). One may use this correlation to obtain a crude estimate for the possible BH masses in globular clusters (although it should be kept in mind that the BHs in galaxies are often hypothesized to have formed through gas collapse and accretion, instead of through collapse of a stellar cluster; e.g., Haehnelt et al. 1997). This yields $M_\bullet \approx 10^3 M_\odot$.

The presence or growth of a BH in a globular cluster affects both the stellar density profile and the stellar dynamics. Observational constraints on BHs in globular clusters can therefore be found from photometric and kinematic studies. So far, no detection of a massive BH has been obtained for any globular cluster, but only few, if any, studies have had sufficient sensitivity to unambiguously detect BHs with masses as low as $M_\bullet \approx 10^3 M_\odot$. On the other hand, advances in observational capabilities and techniques are now making it possible to study BH masses down to this limit, so this is becoming a more active area of interest.

The globular cluster M15 (NGC 7078) at a distance of 10 kpc is one of the densest globular clusters in our Galaxy. The presence of a bright X-ray source (Hut et al. 1992) and several millisecond pulsars (Phinney 1993) are manifestations of its extreme density, which makes M15 one of the best a priori candidates to search for evidence of a central BH. For this reason, M15 has been intensively observed in the past decade using a variety of techniques, and it is also the subject of several ongoing studies.

2   M15 Photometry

Core-collapsed clusters have stellar surface density profiles that rise all the way into the center. Such clusters make up $\sim 20\%$ of all globular clusters in our Galaxy, and stand in marked contrast to King-model clusters, which show flat central cores and are modeled as tidally-truncated isothermal systems. At ground-based resolution M15 has long been known as the proto-typical core-collapsed cluster (Djorgovski & King 1986; Lugger et al. 1987). Studies with the Hubble Space Telescope (HST) in the past decade have provided significantly higher spatial resolution than ground-based data, but have also not provided any evidence for a homogeneous core in M15 (despite early claims to the contrary; Lauer et al. 1991).

Guhathakurta et al. (1996) used the HST/WFPC2 to image M15. Individual stars were resolved well below the main sequence turnoff even in the dense central few arcsec. The projected surface number density profile between 0.3″ (0.017 pc) and 6″ (after correction for the effects of incompleteness and photometric bias/scatter) was found to be well represented by a power law $N(R) \propto R^{-0.82\pm0.12}$. While the density profile cannot be measured reliably at radii $\lesssim 0.3''$, mainly because of small-number statistics and uncertainties in the position of the cluster center, there is no evidence from non-parametric studies that the density profile would flatten at smaller radii.

Sosin & King (1997) obtained even higher spatial resolution data with the HST/FOC. They find $N(R) \propto R^{-0.70\pm0.05}$ for turnoff stars, consistent with the Guhathakurta et al. analysis, and show in addition that the power-law slopes are slightly different for stars of different masses. This is qualitatively consistent with the mass segregation predicted in a cluster in which two-body relaxation has been important.

Bahcall & Wolf (1976, 1977) constructed detailed models for the equilibrium stellar density distribution of a globular cluster in which a BH has been present for much longer than the two-body relaxation time. For a cluster of equal-mass stars one expects $N(R) \propto R^{-3/4}$, in surprisingly good agreement with the star count profile for M15. However, the observed profile can be explained equally well as a result of core-collapse (Grabhorn et al. 1992; but note that the predictions from Fokker-Planck models are quite uncertain).

Hence, the star count profile by itself yields only limited insight. An additional problem is that photometric studies cannot determine whether light follows mass, and what the abundance and distribution of dark remnants are. Kinematical studies are therefore essential to gain further insight.

3   M15 Kinematics

M15 has been the subject of many ground-based kinematical studies. Observational strategies have focused primarily on the determination of velocities of individual stars, using either spectroscopy with single apertures, long-slits or fibers (Peterson, Seitzer & Cudworth 1989; Dubath & Meylan 1994; Dull et al. 1997; Drukier et al. 1998), or using imaging Fabry-Perot spectrophotometry (Gebhardt et al. 1994, 1997, 2000). Integrated light measurements using single apertures have also been attempted (Peterson et al. 1989; Dubath, Meylan & Mayor 1994), but are only of limited use; integrated light spectra are dominated by the light from only a few bright giants, and as a result, inferred velocity dispersions are dominated by shot noise (Zaggia et al. 1992; Dubath et al. 1994).

Line-of-sight velocities are now known for ~ 1800 M15 stars, as conveniently compiled by Gebhardt et al. (2000). The projected velocity dispersion profile inferred from this sample is shown in Fig. 1. It increases monotonically inwards from $\sigma = 3 \pm 1$ km s^{-1} at $R = 7$ arcmin, to $\sigma = 11 \pm 1$ km s^{-1} at $R = 24''$. The velocity dispersion is approximately constant at smaller radii, and is $\sigma = 11.7 \pm 2.8$ km s^{-1} at the innermost available radius $R \approx 1''$ (Gebhardt et al. 2000). The figure also shows the predictions of spherical dynamical models for M15 in which the velocity distribution is isotropic and the stellar population has a mass-to-light ratio $M/L = 1.7$ (in the V-band) that is independent of radius. Different curves correspond to different BH masses. A BH causes the velocity dispersion to rise in Keplerian fashion as $\sigma \propto R^{-1/2}$ towards the center of the cluster. The best-fitting model of this type has $M_\bullet \approx 2000\, M_\odot$. However, it should be noted that the data can be fit equally well with a model in which the M/L of the stellar population increases inwards to a value of $M/L \approx 3$ in the center. This would not be implausible, since mass segregation would tend to concentrate heavy dark remnants to the center of the cluster.

Phinney (1993) used an interesting alternative argument to constrain the mass distribution of M15. There are two millisecond pulsars in M15 at a distance $R = 1.1''$ from the cluster center that have a negative period derivative $\dot{P}$. This must be due to acceleration by the mean gravitational field of the cluster, since the pulsars are expected to be spinning down intrinsically (positive $\dot{P}$). The observed $\dot{P}$ values place a strict lower limit on the mass enclosed within a projected radius $R = 1.1''$. Combined with the observed light profile this yields $M/L > 2.1$ for the total mass-to-light ratio within $R \leq 1.1''$, with a statistically most likely value of $M/L \approx 3.0$. These results

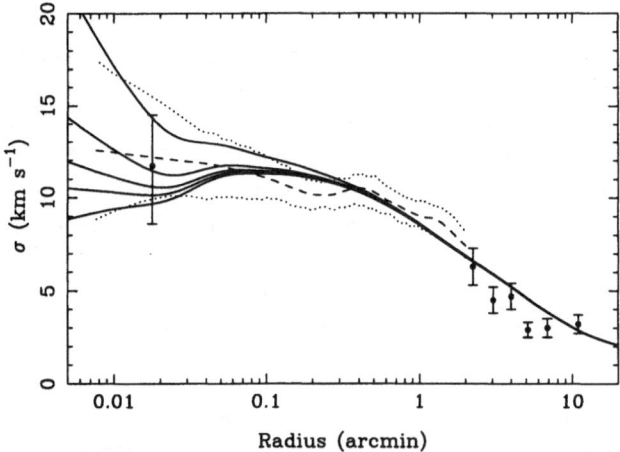

Fig. 1. Projected velocity dispersion profile of M15. Binned data are shown as dots with error bars, while a non-parametric estimate of the dispersion profile in the central arcmin is shown as a dashed line bounded by two dotted lines (the latter representing the 68.3% confidence region). Solid curves are predictions of isotropic models with BHs that have $M_\bullet = 0, 500, 1000, 2000$, and $6000\,M_\odot$, respectively. (Figure reproduced from Gebhardt et al. 2000).

are consistent with the stellar kinematical analysis, and also indicate that there must be some increase in M/L towards the cluster center (since at large radii $M/L \approx 1.7$). However, also the pulsar data cannot discriminate whether this is due to mass segregation or a central BH.

It has been known for some time that M15 has a net global rotation amplitude of $V \approx 2$ km s^{-1}. This is somewhat surprising given the short relaxation time, which tends to isotropize the velocity distribution. However, more recent work (Peterson 1993; Gebhardt et al. 2000) has revealed two even more puzzling facts. First, the position angle of the projected rotation axis in the central region is $\sim 100°$ different from that at larger radii. Second, the rotation amplitude increases to $V = 10.4 \pm 2.7$ km s^{-1} for $R \lesssim 3.4''$, so that $V/\sigma \approx 1$ in this region. Although the central increase in rotation amplitude may have something to do with the presence of a BH, neither of these observations fits naturally in any current theory of globular cluster structure.

4   Ongoing Studies and Future Prospects

To make further progress in our understanding of the structure of M15, and in particular to determine whether it harbors a central BH or not, it is of crucial importance to obtain more stellar velocities close to the center. However, velocity determinations at $R \lesssim 2''$ are very difficult due to severe crowding and the presence of a few bright giants in the central arcsec. Gebhardt et al. (2000) used an Imaging Fabry-Perot spectrophotometer with adaptive optics on the CFHT, and obtained FWHM values as small as $0.09''$. However, the Strehl ratio was only $\lesssim 6\%$, so that even in these observations the light from the fainter turnoff and main-sequence stars in the central arcsec was overwhelmed by the PSF wings of the nearby giants. As a result, there are only 5 stars in the central arcsec with known velocities.

In an attempt to improve this situation my collaborators and I initiated two new observational studies. In the first (Guhathakurta et al., in progress) we used the HIRES echelle spectrograph on the Keck I telescope in multislit mode. The seeing FWHM was $0.7''$ and the slit width $0.5''$. These observations will not allow us to spatially resolve stars in the central arcsec spatially, but the high spectral resolution (as compared to Fabry-Perot spectrophotometry) may allow us to resolve stars spectrally. In the second, more ambitious study we are using the STIS long-slit spectrograph on HST to map the center of M15 spectroscopically (Cycle 8, van der Marel et al., in progress). We are stepping a $0.1''$-wide slit across the center in 23 steps of $0.1''$. This will yield a spectrum of each $0.1'' \times 0.1''$ cell in a rectangular grid on the center of M15. The spectra are taken around the Mg b triplet. The expected signal-to-noise ratio should be sufficient to extract stellar velocities using cross-correlation techniques for several tens of faint stars in the central few arcsec. This will improve our knowledge of the velocity dispersion profile and will put new constraints on the possible presence of a BH. Also, the central escape velocity of M15 in the absence of a BH is expected to be ~ 40 km s^{-1} (e.g., Webbink 1985); any (non-binary) stars found to have velocities exceeding this value will provide additional and independent evidence for a central mass concentration.

More progress in the near future may come from the availability of stellar proper motion measurements with HST. Several groups are pursuing this, both for M15 and for other clusters. The positional accuracy that can be achieved with the HST/WFPC2 is ~ 0.01 PC pixel (0.5 mas). For a 5 year baseline, motion over this distance corresponds to 5 km s^{-1} (at the distance of M15). This opens up the exciting prospect of having all three velocity coordinates for a large sample of stars. Whether proper motions can be determined all the way into the central arcsec remains to be seen though, since the severe crowding will complicate positional measurements in that region.

5   Conclusions and Acknowledgements

Globular clusters could have central BHs, and M15 is the best candidate so far. The available photometry and kinematics are consistent with a BH of mass $M_\bullet \approx 2000\, M_\odot$, although such a BH is not unambiguously required by the data. Ongoing studies should shed more light on the structure of M15 in the coming years.

I am grateful to my collaborators, Raja Guhathakurta, Ruth Peterson, Pierre Dubath, Karl Gebhardt and Tad Pryor for many stimulating discussions on this subject.

References

1. Bahcall, J. N., & Wolf R. A. 1976, ApJ, 209, 214
2. Bahcall, J. N., & Wolf R. A. 1977, ApJ, 216, 883
3. Djorgovski, S., & King, I. 1986, ApJ, 305, 61
4. Drukier, G. A., Slavin, S. D., Cohn, H. N., Lugger, P. M., Berrington, R. C., Murphy, B. W., Seitzer, P. O. 1998, AJ, 115, 708
5. Dubath, P., & Meylan, G. 1994, A&A, 290, 104
6. Dubath, P., Meylan, G., & Mayor, M. 1994, ApJ, 426, 192
7. Dull, J. D., Cohn, H. N., Lugger, P. M., Murphy, B. W., Seitzer, P. O., Callanan, P. J., Rutten, R., Charles, P. 1997, ApJ, 481, 267
8. Gebhardt, K., Pryor, C., Williams, T. B., Hesser, J. E., 1994, AJ, 107, 2067
9. Gebhardt, K., Pryor, C., Williams, T. B., Hesser, J. E., & Stetson, P. B. 1997, AJ, 113, 1026
10. Gebhardt, K., Pryor, C., O'Connell, R., Williams, T. B., & Hesser, J. E. 2000, AJ, in press [astro-ph/9912172]
11. Genzel, R., Eckart, A., Ott, T., & Eisenhauer, F. 1997, MNRAS, 291, 219
12. Grabhorn, R. P., Cohn, H. N., Lugger, P. M., & Murphy, B. W. 1992, ApJ, 392, 86
13. Guhathakurta, P., Yanny, B., Bahcall, J. N., & Schneider, D. P. 1996, AJ, 111, 267
14. Haehnelt, M. G., Natarajan, P., & Rees, M. J. 1998, MNRAS, 300, 817
15. Hut, P. et al., 1992, PASP, 105, 981
16. Kormendy, J., & Richstone, D. 1995, ARA&A, 33, 581
17. Lauer, T. R. 1991, ApJ, 369, L45
18. Lee, M. H. 1993, ApJ, 418, 147
19. Lee, H. M. 1995, MNRAS, 272, 605
20. Lugger, P. M., Cohn, H., Grindlay, J. E., Bailyn, C. D., & Hertz, P. 1987, ApJ, 320, 482
21. Magorrian, J., et al. 1998, AJ, 115, 2285
22. Meylan, G., & Heggie, D. C. 1997, A&AR, 8, 1
23. Miyoshi M., et al. 1995, Nature, 373, 127
24. Peterson, R. C. 1993, in 'Structure and Dynamics of Globular Clusters', eds., Djorgovski & Meylan, ASP Conference Series, Vol. 50, p. 65
25. Peterson, R. C., Seitzer, P., & Cudworth, K. M. 1989, ApJ, 347, 251
26. Phinney, E. S. 1993, in 'Structure and Dynamics of Globular Clusters', eds., Djorgovski & Meylan, ASP Conference Series, Vol. 50, p. 141
27. Quinlan, G. D., & Shapiro, S. L. 1987, ApJ, 321, 199
28. Rees, M. J. 1984, ARA&A, 22, 471
29. Sosin, C., & King, I. R. 1997, AJ, 113, 1328
30. van der Marel, R. P. 1999, AJ, 117, 744
31. Webbink, R. F. 1985, in 'Dynamics of Star Clusters', Proc. IAU Symp. 113, ed. Goodman & Hut, p. 541
32. Zaggia, S. R., Capaccioli, M., Piotto, G., & Stiavelli M. 1992, A&A, 258, 302

Merger Rates of Black-Hole Binaries: Prospects for Gravitational-Wave Detectors

Simon Portegies Zwart[1] and Stephen L.W. McMillan[2]

[1] Hubble Fellow
   Department of Astronomy, Boston University,
   725 Commonwealth Ave., Boston, MA 02215, USA
[2] Department of Physics, Drexel University,
   Philadelphia, PA 19104, USA

Abstract. Mergers of black-hole binaries are expected to release large amounts of energy in the form of gravitational radiation. However, binary evolution models predict merger rates too low to be of observational interest. In this paper we explore the possibility that black holes become members of close binaries via dynamical interactions with other stars in dense stellar systems. In star clusters, black holes become the most massive objects within a few tens of millions of years; dynamical relaxation then causes them to sink to the cluster core, where they form binaries. These black-hole binaries become more tightly bound by superelastic encounters with other cluster members, and are ultimately ejected from the cluster. The majority of escaping black-hole binaries have orbital periods short enough and eccentricities high enough that the emission of gravitational waves causes them to coalesce within a few billion years. We predict a black-hole merger rate of 10^{-8} to 10^{-7} per year per cubic megaparsec, implying gravitational-wave detection rates substantially greater than the corresponding rates from neutron star mergers. For the first generation Laser Interferometer Gravitational-Wave Observatory (LIGO-I), we expect about one detection during the first two years of operation. For its successor LIGO-II, the rate rises to roughly one detection per day. There is about an order of magnitude uncertainty in these numbers.

1 Introduction

Globular clusters contain about one hundred times more low-mass X-ray binaries (LMXBs) per unit mass than does the Galaxy as a whole—the Galaxy, with a mass of $2 \times 10^{11}\,M_\odot$, contains about 100 LMXBs, whereas the Galactic globular cluster population, with a total mass of just $2 \times 10^8\,M_\odot$, contains at least 10. All known cluster LMXBs have neutron stars as primaries.

One might seek an explanation for this discrepancy in LMXB numbers in the obvious population differences between globular clusters and the Galactic disc. The disc contains a mixture of stellar populations, with broad ranges in age and metallicity, while all stars in a given globular cluster have essentially the same age and initial composition. Conceivably, a globular cluster might experience a characteristic "LMXB-rich" epoch as its component stars evolved. This hypothesis, however, is not widely accepted.

A more likely explanation for the excess of LMXBs in globular clusters lies in the radically different dynamics of cluster stars compared to stars in the Galactic disc. The mean stellar density in the disc is about 0.1 star per cubic parsec, with relatively little variation from place to place. Globular clusters, on the other hand, exhibit a huge spread in densities, ranging from values close to the density in the disc near the cluster tidal radius, to tens of millions of stars per cubic parsec in the densest cluster cores. These density differences may be responsible for the higher birthrate of LMXBs in globular clusters relative to the Galactic disc: dynamical interactions favor the formation of LMXBs.

For a cluster age of ∼10 Gyr, neutron stars are more than twice as massive as other cluster members. Dynamical friction causes them to sink to the center of the cluster potential well, where stellar densities are higher and encounters are much more common. Once in the core, close encounters with other stars may lead to two-body tidal capture (Fabian et al. 1975) or to three-body exchange interactions (Phinney & Sigurdsson 1991). In either case, the neutron star gains a low-mass companion, which later evolves to become the donor in an LMXB. High kick velocities imparted to newborn neutron stars cause the majority to be ejected from their parent clusters upon formation (Davies & Hansen 1998). Only about 20% of neutron stars are retained by globular clusters, yet cluster LMXBs still greatly outnumber the population in the Galactic disc. Mass segregation and tidal capture or exchange are evidently very efficient processes.

Given this reasoning, it is all the more striking that no black-hole X-ray binaries are observed in globular clusters. Black holes do not receive a kick upon formation in a supernova (White & van Paradijs 1996), so hardly any escape promptly. Black holes are also considerably more massive than neutron stars, causing them to sink in the cluster core even more rapidly. In equipartition, the black holes' velocity dispersion is $v \propto m^{-1/2}$. Thus, the cross section for a dynamical interaction, which is dominated by gravitational focusing, is

$$\sigma \propto \frac{m}{\sqrt{v}} \propto m^{5/4}. \tag{1}$$

Hence, for a black hole mass of $10\,M_\odot$, we would naively expect that globular clusters should contain almost an order of magnitude more LMXBs with black holes than with neutron stars. However, none are found. The explanation for this discrepancy is as follows.

2   Black-Hole Formation

The initial mass function of globular clusters is well described by a Scalo (1986) distribution, with lower and upper limits of $0.1\,M_\odot$ and $100\,M_\odot$. This IMF has a mean mass $\langle m \rangle \sim 0.5\,M_\odot$ and leads to the formation of about 5×10^{-4} black holes per star. A $10^6\,M_\odot$ star cluster thus produces about

1000 black holes. Black holes resulting from stellar evolution are generally quite massive objects: known black-hole masses range from 6 to 10 $M_\odot$. For clarity we adopt a black-hole mass of $m_{bh} = 10\,M_\odot$; the precise value is not crucial to our discussion, so long as it significantly exceeds $\langle m \rangle$.

As with neutron stars, dynamical friction causes the black holes to sink to the cluster core. The mass segregation time scale is $\sim \langle m \rangle / m_{bh}$ half-mass relaxation times, or about 10^8 yr for $\langle m \rangle = 0.5\,M_\odot$ and a cluster relaxation time of 10^9 years (see Kulkarni et al. 1992 and Sigurdsson & Hernquist 1992 for details). As mass segregation proceeds and the cluster core contracts, binaries are formed, providing the energy needed to support the core against further gravothermal collapse (Heggie 1975). The black holes will preferentially form binaries with one another, both because it is energetically favorable for them to do so, and because of the generally larger black-hole interaction cross sections. Subsequently, the black-hole binaries evolve via dynamical encounters with other cluster components. On average, each encounter between a black-hole binary and a single black hole hardens the binary (increases its binding energy) by about 20%. Two-thirds of the energy released goes into binary recoil, the rest into recoil of the other black hole involved in the interaction. The hardening process continues until the recoil velocity exceeds the clusters' escape speed and the black-hole binary is ejected from the cluster.

A binary can release enough energy to eject itself from the cluster once its binding energy exceeds ~ 1000 times the mean kinetic energy of cluster stars. By this time the binary has typically experienced some 40–50 hard encounters. The recoil energetics imply that, on average, a black-hole binary is ejected after its *previous* encounter has already ejected a single black hole. Thus, for each ejected black-hole binary one expects two single black holes to be ejected. There are two possible dynamical scenarios for binary ejection: (1) there will be at most one or two black-hole binaries in the core at any given time, and a new binary can form only after these are ejected; or (2) the core is able to support a large population of black-hole binaries. In the former case, the ejection process takes considerably longer, as the binaries are ejected sequentially. In the latter, the binaries may be ejected more or less simultaneously. Our simulations are not sufficiently detailed to discriminate between these alternatives.

In order to eject a black-hole binary following an encounter with a low-mass cluster member, the binding energy of the black-hole binary must exceed $\sim 4 \times 10^4$ kT. However, by this time, the black-hole binary has shrunk to such a small orbital separation that it likely merges due to emission of gravitational wave radiation before another encounter takes place. On the other hand, the black-hole binary easily ejects low mass stars. The black-hole binary starts to eject low-mass stars as soon as its binding energy exceeds ~ 25 kT. At least 20 low-mass stars are ejected for each single black hole.

3   Characteristics of Ejected Binaries

The energy of an ejected binary and its orbital separation are coupled to the dynamical characteristics of the star cluster. For a cluster in virial equilibrium, we have

$$kT = \frac{2E_{\text{kin}}}{3N} = \frac{-E_{\text{pot}}}{3N} = \frac{GM^2}{6Nr_{\text{vir}}}, \tag{2}$$

where M and N are the total cluster mass and number of stars, respectively, and r_{vir} is the virial radius. A black-hole binary with semi-major axis a has

$$E_b = \frac{Gm_{\text{bh}}^2}{2a}, \tag{3}$$

and therefore

$$\frac{E_b}{kT} = 3N \left(\frac{m_{\text{bh}}}{M}\right)^2 \frac{r_{\text{vir}}}{a}. \tag{4}$$

We can thus compute the properties of black-hole binaries produced by globular clusters of given masses and virial radii. These cluster parameters are assumed to be distributed as independent gaussians with means and dispersions of $\log M = 5.5 \pm 0.5$ and $\log r_{\text{vir}} = 0.5 \pm 0.3$, respectively (Djorgovski & Meylan 1994). A recent parameter-space survey of cluster initial conditions (Takahashi & Portegies Zwart 2000) finds that typical globular clusters which have survived for a Hubble time have lost $\gtrsim 60\%$ of their initial mass and have expanded by about a factor of three. We correct for this by changing the adopted distributions to $\log M = 6.0 \pm 0.5$ and $\log r_{\text{vir}} = 0 \pm 0.3$.

4   Production of Gravitational Radiation

An approximate formula for the merger time of two stars due to the emission of gravitational waves is given by Peters & Mathews (1963):

$$t_{\text{mrg}} \approx 150\,\text{Myr} \left(\frac{M_{\odot}}{m_{\text{bh}}}\right)^3 \left(\frac{a}{R_{\odot}}\right)^4 (1 - e^2)^{7/2}. \tag{5}$$

Here e is the orbital eccentricity of the black-hole binary. About 90% of the black-hole binaries formed in the cores of star clusters merge within a Hubble time due to gravitational radiation. This fraction is based on the assumption that the binary binding energies are distributed flat in $\log E_b$ between $1000\,kT$ and $10000\,kT$, that the eccentricities are thermal, independent of E_b (these assumptions are supported by detailed N-body simulations of smaller systems), and that the universe is $15\,\text{Gyr}$ old (Jha et al. 1999). The specific contribution to the total merger rate of black-hole binaries from globular clusters is then about 0.04 per star cluster per million years.

4.1   Merger Rate in the Local Universe

We estimate the number density of globular clusters in the universe to be

$$\phi_{GC} \approx 8.4\, h^3 \, \text{Mpc}^{-3}, \tag{6}$$

where $h = H_0/100 \, \text{km}\,\text{s}^{-1}\,\text{Mpc}^{-1}$. Combining the specific number density of globular clusters with their contribution to the black-hole merger rate results in a total rate density of black-hole mergers in the universe of

$$\mathcal{R}_{GC} \approx 3.2 \times 10^{-7} h^3 \, \text{yr}^{-1} \, \text{Mpc}^{-3}. \tag{7}$$

We note that this figure is larger than the current best estimates of the neutron-star merger rate $\mathcal{R} \sim 2 \times 10^{-7} \, h^3 \, \text{yr}^{-1} \, \text{Mpc}^{-3}$ (Narayan et al. 1991; Phinney 1991; Portegies Zwart & Spreeuw 1996) .

4.2   LIGO Observations

The current best estimate of the maximum distance within which LIGO-I can detect an inspiral event is

$$R_{\text{eff}} \approx 18 \, \text{Mpc} \, \left(\frac{M_{\text{chirp}}}{M_\odot} \right)^{5/6} \tag{8}$$

(K. Thorne, private communication). Here, the "chirp" mass for a binary with component masses m_1 and m_2 is $M_{\text{chirp}} = (m_1 m_2)^{3/5}/(m_1 + m_2)^{1/5}$. For neutron-star spiral-in, $m_1 = m_2 = 1.4 \, M_\odot$, so $M_{\text{chirp}} \approx 1.22 \, M_\odot$, $R_{\text{eff}} \approx$ 21 Mpc. For black-hole binaries with $m_1 = m_2 = m_{\text{bh}} = 10 \, M_\odot$, we find $M_{\text{chirp}} \approx 8.71 \, M_\odot$, $R_{\text{eff}} \approx$ 109 Mpc, and a LIGO-I detection rate of about $1.7 \, h^3$ per year. For $h \sim 0.65$ (Jha 1999), this results in about one detection event every two years. LIGO-II should become operational by 2007, and is expected to have R_{eff} about ten times greater than LIGO-I, resulting in a detection rate 1000 times higher: roughly one event per day.

5   Discussion

Black-hole binaries ejected from galactic nuclei, the most massive globular clusters (masses $\gtrsim 10^6 \, M_\odot$), and globular clusters which experience core collapse soon after formation tend to be very tightly bound, have high eccentricities, and merge within a few million years of ejection. These mergers therefore trace the formation of dense stellar systems with a delay of a few Gyr (the typical time required to form and eject binaries), making these systems unlikely candidates for LIGO detections, as the majority merged long ago. This effect may reduce the current merger rate by an order of magnitude, but more sensitive future gravitational wave detectors may be able to see some of these early universe events. In fact, we estimate that the most massive globular clusters contribute about 90% of the total black-hole merger rate. While their black-hole binaries merge promptly upon ejection, the longer relaxation times of these clusters mean that binaries tend to be ejected much later than in lower-mass systems. Consequently, we have retained these binaries in our merger rate estimate.

We have assumed that the mass of a stellar black hole is $10\,M_\odot$. Increasing this mass to $18\,M_\odot$ decreases the expected merger rate by about 50%; higher-mass black holes tend to have wider orbits. However, the larger chirp mass increases the signal to noise, and the distance to which such a merger can be observed increases by about 60%. The detection rate on Earth therefore increases by about a factor of three. For $6\,M_\odot$ black holes, the detection rate decreases by a similar factor. For black-hole binaries with component masses $\gtrsim 12\,M_\odot$, the first generation of detectors will be more sensitive to the merger itself than to the spiral-in phase that precedes it (Flanagan & Hughes 1998). Since the strongest signal is expected from black-hole binaries with high-mass components, it is critically important to improve our understanding of the merger waveform. Even for lower-mass black holes (with $m_{bh} \gtrsim 10\,M_\odot$), the spiral-in signal comes from an epoch when the holes are so close together that the post-Newtonian expansions used to calculate the wave forms are unreliable. The wave forms of this "intermediate binary black-hole regime" (Brady et al. 1998) are only now beginning to be explored. Finally we stress that the black-hole binaries are highly eccentric, which affects their gravitational wave signals and also influences their detectability.

Acknowledgments We thank Piet Hut, Jun Makino and Kip Thorne for insightful comments on this work. This work was supported by NASA through Hubble Fellowship grant HF-01112.01-98A awarded (to SPZ) by the Space Telescope Science Institute, which is operated by the Association of Universities for Research in Astronomy, Inc., for NASA under contract NAS 5-26555, and by ATP grant NAG5-6964 (to SLWM). SPZ is grateful to Drexel University and Tokyo University for their hospitality and for the use of their GRAPE systems. Part of the calculations are performed on the SGI/Cray Origin2000 supercomputer at Boston University.

References

1. Brady, P., Creighton, J., Thorne, K., 1998, Phys. Rev. D. 57, 1111
2. Davies, M., Hansen, B., 1998, MNRAS, 301, 15
3. Djorgovski, S., Meylan, G., 1994, AJ 108, 1292
4. Fabian, A.C., Pringle, J.E., Rees, M.J., 1975, MNRAS 172, 15
5. Flanagan, É. É., Hughes, S. A. 1998, Phys. Rev. D 57, 4535
6. Heggie, D. C., 1975, MNRAS 173, 729
7. S. Jha, P., Garnavich, R. Kirshner, P. Challis, A. Soderberg, L. Macri, J. Huchra, P. Barmby, E. Barton, P. Berlind, W. Brown, N. Caldwell, M. Calkins, S.

Kannappan, D. Koranyi, M. Pahre, K. Rines, K. Stanek, R. Stefanik, A. Szent-gyorgyi,P. Vaisanen, Z. Wang, J. Zajac, A. Riess, A. Filippenko, W. Li, M. Modjaz, R. Treffers, C. Hergenrother, E. Grebel, P. Seitzer, G. Jacoby, P. Benson, A. Rizvi, L. Marschall, J. Goldader, M. Beasley, W. Vacca, B. Leibundgut, J. Spyromilio, B. Schmidt, P. Wood, to appear in ApJS, (astro-ph/9906220)

8. Kulkarni, S. R., Hut, P., McMillan, S. L. W., 1993, Nature, 364, 421
9. Narayan, R., Piran, T., Shemi, A., 1991, ApJ 379, L17
10. Phinney, E. S., 1991, ApJ 380, L17
11. Phinney, E. S., Sigurdsson, S., 1991, Nat 249, 220
12. Peters, P. C., Mathews, J., 1963, Phys. Rev. D, 131, 345,
13. Portegies Zwart, S. F., Spreeuw, F., 1996, A&A 312, L670
14. Scalo, J. M., 1986, Fund. of Cosm. Phys., 11, 1
15. Sigurdsson, S., Hernquist, L., 1993, Nature, 364, 423
16. Takahashi, K., Portegies Zwart, S. F., 2000, ApJ submitted (astro-ph/9903366)
17. White, N. E., van Paradijs, J. A., 1996, ApJ 473, L25

Central Black Holes and Galaxy Evolution

Karl Gebhardt

UCSC/Lick Observatory, Santa Cruz, Ca 95064, USA

Abstract. The study of supermassive black holes in the centers of galaxies has recently progressed from mere detection to using their correlations with galaxy properties to infer formational and evolutionary histories of both the galaxy and the black hole itself. In this paper, I present two recent conclusions concerning the black-hole surveys and galaxy models: 1) the ratio of black-hole mass to the brightness of the bulge component is now a factor of 5-10 smaller than previous estimates. This decrease is due mainly to the inclusion in our sample of galaxies not formerly thought to contain black holes; 2) The stellar orbital structure of the stars near the galaxy center show strong tangential velocity anisotropy. Theoretical models which best match this trend are black-hole binary/merger models.

1 Black-Hole Studies

Kormendy (1993) and Kormendy & Richstone (1995) first studied possible correlations between the mass of a central black hole and the luminosity of the bulge component of the host galaxy. They found that the black-hole mass was about 0.5% of the host luminosity in solar units. Many studies, subsequently, followed up on this observation by using larger datasets (Magorrian et al. 1998) and different analysis techniques (van der Marel 1999). However, all of these studies suffer from possible biases in their assumptions—velocity isotropy, adiabatic models, and so forth. In order to understand both whether such a relationship exists and how to explain it, we need to obtain more reliable mass estimates and use general models that are free from assumption biases.

The techniques that have been used most often for black-hole studies are stellar dynamics, gas dynamics, masers, proper motions, and reverberation maps. Of these, the stellar dynamical analysis is the most useful since it can be applied to any system; other techniques can only be applied in particular galaxies under special circumstances. The great advantage of using stellar dynamics—other than applicability—is that one obtains the black hole mass and the stellar orbital distribution in the galaxy simultaneously. However, stellar dynamics are normally the most difficult to use since they require high signal-to-noise observations and detailed modeling.

2 Stellar Dynamical Models

I use 3-integral axisymmetric dynamical models to measure both the black hole mass and the stellar orbital distribution. Based on Schwarzschild's method of building a library of characteristic orbits and then finding the best linear superposition of weights to match the observations, this technique is presented in Gebhardt et al. (2000) and Richstone et al. (2000) and is an extension of the spherical maximum-entropy models of Richstone & Tremaine (1988). A very similar axisymmetric model has been developed by Cretton et al. (2000). Even more general models could include triaxial distributions, but these will not be discussed here (see Merritt 1999 for a review of the current understanding of whether galaxies may be triaxial or axisymmetric).

Briefly, I outline these models in five steps: 1) use the surface brightness profile in two dimensions to infer the luminosity density; 2) determine the mass density from the luminosity density (generally using constant mass-to-light ratio); 3) define the potential from the density profile and any other sources—i.e., a central black hole and a dark halo; 4) run a representative set of orbits in the galaxy potential; 5) solve for the set of orbital weights which best matches the observations, and provide an estimate of the goodness-of-fit. The first step requires a deprojection which is only unique for edge-on configurations. Thus to ensure the most generality one should allow for a family of deprojections. I then vary any of the above parameters (black-hole mass, M/L profile, etc.) and find the model which has the best goodness-of-fit.

The two most important requirements for these models are to obtain two-dimensional kinematic coverage and to use the full line-of-sight velocity distribution (LOSVD). Since these models are axisymmetric, two-dimensional coverage is essential if one wishes to obtain an un-biased estimate of the stellar orbital distribution. The effects from using only one-dimensional kinematic coverage can be quite dramatic in terms of measuring the anisotropies (see Richstone et al. 2000 for details). In addition, we have now advanced both in S/N and data analysis techniques to the point where there is no need to parameterize the velocity profile in terms of moments. The most common analysis involved only the first two moments of the LOSVD (velocity and velocity dispersion), and more recently have included higher order moments in terms of Gaus-Hermite expansions. However, any parameterization limits our ability to measure the black hole and possibly creates biases in unknown ways. For example, the most important part of the LOSVD for measuring the black hole mass are its wings. As we explore regions closer to the black hole we expect a power-law profile in the LOSVD at large velocities. Gaus-Hermite polynomials do not allow a power-law tail and as such will bias the results. In addition, the shape of the LOSVD directly determines the orbital structure and, again, parameterization will be a limitation. Thus, it is prudent to use the full LOSVD when possible.

The data requirements for 3-integral models are expensive. In order to measure the black-hole mass accurately, we require HST observations, since in almost all cases the radius at which the black hole begins to dominate the potential is less than $0.2''$. Ground-based observations are also essential to complete the velocity and spatial coverage, because to infer the orbital structure we require both high S/N spectra to measure the shape of the LOSVD and spectra along many position angles. For each galaxy modeled, I generally use HST/STIS observations and spectra along 3-4 position angles from the ground.

3   Results

Figure 1 plots the galaxies with reliable black-hole estimates. I include all of the mass estimates for those studies which do not use stellar dynamical analysis. For those galaxies which use stellar kinematics, I only include those that have both HST spectral observations and have been modeled with 3-integrals.

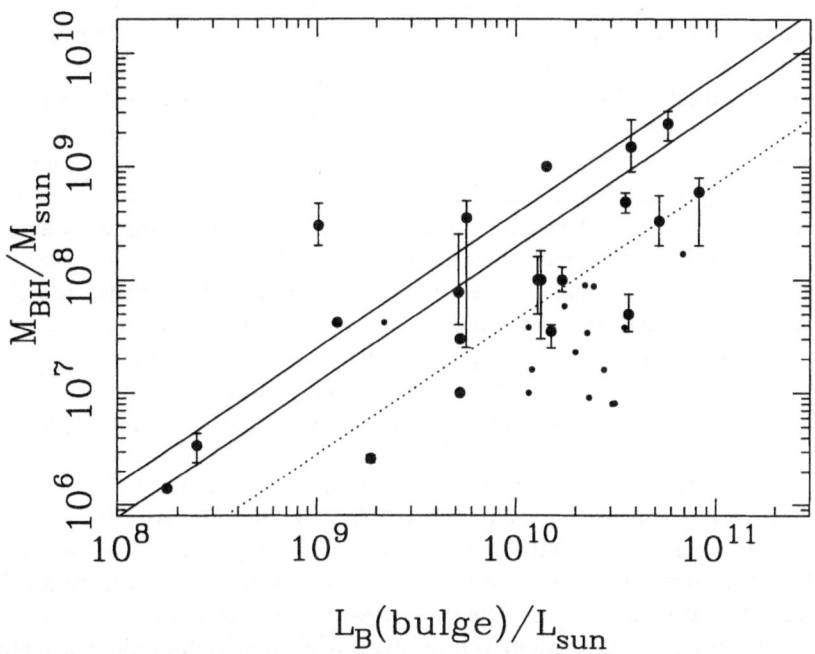

Fig. 1. Black hole mass versus luminosity of the bulge component. The points come from a variety of sources and techniques. The error bars represent the 68% confidence as quoted by the source. The small points without error bars come from reverberation mapping.

I include several lines on Fig. 1; the top solid line is the relation found in Magorrian et al. (1998), the next line down is that from Kormendy & Richstone (1995), and the bottom dotted line is the prediction from comparing the quasar density to the galaxy density (see Richstone et al. (1998) for details). The main point from this figure is that the relation between the black-hole to galaxy mass has decreased with the inclusion of the newer data and models. The two reasons for this are: 1) galaxies tend to have slight radial-velocity anisotropy at most of their radii (see next section) and thus the previous isotropic models overestimate the black hole mass, and 2) the sample now contain galaxies which were not previously suspected to contain black holes and thus we are not biased towards those galaxies with relatively large black-hole masses. The sample of galaxies with accurate black hole mass measurements will grow by another 20-30 galaxies over the next year, and any correlations, if they exist, will become much more apparent.

4   Stellar Orbital Structure

The stellar orbital distribution provides a powerful observation to measure the dominant processes which govern the formation and evolution of both the galaxy and black hole itself. Fig. 2 presents histograms of the radial-to-tangential motion in the central regions and in an average region (from 0 to the half-light radius) for the current sample of galaxies with 3-integral models. These histograms show that most galaxies have large tangential anisotropy near their centers, but the orbits are slightly radial in the main body of the galaxy. Fig. 2 represents the results along the major axes of the galaxies, however the results are similar along *any* position angle. Tangential anisotropy is a likely consequence of having a point mass in the center since those stars that pass near the center (i.e., those on radial orbits) are either scattered out of the central regions or destroyed by the black hole. However, theoretical models predict a range of central anisotropies. Three models that have been studied are adiabatic BH growth (Quinlan et al. 1995), BH infall models (Nakano & Makino 1999), and BH binary/merger models (Quinlan & Hernquist 1997). The models predict a different value of the central aniso-tropy parameter, $\beta(= 1 - v_t^2/v_r^2)$: BH infall models have $0 > \beta > -0.3$, adiabatic models have $\beta \approx -0.3$, and BH binary models have $\beta \approx -1.0$. The models that agree best with the observed anisotropies are the BH binary models. However, the models which have been studied have all used a limited set of initial conditions. The most restrictive is that they start with an isotropic velocity distribution. While this initial condition is likely a realistic starting point based on merger simulations, a detailed comparison requires more general theoretical models and more observations.

5   Conclusions

All galaxy models which include HST spectral observations require a super-massive black hole in their core. We have now explored those galaxies which had shown no evidence for having a black hole and as such have not pre-selected the current sample. Thus, it appears that central black holes are a common feature in galaxies. Given their special location at galaxy centers

Galaxies with 3−Integral Models

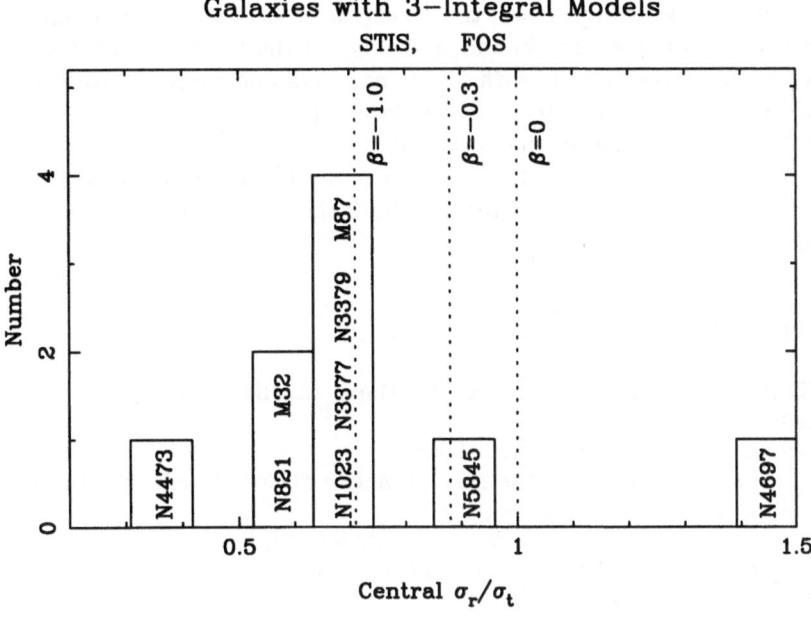

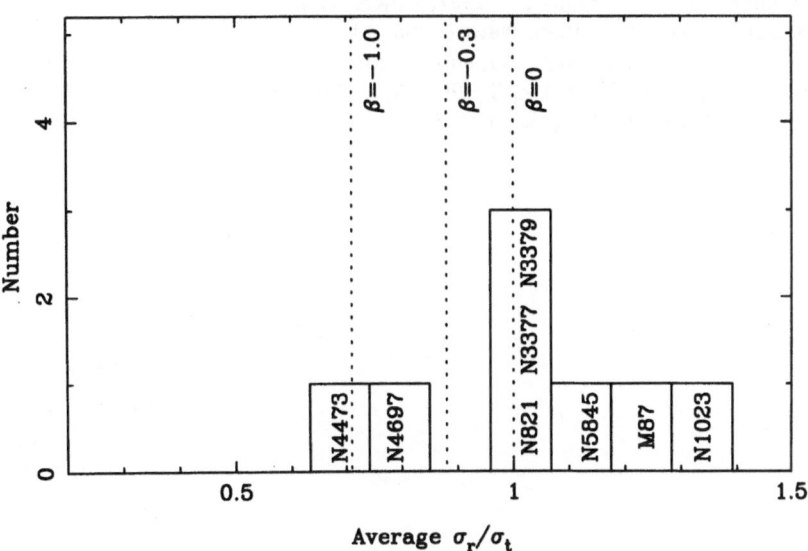

Fig. 2. Histograms of the radial to tangential dispersion in the central (top plot) and the average (bottom plot) regions. The tangential dispersion is defined as $\sqrt{(\sigma_\theta^2 + \sigma_\phi^2)/2}$.

and their relatively high mass, these central objects will have a significant influence on galaxy evolution and, possibly, on the formation process itself. As we improve the sample with larger numbers and more accurate masses we will then be able to determine the dominant evolutionary processes. The present evidence suggests that black-hole binary mergers are the most important events which shaped the central structure. Over the next few years we will see significant improvements in both the observations and theoretical models.

References

1. Cretton, N., de Zeeuw, P.T., van der Marel, R., Rix, H.-W. 1999, ApJS, 124, 383
2. Gebhardt, K., et al. 2000, AJ, March
3. Kormendy, J. 1993, in "The Nearest Active Galaxies", eds. J. Beckman, L. Colina, & H. Netzer (Madrid), 197
4. Kormendy, J., & Richstone, D. 1995, ARA&A, 33, 581
5. Magorrian, J. et al. 1998, AJ, 115, 2285
6. Merritt, D. 1999, PASP, 111, 129
7. Nakano, T., & Makino, J. 1999, ApJ, 510, 155
8. Quinlan, G., Hernquist, L., & Sigurdsson, S. 1995, ApJ, 440, 554
9. Quinlan, G., & Hernquist, L. 1997, NewA, 2, 533
10. Richstone, D., et al. 1998, Nature, 395, 14
11. Richstone, D., et al. 2000, AJ, submitted
12. Richstone, D., & Tremaine, S. 1988, ApJ, 327, 82
13. van der Marel, R. 1999, AJ, 117, 744

X-Ray Candidates for a Population of Nuclear Cores in Local Group Galaxies

E.J.A. Meurs and Z. Zang

Dunsink Observatory, Castleknock, Dublin 15, Ireland

Abstract. Because of their closeness, the galaxies of the Local Group are an interesting sample for finding weak manifestations of nuclear activity (as is known for M31 and the Milky Way). Archival ROSAT PSPC and HRI observations have been examined for X-ray sources associated with the centres of several Local Group members. Detections will be considered candidate massive Black Holes. Besides four relatively luminous systems (mostly known already), two smaller galaxies contain candidate central cores. These are however in Irregular systems without well-defined nuclei. So far, and also given observational results in other wavebands, only those four present good candidates for massive nuclear Black Holes within the Local Group.

Many nuclei of galaxies exhibit some level of activity. Very low level AGNs were recognized in some of the nearest galaxies: the Milky Way ($10^{35} erg\ s^{-1}$, Genzel & Townes 1987), M31 ($2.1\ 10^{37} erg\ s^{-1}$, Trinchieri & Fabbiano 1991), M81 ($1.7\ 10^{40} erg\ s^{-1}$, Elvis & Van Speybroeck 1982). Useful diagnostics for the presence of active cores are found at X-rays. Examples that illustrate this are the many QSOs and AGNs detected at high energies.

The ROSAT Public Archive allows to check several of the very nearest galaxies (members of the Local Group) for high energy emission. X-ray emission from their centres may be detected till very low luminosity levels and the spatial resolution that can be attained for such close-by objects will be helpful for determining the nature of any source coincident with these galaxies. The ROSAT PSPC and HRI frames are analysed according to standard procedures. In only a few instances sources are detected in the central areas of the galaxies. For undetected centres upper limits are determined.

We leave out a few major members that have already been discussed extensively in the literature (M31, M33, LMC, SMC and the Milky Way itself). Of the remaining Local Group galaxies, many are Irregulars that do not contain well-defined centres.

By far the strongest case is M32, for which we re-analysed PSPC and HRI observations that had already been reported in the literature. Eskridge et al. (1996; PSPC) suggested either a low-level AGN or a collection of stellar sources. Loewenstein et al. (1998; HRI) interpret the higher resolution data, in conjunction with ASCA observations, as a low-mass X-ray binary on the basis of strong variability on short timescales and an apparent offset from the optical nucleus. Re-analysing these data we find that the M32 source may be slightly extended and that an offset from the centre is not well established. At the present stage a nuclear source remains thus an interesting possibility.

Only in two other cases we find a ROSAT source close to the optical centres: NGC6822 and WLM. These are both Irregular systems and it is doubtful whether optical positions are close to their actual centres or would refer to proper nuclei at all. From Hardness Ratios or inferred luminosities we cannot find clear indications for being low-level nuclei.

Low-level activity is well-established only for the Milky Way and M31. M32 remains a promising candidate. M33 features a strong source (M33 X-8) very close to its nucleus. A period was suggested that could imply a stellar binary (Dubus et al. 1998), but widely and fairly regularly spaced observations could produce spurious periods and we keep this source as a possible core. The LMC and SMC have many sources but it is perhaps not sufficiently clear where their centres are located for deciding anything and we do not attempt to derive upper limits in these two cases.

Some of the upper limits achieved are at quite low luminosities thanks to the closeness of the galaxies. All the Spirals have certain or possible nuclear sources; these are the three most luminous stellar systems in the Local Group. The most luminous Elliptical (M32) also has a possible nuclear source.

Masses for the central condensed objects have been determined for the Milky Way ($2.6\ 10^6\ M_\odot$, Genzel et al. 1997) and M32 ($3.4\ 10^6\ M_\odot$, Van der Marel et al. 1998). In both cases the accretion rate is extremely low and well into the expected Advection Dominated Accretion Flow regime. If the X-ray source in M32 were offset from the nucleus, then the true nucleus would be simmering at an even lower fuelling rate.

Central high-energy sources are thus found among the most luminous members of the Local Group. These are candidate massive Black Holes, for which confirmation can be found from data at other wavelengths in the cases of the Milky Way and M31. If also M32 and M33 have massive Black Holes in their nuclei, there could be at least four massive Black Holes immediately around us. For the moment one could conclude that the 20 or so smaller and/or Irregular and dwarf-Spheroidal systems do not seem to possess massive Black Holes, but the sample of M32 shows that the activity level can be so low that a non-detection may not necessarily imply no Black Hole.

References

1. Dubus et al. 1997, ApJ 490, L47
2. Elvis & Van Speybroeck 1982, ApJ 257, L51
3. Eskridge et al. 1996, ApJ 463, L59
4. Genzel & Townes 1987, ARA&A 25, 377
5. Genzel et al. 1997, MNRAS 291, 219
6. Loewenstein et al. 1998, ApJ 497, 681
7. Van der Marel et al. 1998, ApJ 493, 613
8. Trinchieri & Fabbiano 1991, ApJ 382, 82

Global X-Ray Emission and Central Properties of Early Type Galaxies

Silvia Pellegrini

Dipartimento di Astronomia, Università di Bologna, via Ranzani 1,
I-40127 Bologna, Italy

Abstract. *Hubble Space Telescope* observations revealed that the central surface brightness profiles of early type galaxies show either *cuspy cores* or *featureless power law* profiles. I find that there is a clear *dichotomy* also in the X-ray properties: cuspy core galaxies span the whole observed range of L_X values (roughly two orders of magnitude in L_X), while power law ones are confined to log L_X (erg s^{-1}) < 41. So, a *global* property, such as L_X, that measures the hot gas content on a galactic scale, turns out to be well linked to a *nuclear* one. This could indicate that a central massive black hole (MBH) plays an important role in determining L_X (i.e., the hot gas content of early type galaxies).

1 Results

I have cross-correlated the sample of early type galaxies with inner profile measured by *HST* with existing catalogs of X-ray emission, collecting a sample of 59 galaxies (see Pellegrini 1999 for more details). Figure 1 plots the $L_X - L_B$ relation for the galaxies in this sample. We note that:

1) The least optically luminous galaxies and the most optically luminous ones are respectively power law and cuspy core galaxies (as found by Faber et al. 1997).

2) The least X-ray luminous galaxies and the most X-ray luminous ones are again respectively power law and cuspy core galaxies, consistent with the known $L_X - L_B$ correlation.

3) At intermediate L_B, where the two families coexist, cuspy core galaxies span the whole range of L_X values, while power law ones are confined below log L_X (erg s^{-1}) = 41.

It looks as if power law galaxies cannot be very X-ray bright, while cuspy core galaxies show L_X values extending from the lowest to the highest observed. What is also surprising is that this trend is sharper than those between L_X and the other basic properties used recently to divide early type galaxies into two families (Faber et al. 1997): the deviation of the isophotal shape from a pure elliptical shape, and the degree of anisotropy in the velocity dispersion tensor. Disky objects are low X-ray emitters, and generally flattened by rotation; boxy and irregular objects show the whole range of L_X, and various degrees of velocity anisotropy [see Pellegrini (1999) for more details].

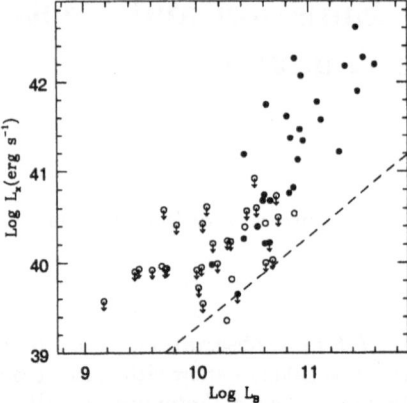

Fig. 1. Power law galaxies (open circles) and cuspy core ones (full circles) in the $L_X - L_B$ plot. Downward arrows show upper limits on L_X. The dashed line is $L_X \propto L_B$.

2   Discussion

Differences of the inner profile shape are not expected to produce large variations of the hot gas content, based on numerical simulations. A higher degree of galactic rotation and of flattening of the mass distribution (likely more important on average among power law galaxies) may have an effect; but do not offer a general explanation of the reduction in L_X, and a difference in L_X as large as observed.

It is now commonly accepted that early type galaxies host central MBHs (Richstone et al. 1998), and the presence of central cusps has been related to their role during galaxy formation and evolution. If a MBH is important also in determining L_X, the conventional approach to the problem of interpreting the X-ray properties of early type galaxies requires revision. One should focus now on the relation between central MBH presence and amount of hot gas. A few solutions to figure out why power law galaxies are found *only* below 10^{41} erg s^{-1}, while cuspy core galaxies span the whole observed scatter in L_X, might involve also other ingredients: the shape of the galaxies (triaxial versus axisymmetric) can have an effect on the way accretion proceeds. Or the very MBH properties (such as its mass) could play a role.

References

1. Faber, S. M., et al. (1997) The centers of early-type galaxies with *HST*. IV. AJ **114**, 1771–1796
2. Pellegrini S. (1999) Global X-ray emission and central properties of early type galaxies. A&A in press
3. Richstone, D., et al. (1998) Supermassive black holes and the evolution of galaxies. Nature **395**, 14–19

A New Monte Carlo Code
for Dynamical Simulations of Galactic Nuclei
Hosting Massive Black Holes

Marc Freitag[1] and Willy Benz[2]

[1] Observatoire de Genève, CH-1290 Sauverny, Switzerland
[2] Physikalisches Institut, Universität Bern, Sidlerstrasse 5
   CH-3012 Bern, Switzerland

Abstract. We have developed a new Monte Carlo code to follow the long-term evolution of a spherical dense galactic nucleus with a central massive black hole (BH). The main physical processes taken into account are 2–body relaxation, collisions, tidal disruptions and BH's growth. Self–gravity, velocity anisotropy and a stellar mass spectrum are properly modeled. This tool enables us to get reliable predictions for the rate and characteristics of star destroying events that induce bright accretion phases. Here, we summarize the design concepts of this code and show the results of test calculations and preliminary investigations.

1   Astrophysical Motivation

The now well established presence of massive black holes (BHs) in the center of some (maybe most) bright galaxies raises many yet unanswered questions. The ones we want to address are:

1. What is the long-term evolution of the structure of the BH engulfing central star cluster? How do the BH and the various physical processes imprint this structure?
2. What are the rates and characteristics of stellar collisions and tidal disruptions by the BH?

The backbone of our approach of these problems is of stellar dynamical nature and consists of a cluster evolution simulation program which will be briefly described in the next section. To add flesh to our study, we need realistic descriptions for the outcome of disruptive events. We are in the process of deriving such prescriptions for stellar collisions based on the results of many thousands of hydrodynamical simulations of these events computed with a "Smoothed Particle Hydrodynamics" (SPH) code [1].

2   The Monte Carlo Code

To get a better understanding of such complex systems as galactic nuclei where many processes interplay, numerical simulations are clearly called for.

Amongst the various schemes described in the literature, we chose to adopt the Monte Carlo method first proposed by Hénon [2] as it is a nice compromise between realism and efficiency. "Realism" means that all the important processes at play can be included in the computations. These are: 2–body relaxation, collisions, tidal disruptions, BH's growth and stellar evolution. To spare development time and facilitate the interpretation of the first sets of simulations, stellar evolution has been left out to be added in a future stage. Furthermore, our code properly tackles self–gravitation of the cluster and arbitrary velocity dispersion and stellar mass spectrum without added difficulty. On the other hand, "efficiency" is required in order to perform many quick simulations with various initial conditions and physical ingredients.

As a complete description of this code will be published in [3], we only briefly outline here the basic concepts underlying its design. It relies on two major assumptions, namely those of spherical symmetry and dynamical equilibrium (age $\gg t_{\text{rel}} \gg t_{\text{dyn}}$). The cluster is realized as a set of thin spherical shells composed of stars sharing the same energy, angular momentum (in modulus) and orbital phase. Together with the BH, these shells produce a smooth, non–relaxing, potential. Their rosette orbits are not explicitly integrated but sampled at each time step by randomly picking a radial position from the proper probability distribution: $dP/dR \propto 1/v_{rad}(R)$. While this approach forbids to track processes on dynamical time scales, it allows to adapt the local time step to the relaxation and/or collision time rather than the much shorter orbital time. The simulation of relaxation proceeds through 2–body gravitational encounters between neighbouring shells. The deflections are tailored to lead to the same diffusion of orbits as caused by genuine relaxation. Collisions are likewise simulated between stars from neighboring shells according to the local collision probability. The Monte Carlo approach ensures a correct sampling of the collision's initial parameters, i.e. stellar masses, relative velocity and impact parameter. Tidal disruptions are the fate of shells entering the so–called loss–cone [4]. These events are detected by mimicking the relaxation–induced random walk of the tip of the star's velocity vector during the many orbits a time step consists of.

3   Code Testing

We first concentrated on the relaxation–induced core collapse of idealized globular clusters either with all stars having the same mass (as in Fig. 1) or with a mass spectrum. These successful simulations ensure that relaxation and the related phenomena of evaporation and mass segregation are properly simulated.

Turning to collisions, we obtained a very good agreement between the analytical predictions for the differential collision rate (as a function of radius and of stellar masses) and the statistics of Monte Carlo runs with cluster evolution inhibited.

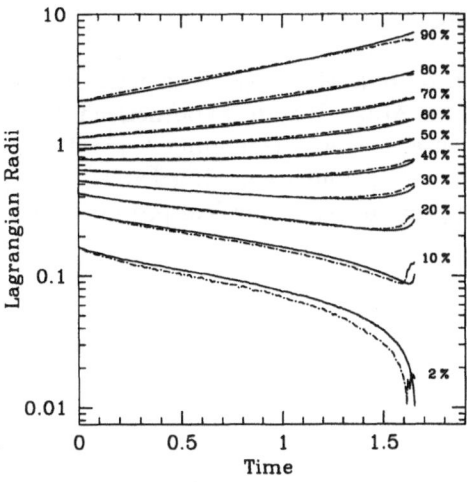

Fig. 1. Simulation of the core collapse of an isolated mono–mass Plummer cluster. We show the evolution of radii of spheres containing the indicated fractions of the cluster's total mass. Our results (*solid lines*) are compared with those of Mirek Giersz (*dash–dotted lines*) [5], [6]. The time unit is $N_*/\ln(0.11N_*)GM_0^{5/2}\,(4E_0)^{-3/2}$ and the length unit $GM_0^2\,(4E_0)^{-1}$, where M_0, E_0 and N_* are the cluster's initial mass, energy and number of stars. Our 512k shells simulation required about 150 CPU hours on a 400 MHz PentiumII computer.

Another test was to reproduce the models for dense galactic nuclei presented in [7]. These include the effects of tidal disruption and completely destructive collisions. Figure 2 shows that the agreement, although not perfect, is satisfying given the important differences between our code and the numerical scheme used in [7]. Also, to assess the consequences of a more realistic prescription for the outcome of collisions, we used the formulae by M. Davies published in [8]. As Fig. 2 testifies, this improvement induces at early times a rather drastic decrease in the amount of gas fed by collisions to the BH.

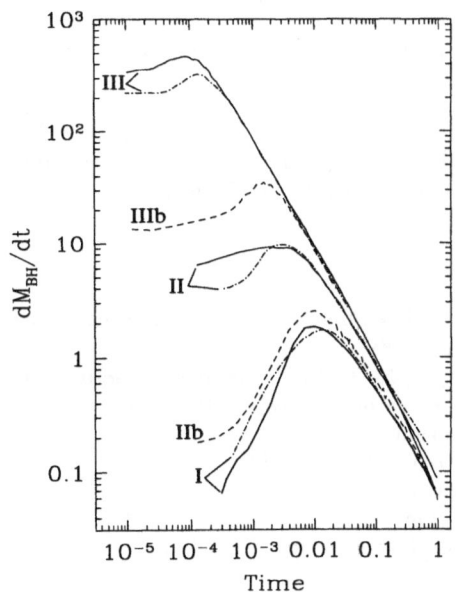

Fig. 2. Evolution of the accretion rate on the central black hole for the three nucleus models of [7]. Models I and II share the same initial conditions but model I does not include stellar collisions while model II treats them as causing complete disruption of stars. Model III starts with a denser star cluster and a larger BH and includes destructive collisions. Our results (*solid lines*) are compared with those of [7] (*dash–dotted lines*). Additionally, we re-computed models II and III with a more realistic treatment of collisions (*dashed lines*, labels IIb and IIIb). For models I and II, the time unit is 1.37×10^{11} yrs and the unit for dM/dt is $2.6 \times 10^{-3}\,M_\odot\mathrm{yr}^{-1}$. In model III, these units are 9.81×10^{11} yrs and $5.8 \times 10^{-3}\,M_\odot\,\mathrm{yr}^{-1}$

4   Preliminary Investigations

Our code allows us to study the evolution of a wide class of galactic central clusters, ranging from the nucleus of the Milky Way to hypothesized AGN models consisting of an extremely dense cluster surrounding a giant BH. Figure 3 shows results obtained in a first set of simulations aimed at exploring this parameter space and disentangling the role each physical process plays.

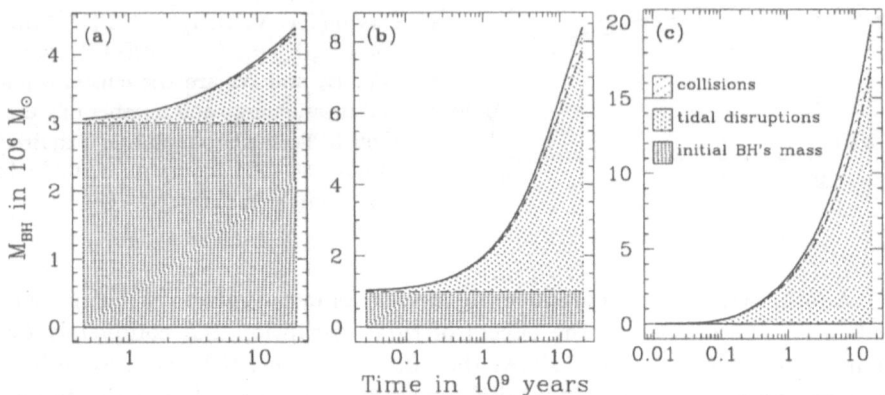

Fig. 3. Growth of the central BH for three nucleus models of increasing initial stellar density. Immediate and complete accretion of the gas released by tidal disruptions and stellar collisions (using formulae from [8]) is assumed. The cluster density profile is initially of the form $\rho(R) = \rho_0 \left(1 + (R/R_b)^2\right)^{-1}$. Initial conditions for (a)/(b)/(c) are respectively: $R_b = 0.2/0.1/0.05\,\mathrm{pc}$, $\rho_0 = 5 \times 10^6/10^8/10^9\ M_\odot\mathrm{pc}^{-3}$, $M_{BH} = 3 \times 10^6/10^6/10^4\ M_\odot$ with a stellar mass spectrum of $dN_*/dM_* \propto M_*^{-2.35}$ between $0.2\ M_\odot$ and $2\ M_\odot$. Model (a) is similar to the nucleus of the Milky Way. Parameters of model (c) were set to demonstrate the growth of a massive BH from a "small" seed. Model (b) is an intermediate case

M. F. wants to thank warmly Mirek Giersz for kindly providing us with simulation data and discussing many aspects of Monte Carlo algorithms.

References

1. Benz, W. (1990) In: Buchler, J. R. (Ed.) The numerical modelling of nonlinear stellar pulsations, Kluwer, Dordrecht, 269–288
2. Hénon, M. (1973) In: Martinet, L., Mayor, M. (Eds.) Dynamical structure and evolution of stellar systems, Geneva Observatory, Sauverny, 183–260
3. Freitag, M., Benz, W. (2000) in preparation
4. Lightman, A. P., Shapiro, S. L. (1977) ApJ **211**, 244–262
5. Giersz, M. (1998) personal communication
6. Giersz, M. (1998) MNRAS **298**, 1239–1248
7. Duncan, M. J. Shapiro, S. L. (1983) ApJ **268**, 565–581
8. Rauch, K. P. (1999) ApJ **514**, 725–745

Luminosity and Mass Functions of Active Galactic Nuclei

Lutz Wisotzki

Hamburger Sternwarte, Gojenbergsweg 112, 21029 Hamburg, Germany,

1 Introduction

Luminosity functions are a standard tool of observational astrophysics. Here I explore the interrelation between luminosity and mass functions for Active Galactic Nuclei. The formal relation relating BH mass and bolometric luminosity is $L_{\mathrm{bol}} = \epsilon M_{\mathrm{BH}}\,c^2/t_{\mathrm{E}}$ with the 'Eddington ratio' $\epsilon \equiv L/L_{\mathrm{E}}$. Using a semi-empirical $M(L)$ relation for AGN, we can (statistically) convert luminosity functions into mass spectra of radiatively accreting black holes.

2 Luminosity Function

To construct a combined LF of low- and high-luminosity Seyferts, I adopt the flux of the broad Hα emission lines as robust luminosity indicator. The following samples were used: (a) The Hα luminosity function of luminous local QSOs and bright Seyfert 1 nuclei using the Hamburg/ESO survey [5]; (b) The 'Dwarf' Seyfert 1 nuclei in nearby galaxies [1]. The combined LF of low-redshift AGN covers over 6 orders of magnitude in luminosity and is still nearly consistent with a single power law (Fig. 1).

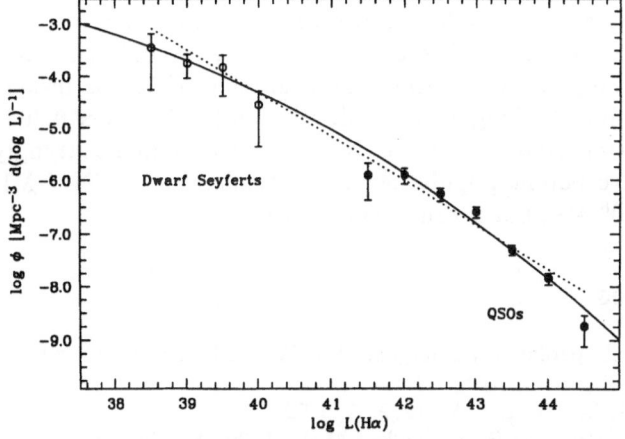

Fig. 1. Hα luminosity function of local AGN, with analytic approximations.

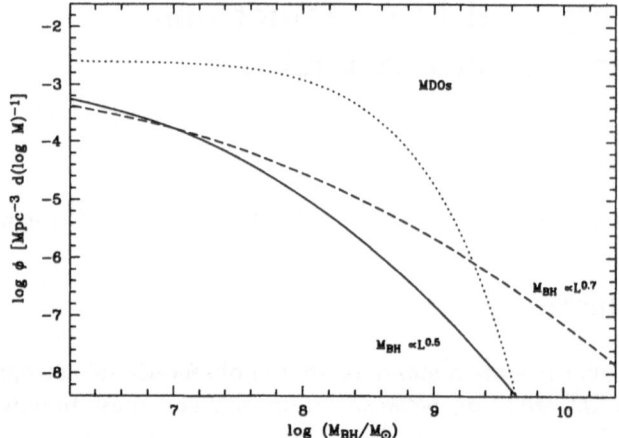

Fig. 2. Reconstructed AGN mass function, using the parametric fit from Fig. 1.

3   The Mass-Luminosity Relation for AGN

If the BLR clouds are gravitationally bound [3], the mass of the black hole is $M_{BH} \approx v^2 R/G$, where v can be obtained from the broad emission lines. The line width is only weakly correlated with luminosity if at all; a power law fit to the above mentioned combined sample gives $v \propto L^{0.045}$. The BLR radius R is ideally obtained from continuum-line reverberation [4]; so far, ~ 20 Seyfert galaxies have been observed, giving $R \propto L^{0.5}$. Combining these relations yields an approximate $M(L)$: $M_{BH} \propto L^{\beta}$ with $0.5 < \beta \lesssim 0.7$.

4   Mass Function

Having adopted a $M(L)$ relation, we can convert the luminosity distribution into a mass function. This is shown in Fig. 2 for two different values of β. For comparison we give the *expected* mass function of MDOs inside of galactic spheroids, using the Magorrian et al. relation [2] $M_{MDO} \propto 0.006\, M_{sph}$, with assumed scatter ± 0.5 dex. The AGN mass function differs strongly from that of its putative parent population. In particular, the AGN / MDO ratio is 1:100 at $\sim 10^8\, M_{\odot}$, but much larger elsewhere.

References

1. Ho L.C., Filippenko A.V., Sargent W.L.W., 1997, ApJS 112, 391
2. Magorrian J., et al., 1998, AJ 115, 2285
3. Peterson, B.M., Wandel A., 1999, ApJ 521, 95
4. Wandel A., Peterson B.M., Malkan M.A., 1999, ApJ in press
5. Wisotzki L., Christlieb N., Bade N., et al., 1999, A&AS submitted

Accreting Black Holes: Modelling Individual Objects and Whole Populations

Ewa Szuszkiewicz[1,2], Luigi Danese[2], John C. Miller[2,3], Pierluigi Monaco[4,2], and Paolo Salucci[2]

[1] Toruń Centre for Astronomy, Nicolaus Copernicus University, ul. Gagarina 11, 87-100 Toruń, Poland
[2] International School for Advanced Studies, SISSA, via Beirut 2-4, 34013 Trieste, Italy
[3] Nuclear and Astrophysics Laboratory, University of Oxford, Keble Road, Oxford OX1 3RH, England
[4] Dipartimento di Astronomia, Università di Trieste, via Tiepolo 11, 34131 Trieste, Italy

Abstract. We are carrying out a programme of non-linear, time-dependent calculations to study non-stationary accretion onto black holes. We plan to use this study as a basis for considering whole populations of accreting objects. Some of the most interesting results are presented.

1 Individual Accreting Black Holes

Accretion discs, which are thought to be present in galactic black hole candidates and active galactic nuclei (AGN), are subject to several types of instability which can be responsible for the variability observed in these objects. The aim of our study is to calculate the observational consequences of each instability and compare these with the actual measurements. We have started our programme of work by investigating the thermal instability driven by radiation pressure, which is relevant for intrinsically bright sources, using the "slim disc" model with vertically-integrated equations [1]. Our main result is that the thermally-unstable discs undergo limit-cycle behaviour with successive evacuation and refilling of the central parts of the disc. Moreover, we found that models predicted to be stable by local analysis do indeed remain stable and stationary. The same applies for models which, according to local analysis, would have a potentially unstable region smaller than the minimum wavelength for unstable perturbations. Systematic studies of this instability for different black hole masses, accretion rates and viscosity parameters are in progress.

2   Populations of Accreting Black Holes

A different type of thermal instability, driven by partial ionization of the disc material, has been found to operate in most of the accretion disc models relevant for AGN (with the exception of the brightest sources for which the discs are completely ionized) [2]. Two important consequences follow: (1) quiescent AGN must appear as quite normal galaxies, and (2) the average mass fuelling rate in many if not all AGN is much lower than implied by their current luminosities. This in turn limits the masses that their central black holes are expected to reach. The physics learned from studying individual

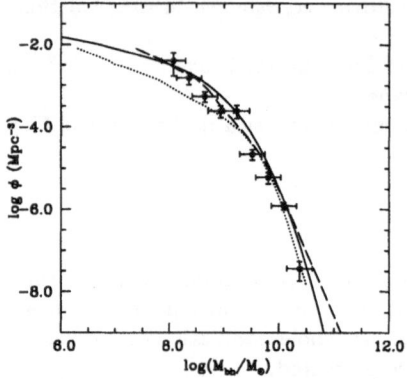

Fig. 1. The comparison between optical (solid line), radio (points with error bars) and theoretical (dashed line) mass functions derived by [3]. The dotted line is the mass function of the relic black holes as predicted by [4].

objects should be consistent with observational characteristics derived from whole populations of accreting black holes. One of the possible ways to constrain accretion scenarios is to construct the mass functions (Fig. 1) derived from investigations of massive dark objects resident in local galaxies and the mass function of the black holes inferred from the past activity of AGN.

References

1. Szuszkiewicz, E., Miller, J. C. (1998) Limit-cycle behaviour of thermally unstable accretion flows on to black holes. MNRAS **298**, 888–896
2. Burderi, L., King, A. R., Szuszkiewicz, E. (1998) Does the thermal disk instability operate in active galactic nuclei? ApJ, **509**, 85–92
3. Salucci, P., Szuszkiewicz, E., Monaco, P., Danese, L. (1999) Mass function of dormant black holes and the evolution of active galactic nuclei . MNRAS **307**, 637–644
4. Cavaliere, A., Vittorini V. (1998) The rise and fall of the quasars in The Young Universe; Galaxy Formation and Evolution at Intermediate and High Redshift, eds. S. D'Odorico, A. Fontana, E. Giallongo, Astron. Soc. Pac. Conf. Ser. 146, 26

BLACK-HOLE FORMATION

Formation and Evolution
of Black-Hole Binaries

Frank Verbunt

Astronomical Institute, Postbox 80.000, 3508 TA Utrecht, the Netherlands

Abstract. A brief overview is given of the circumstances which allow a black hole to be formed in a binary. The importance of mass loss before the supernova explosion from the progenitor of the black hole is explained, as are the various ways in which this may happen and the properties of the resulting X-ray binaries. X-ray binaries in globular clusters are also discussed. We indicate recent interesting developments in theory, i.e. studies of mass transfer from a main-sequence donor, of the evolution of stellar cores without envelopes, of low-mass binaries with $\geq 2M_\odot$ donors; as well as in observations, e.g. the discovery by *Beppo*SAX of many soft X-ray transients with neutron stars.

1 Introduction: The Problem

Black holes by definition are not observable directly, and our knowledge of their existence derives from the observation of radiation emitted by gas streaming towards them. In a binary strong X-ray emission indicates a compact donor, i.e. a neutron star or a black hole; if a radial-velocity study of the binary indicates that the compact object has a mass in excess of the highest mass deemed acceptable for a neutron star, $> 3M_\odot$ say, we conclude that the compact object must be a black hole.

That a compact star can be found in a binary is a surprise, for the following reason. Consider a primordial binary, with stars of masses M and m, where $M > m$. The more massive star will evolve first, and provided it is sufficiently massive, evolve into a supernova which may leave a compact star. Write the mass loss during the supernova as ΔM, and assume for simplicity that a) the pre-supernova orbit is circular and b) the instantaneous velocity of the compact object is equal to the velocity V of the pre-supernova star. Simple newtonian equations then give an orbital eccentricity e and (change in) velocity of the binary center of mass v as

$$e = \frac{\Delta M}{M + m - \Delta M} \quad \text{and} \quad v = eV. \tag{1}$$

The supernova explosion unbinds the orbit if $e > 1$, i.e. if more than half of the total mass of the binary is lost. Because it is the more massive star that explodes first, one expects the binary to be dissolved, unless the initial masses of the two binary components are close: $M - 2M_c < m < M$, where

$M_c \equiv M - \Delta M$ is the mass of the compact object. The problem is severe for a neutron star $(M_c = 1.4 M_\odot)$; somewhat less so for a black hole $(M_c = 7 M_\odot)$.

To keep a binary after the explosion, various solutions have been envisaged. First, the compact star can obtain an extra velocity, the so-called kick velocity, when formed; that such is the case is indicated by the high spatial velocities of radio pulsars. If the kick is in a direction opposite to the orbital velocity, it may save a binary which without the kick would have dissolved. Second, the more massive star may lose much of its mass before exploding, in the form of a stellar wind, by transferring it to its companion, or when the companion violently enters its envelope and expels it in a spiral-in. We will discuss these possibilities below in discussing the scenarios for the formation of high-mass X-ray binaries (Sect. 2), and low-mass X-ray binaries (Sect. 3). The formation of X-ray binaries in globular clusters is discussed in Sect. 4. A more extended overview of the topics reviewed here can be found in e.g. Verbunt (1993); some recent developments are discussed by Kalogera, Langer and Wellstein in these proceedings.

2   High-Mass X-Ray Binaries

Mass transfer in a binary can occur when one of the two components expands to fill its Roche lobe, the innermost equipotential surface that surrounds both stars. When mass transfer is initiated by expansion on the main sequence, it is referred to as case A; when on the first ascent of the giant branch (after core hydrogen exhaustion) as case B; and when on the second ascent of the giant branch (after core helium exhaustion) as case C. The radius of the Roche lobe of a binary star is proportional to the semi-major axis of the binary. Thus, the relative frequencies of binaries in which mass transfer is initiated in cases A, B and C is proportional to the relative frequencies of the initial semi-major axes of binaries. It is often assumed that binary orbital periods P_b have a distribution which is flat in $\log P_b$; this implies that the distribution of semi-major axes is flat in $\log a$; and this in turn implies that most binaries will initiate mass transfer in the evolutionary state where the expansion is most pronounced, i.e. case B is most frequent. For this reason most studies of the formation of X-ray binaries assume case B.

2.1   Conservative Mass Transfer and Be/X-Ray Binaries

The total angular momentum of a binary of stars with masses M and m, radii R and r, and rotational velocities Ω and ω can be written as the sum of the orbital angular momentum and the spin angular momenta of the stars:

$$J = Mm\sqrt{\frac{Ga}{M+m}} + R_g^{\,2} M R^2 \Omega + r_g^{\,2} m r^2 \omega, \qquad (2)$$

where R_g and r_g are the gyro radii of the two stars. The mass transfer in the binary is called conservative when total mass $M + m$ and total angular momentum J are conserved. In most cases the spin momenta are negligible. The semi-major axis then changes with M (and m) according to

$$Mm\sqrt{a} = \text{constant} \qquad (3)$$

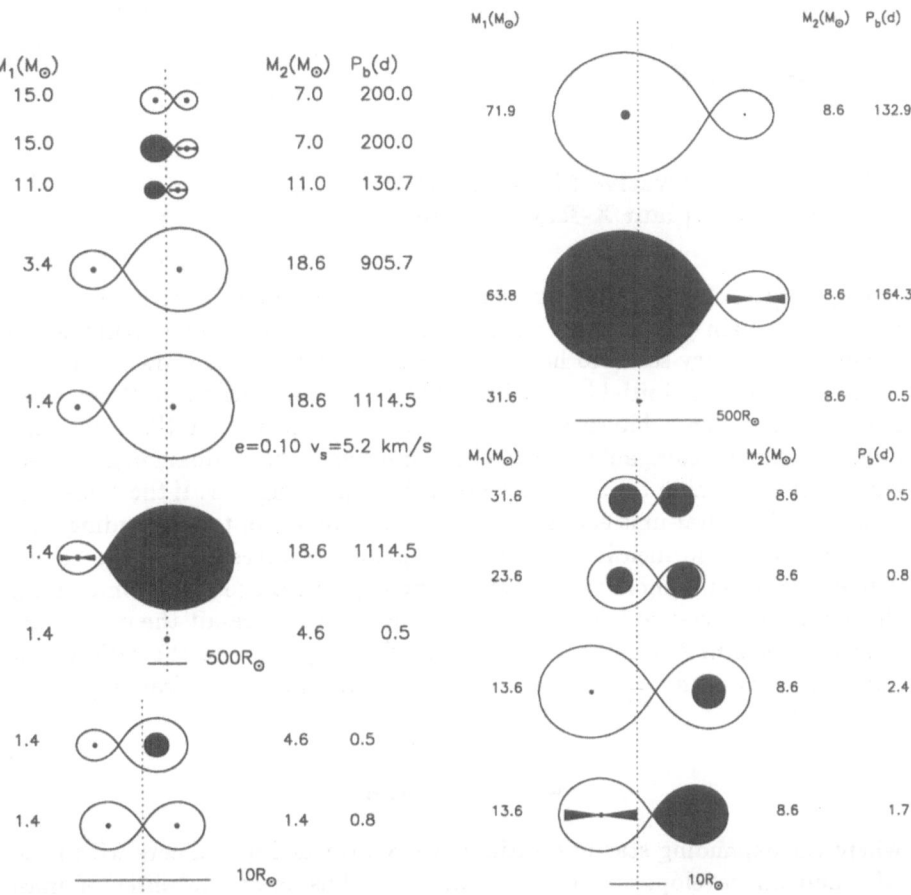

Fig. 1. Conservative (left) and non-conservative (right) evolution of a primordial binary into a high-mass X-ray binary. For each evolutionary state the masses of both stars and the binary period are indicated. For details see text. Note that the evolution before and after spiral-in are shown at different scales

The conservative evolution of a binary leading to a Be/X-ray binary is illustrated in Fig. 1 (left). The more massive star evolves first, expands to fill its Roche lobe, and transfers almost all of the mass outside its helium core to its companion. In this process, the orbit shrinks a little at first, and then expands a lot, according to Eq. 3. The helium core continues to evolve until it explodes as a supernova and leaves a neutron star. The mass loss induces an eccentricity in the orbit, and a velocity of the binary as a whole, according to Eq. 1. The companion of the neutron star has accreted a lot of angular momentum with the transferred mass, and thus is expected to rotate rapidly. Such a rapidly rotating star loses mass in a wind mainly in its equatorial plain: it is a Be star. The neutron star catches some of this wind mass and the accretion energy is emitted as X-rays. This type of wide X-ray binary is the most common type of X-ray binaries with a high-mass donor. None of the Be X-ray binaries discovered so far harbours a black hole.

2.2   Non-Conservative Mass Transfer and Supergiant X-Ray Binaries

X-ray binaries formed via conservative mass transfer in case B are always wide. To explain the close high-mass X-ray binaries with supergiants, a different evolutionary scenario has been devised, which is also shown in Fig. 1. In this scenario the initial binary has a more extreme mass ratio. When mass transfer stars, the orbit shrinks so rapidly (see Eq. 3) that the expanding donor engulfs its companion. The orbital motion of the companion is braked, causing it to spiral in towards the core of the expanding star. If the energy released by the spiral-in is enough to expel the envelope of the expanding star, a close binary is formed in which the companion revolves around the helium core of the initially more massive star. The physical processes accompanying the spiral-in process are not well understood, and as a result the outcome is highly uncertain. A simple consideration of energy leads to the ratio of the semi-major axes immediately before (a_i) and after (a_f) the spiral in given by

$$\frac{a_f}{a_i} = \frac{\lambda \alpha}{2} \frac{M M_{co} R}{m M_e a_i} \,, \tag{4}$$

where the expanding star has initial mass M, divided over a core with mass M_{co} and an envelope of mass M_e; and a radius R at the onset of mass transfer. The mass of the companion is m; λ measures the effective radius (in terms of binding energy) of the envelope of the expanding star, and thus can be computed from a stellar model; α absorbs our ignorance of the spiral-in physics, and must be guessed. Again, the helium core is assumed to continue its evolution until it collapses to a neutron star or a black hole. Once the companion develops a strong wind or expands to fill its Roche lobe, the compact star accretes matter and turns into a bright X-ray source. Such X-ray binaries with a supergiant donor are rare, only a few are known in our galaxy, and a few in the Magellanic Clouds. Three of these harbour black holes: Cygnus X-1 in our galaxy and LMC X-1 and LMC X-3 in the Large Magellanic Cloud.

2.3 Recent Developments

Two interesting developments have occurred in our thinking about the formation of high-mass X-ray binaries in recent years. The first of these concerns the evolution of the helium core of a star which has lost its envelope. In early computations of binary evolution it has mostly been assumed that such an unwrapped core evolves pretty much in the same way as it would have done inside the whole star. By computing the evolution of unwrapped cores explicitly several authors (in particular Woosley et al. 1993, 1995) have shown that this assumption is not correct. In particular, even the cores of very massive stars, which would have evolved into a black hole inside the full star, evolve into a neutron star instead when the star loses its envelope at an early evolutionary stage. This explains why no Be/X-ray binary (formed via case B mass transfer) contains a black hole (Brown et al. 1996). An important consequence is that one can no longer transfer conclusions about the progenitor mass of a black hole from single-star evolution to binary evolution or vice versa. The presence of black holes in close binaries can only come about via a spiral-in initiated by case C mass transfer, when the core of the mass-losing star has evolved far enough before it loses its envelope.

The second development is the realization that case A mass transfer may lead to supergiant binaries, which explains that a neutron star can be accompanied by a very massive donor (Wellstein & Langer 1999; see also their paper in these proceedings). As a result, one can no longer conclude from the binary Wray 977, in which a $48 M_\odot$ star transfers mass to a neutron star, that the progenitor of the neutron star had an initial mass higher than $48 M_\odot$; it may be as low as $25 M_\odot$. An interesting question regarding the systems arising from case A mass transfer is whether the mass donor to the compact star rotates rapidly or not: the donor in Wray 977 does not rotate rapidly, even though almost half of its mass may have been accreted from its companion. If this slow rotation is generally the case, it may be used to discriminate between systems evolved via case A and case B evolution; more research is needed into this question, however.

2.4 Continued Evolution

The continued evolution of a high-mass X-ray binary in a close orbit has been studied by Bagot (1996), who builds on earlier work by e.g. Hut (1980), and shows that the evolution of the semi-major axis – which can be measured from extended timing of the X-ray pulses – cannot be computed from Eq. 3, but requires the fuller Eq. 2. These close high-mass X-ray binaries are close to the Darwin instability (i.e. the stellar spin momentum is more than a third of the orbital angular momentum), and a spiral-in will follow, which almost certainly leads to a merger in the case of a neutron star. However, if the mass recipient is a black hole, a spiral-in may be avoided, and the donor can evolve into a compact star, most likely a neutron star. We thus anticipate the discovery of a young radio pulsar in orbit around a black hole!

The continued evolution of a Be/X-ray binary may lead to a binary consisting of two neutron stars, via a spiral-in. Several such binaries with two neutron stars are known now, and are thought to have formed this way (for a recent discussion, see Bagot 1997).

3   Low-Mass X-Ray Binaries

In low-mass X-ray binaries a neutron star or black hole accretes matter from a low-mass companion. In Fig. 2 I illustrate the galactic distribution of low-mass X-ray binaries with neutron stars and with black holes: near the Sun, where observational selection effects are least, the black-hole binaries appear more common than neutron-star binaries. The origin of such a binary is thought to be analogous to the scenario shown in Fig. 1 (right), with the crucial difference that the initially less massive star – now the donor – had a mass $\leq 1\,M_\odot$. The spiral-in with such a low-mass donor can easily lead to a merger into a single star, and to avoid this the spiral-in should be initiated by mass transfer at as late an evolutionary state as possible, e.g. case C; i.e. the initial binary must be rather wide. This may explain, incidentally, why the fraction of black holes in low-mass X-ray binaries is relatively high: the core of the progenitor has evolved mainly with its envelope still present.

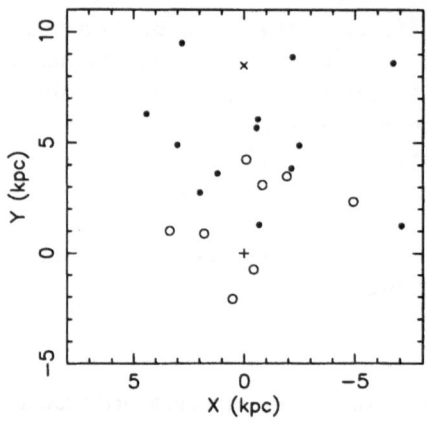

Fig. 2. The spatial distribution in our galaxy of low-mass X-ray binaries with black holes (o) and with neutron stars (•). The Sun (+) and galactic center (×) are also indicated. For black holes we take only those with dynamical mass estimates (Bailyn et al. 1998, distances from the review by Chen et al. 1997), for neutron stars the bursters (distances from Van Paradijs & White 1995)

The initial orbit cannot be too wide, however, because the initially more massive star must reach its Roche lobe for mass transfer and spiral-in to be initiated. It is not at all obvious that the lower limit to the initial orbital size required by the survival of spiral-in is compatible with the higher limit required by the condition that the primary reaches its Roche lobe (e.g. Portegies Zwart et al. 1997). This question has been investigated in detail by Kalogera & Webbink (1996, 1998), who confirm earlier suggestions that the velocities acquired by neutron stars when they are formed in an asymmetric collapse may play a decisive role in the formation of low-mass X-ray binaries with neutron stars. Whether kicks are important also in the formation of black-hole binaries is less obvious (Kalogera 1999, also her contribution to these proceedings); analysis of the spatial motion of black hole binaries in terms of Eq. 1 indicates that black holes do not acquire kicks at formation (Nelemans et al. 1999; these proceedings).

A recent census of O-star binaries does find binaries with long orbital periods (more than a year) and extreme mass ratios, suitable progenitors for low-mass X-ray binaries (Mason et al. 1998). The wider binaries appear to have mass ratios which are extremer than the close binaries; the consequence of this for population synthesis studies – which so far assumed the same mass ratio distribution for all binary periods (e.g. Portegies Zwart & Verbunt 1996) – will have to be investigated.

The evolution of low-mass X-ray binaries is driven by mass transfer. The rate of mass transfer $\dot{M}$ is derived observationally from the X-ray luminosity L_x, according to $L_x \simeq GM\dot{M}/R$; because most X-ray sources show variability on time scales of decades (i.e. the period during which X-ray observations have been made) it is not clear in how far we can connect the observed mass transfer rates with an evolutionary time scale. Changes in sign of the derivative of the orbital period have been determined for more than one low-mass X-ray binary: this proves that for these systems, at least, the mass transfer rate varies due to processes acting on short time scales, perhaps stellar activity cycles.

The mass transfer may be the consequence of loss of angular momentum from the binary, which drives the two stars together; this is only efficient in binaries with periods less than about half a day. The mass transfer rate $\dot{M}$ is proportional to the loss of angular momentum $\dot{J}$; for a donor mass M and an orbital angular momentum J one roughly has $\dot{M}/M \simeq \dot{J}/J$. Gravitational radiation then leads to $\dot{M} \simeq 10^{-10} M_\odot/\text{yr}$ for a main-sequence donor. In view of the above remark about variability, it is not clear whether the higher observed mass transfer rates require higher loss of angular momentum; if they do, magnetic braking is a possible mechanism. Whereas cataclysmic variables are known with orbital periods shorter than 2 hrs and longer than 3 hrs, very few such systems have been found between 2 and 3 hrs. This so-called period gap was one of the most potent arguments in favour of magnetic braking. However, it appears possible that the period gap is an observational selection effect, in which case there is no proven need for magnetic braking as a driver of binary evolution (Verbunt 1997).

In wider binaries the mass transfer is driven by the expansion of the donor on its first ascent of the giant branch. For a donor radius R and its time derivative $\dot{R}$, we roughly have $\dot{M}/M \simeq \dot{R}/R$. This implies that the mass transfer rate increases with the orbital period, since large giants expand faster. The mass transfer continues until the giant has transferred all of its envelope to the compact star, and leaves its core in a wide, circular orbit around the neutron star or black hole. The many millisecond radio pulsars now known in circular orbits around undermassive white dwarfs (reviewed by Bailes 1996) attest to the basic correctness of this scenario (e.g. Phinney 1992).

Three recent developments are worth mentioning. First, it has been found that several low-mass X-ray binaries have donors with masses that are not as low (viz. $\leq 1 M_\odot$) as generally assumed for low-mass X-ray binaries. For example, the black-hole binary GRO J 1655 − 40 has a donor with a mass of about $2.3 M_\odot$ (Orosz & Bailyn 1997). It would appear that the donor must be a subgiant to fill its Roche lobe in the 2.6 d orbit. However, Regős et al. (1998) note that accurate radius determinations of main-sequence stars in double-lined eclipsing binaries (Andersen 1991) show that stars with masses in the range 2-4 $M_\odot$ expand sufficiently on the main-sequence to explain mass transfer from a main-sequence star in GRO J 1655 − 40.

Secondly, Brown et al. (1999) remark that the observation of black-hole binaries in low-mass systems with evolved donors implies that there are many times more – in the ratio of the main-sequence life time to the giant life time, i.e. a factor ~ 100 – black-hole binaries with an unevolved companion which does not fill its Roche lobe. This has obvious consequences for the estimated birth rate of black-hole binaries.

And finally, the Wide Field Camera on board of the *Beppo*SAX X-ray satellite has discovered relatively dim X-ray transients, with peak luminosities $\leq 10^{37}$ erg/s, thanks to its unique combination of a large field of view and small angular separation. Most of these dim transients are bursters, i.e. neutron stars, which confounds the recent speculations that the vast majority of X-ray transients with low-mass donors are black-hole systems (Heise 1998, Heise et al. 1999).

4   Globular Cluster X-Ray Sources

Whereas the system of globular clusters contains only 0.1% of the mass of our Galaxy, it contains no less than 12 bright X-ray sources, about 10% of the total number of such sources in our Galaxy. It is thought that the X-ray sources in globular-cluster are low-mass X-ray binaries, formed via two processes that occur only in the dense cluster cores: tidal capture and exchange encounters (see the review by Hut et al. 1992). Tidal capture occurs when a neutron star transfers some of its kinetic energy to tides in another star during a close passage, and enough tidal energy is dissipated to bind the neutron star in orbit around its captor. An exchange encounter occurs when a neutron star ejects one of the stars in a binary in a close encounter, and takes its place. These processes are efficient especially in globular clusters with dense cores, where the average distance between stars is relatively small.

The Wide Field Camera on board of *Beppo*SAX has contributed to our knowledge of X-ray sources in globular clusters by increasing the number of sources from which an X-ray burst has been detected to 11 out of the 12 known (in 't Zand et al. 1998, 1999). Thus, we now know that 11 of the 12 bright X-ray sources in globular clusters contain a neutron star; whether the twelfth system contains a neutron star or a black hole is not as yet known. An X-ray burst has also been detected from NGC 6626, which is remarkable because this cluster does not contain a bright X-ray source; probably the burst comes from a neutron star accreting at a very low ('quiescent') rate from a companion (Gotthelf & Kulkarni 1997). The dearth of black-hole binaries in globular clusters may be related to the dynamical evolution of the cluster in its early stages, during which virtually all black holes were kicked from the cluster via close encounters (Portegies Zwart & McMillan 2000; these proceedings).

Two more orbital periods have been discovered in recent years, viz. a 20.6 minute (or its Nyquist alias 13 minute) period for the source in NGC 6712 (Homer et al. 1996) and a 12.36 hr period for the source in Terzan 6 (in 't Zand et al. 2000). Two of five orbital periods for globular cluster sources now known are shorter than the period of any X-ray binary in the Galactic disk, and imply that the mass donor is a white dwarf. The two periods in excess of half a day imply a donor which is expanding onto the subgiant branch; the fifth period is compatible with a mass donor on the main sequence. It has been suggested recently that exchange encounters are much more efficient in forming X-ray binaries than tidal capture; it appears to me, however, that the observed preference for short orbital periods clearly favours tidal capture as the formation mechanism. Circularization of the orbit following tidal capture requires dissipation of (almost) enough energy to destroy the companion of the compact object; to avoid such destruction, energy must be exchanged between the tides of the companion and the orbital motion, possibly in a chaotic manner (Mardling 1995).

5   The Future

Recent years have shown remarkable and interesting progress in the field of the evolution and formation of X-ray binaries. In the near future we may expect the new work to continue. Progress will, I think, be made along the following lines.

The study of mass loss from the binary and of the loss of angular momentum that it implies will be done from first principles, and replace the somewhat arbitrary parametrization of these processes that are currently used.

Further population synthesis studies will be used to obtain statistical arguments on the importance of the various cases of binary evolution. Detailed studies – observationally and theoretically – of individual systems (like those of Wray 977 and GRO J 1655−40) will provide more direct knowledge on some of the physical processes involved.

With the 8m-class optical telescopes and the sensitive XMM X-ray satellite it will become possible to discriminate between high-mass and low-mass X-ray sources in M 31, making a nearby galaxy available for study of its population of X-ray sources.

References

1. Andersen, J. 1991, A&A Review, 3, 91
2. Bagot, P. 1996, A&A, 314, 576
3. Bagot, P. 1997, A&A, 322, 533
4. Bailes, M. 1996, in S. Johnston, M. Walker, M. Bailes (eds.), Pulsars: problems and progress, ASP Conference Series 105, ASP, San Francisco, p. 3
5. Bailyn, C., Jain, R., Coppi, P., Orosz, J. 1998, ApJ, 499, 367
6. Brown, G., Lee, C.-H., Bethe, H. 1999, New Astronomy, 4, 313
7. Brown, G., Weingartner, J., Wijers, R. 1996, ApJ, 463, 297
8. Gotthelf, E., Kulkarni, S. 1997, ApJ (Letters), 490, 161
9. Heise, J. 1998, Nucl.Phys.B Proc., 69, 186
10. Heise, J., in 't Zand, J., Smith, M., Muller, J., Ubertini, P., Bazzano, A., Cocchi, M. 1999, in: The extreme universe, Astrophys.Lett.Com. 38, 297
11. Homer, L., Charles, P., Naylor, T., van Paradijs, J., Aurière, M., Koch-Miramond, L. 1996, MNRAS, 282, L37
12. Hut, P. 1980, A&A, 92, 167
13. Hut, P., McMillan, S., Goodman, J., Mateo, M., Phinney, S., Pryor, C., Richer, H., Verbunt, F., Weinberg, M. 1992, PASP, 104, 981
14. in 't Zand, J., Bazzano, A., Cocchi, M., et al. 2000, A&A 355, 145
15. in 't Zand, J., Verbunt, F., Heise, J., Muller, J., Bazzano, A., Cocchi, M., Natalucci, L., Ubertini, P. 1998, A&A, 329, L37
16. in 't Zand, J., Verbunt, F., Strohmayer, T., et al. 1999, A&A, 345, 100
17. Kalogera, V. 1999, ApJ, 521, 723
18. Kalogera, V., Webbink, R. 1996, ApJ, 458, 301
19. Kalogera, V., Webbink, R. 1998, ApJ, 493, 351
20. Mardling, R. 1995, ApJ, 450, 722,732
21. Mason, B., Gies, D., Hartkopf, W., Bagnuolo, W., ten Brummelaar, T., McAlister, H. 1998, AJ, 115, 821
22. Nelemans, G., Tauris, T., van den Heuvel, E. 1999, A&A, 352, L87
23. Orosz, J., Bailyn, C. 1997, ApJ, 477, 876
24. Phinney, E. 1992, Phil. Trans. R. Soc. London A, 341, 39
25. Portegies Zwart, S., McMillan, S. 2000, ApJ, 528, L17
26. Portegies Zwart, S., Verbunt, F. 1996, A&A, 309, 179
27. Portegies Zwart, S., Verbunt, F., Ergma, E. 1997, A&A, 321, 207
28. Regös, E., Tout, C., Wickramasinghe, D. 1998, ApJ, 509, 362
29. van Paradijs, J., White, N. 1995, ApJ (Letters), 447, 33
30. Verbunt, F. 1993, ARA&A, 31, 93
31. Verbunt, F. 1997, MNRAS, 290, L55
32. Wellstein, S., Langer, N. 1999, A&A, 350, 148
33. Woosley, S., Langer, N., Weaver, T. 1993, ApJ, 411, 823
34. Woosley, S., Langer, N., Weaver, T. 1995, ApJ, 448, 315

Empirical Lower Mass Limit for Black-Hole Formation in a Massive Binary

Lex Kaper[1] and Anatol Cherepashchuk[2]

[1] Astronomical Institute, University of Amsterdam,
Kruislaan 403, 1098 SJ Amsterdam, The Netherlands
[2] Sternberg State Astronomical Institute, Moscow State University,
119899, Universitetskij pr., 13, Moscow, Russia

Abstract. Observations of massive binaries provide important constraints on the evolutionary fate of massive stars. Although stellar evolution in a binary system is different from single-star evolution, practically the only way to derive information on the progenitor masses of neutron stars and black holes is by studying massive binaries and their descendants. Wolf-Rayet binaries show a rather continuous mass distribution, while there might be a gap between the neutron-star and black-hole masses. Wray 977 (GX301-2) hosts the most massive OB star with X-ray pulsar companion, and sets an empirical lower mass limit for black-hole formation in a massive binary. HD153919 is the OB supergiant with earliest spectral type, and potentially the most massive OB star in a high-mass X-ray binary, but the nature of the X-ray source (4U1700-37) is not clear. Although it might be a low-mass black hole, new observations suggest it is a neutron star.

1 The Evolution of Massive Binaries

Massive stars end their lives as neutron stars or black holes. The final evolution of massive stars, especially the formation of a compact object during a supernova explosion, is, however, poorly understood. It is commonly believed that the most massive stars form black holes, while massive stars with a mass below a certain limit form neutron stars. Mass loss in the form of a stellar wind strongly affects the evolution of a massive star. It explains the transition of OB-type stars into Wolf-Rayet stars, the "naked" helium cores that are exposed after the mantle is lost due to phases of strong mass loss. The mass-loss history of the star has to be known accurately for the calculation of the mass of the remaining helium core (or the pre-supernova iron core), which sets its ultimate fate. Timmes et al. [12] predict a bimodal distribution of neutron-star masses, due to the difference in pre-supernova structure of stars above and below 19 $M_\odot$. For a given core mass, the neutron-star equation of state (Srinivasan, these proceedings) finally determines whether the compact remnant can remain as a neutron star or has to collapse into a black hole. It has been argued that the mass-loss rate of very massive stars is so high that the remaining helium core will form a neutron star, and not a black hole (cf. Maeder [9], Woosley et al. [16]).

The masses of neutron stars and black holes, determined from radial-velocity studies and/or pulse-timing analyses of binary radio pulsars, high-mass X-ray binaries, and soft X-ray transients, indicate that the neutron-star masses have a very narrow spread around 1.4 $M_\odot$, while the average black-hole mass is about 7 $M_\odot$, i.e. significantly higher than the maximum neutron-star mass of about 3.2 $M_\odot$ (cf. Charles [4]). Is this "gap" in mass due to observational bias, or is the formation mechanism of neutron stars and black holes fundamentally different?

Observations of massive binary systems can provide valuable constraints on the properties of the progenitors of neutron stars and black holes. A drawback is that the evolution of a massive star in a close binary system is different from that of a single star. The initially most massive star, the primary, evolves fastest. As soon as the core-hydrogen burning ends, the core will contract. When hydrogen-shell burning commences, the star expands and becomes a supergiant until the outer layers reach the critical Roche lobe initiating a phase of mass transfer to the less-massive secondary star. If no mass is lost from the system, this scenario is called conservative case B evolution. In case A, the mass transfer already starts when the primary is still undergoing core-hydrogen burning (cf. Van den Heuvel [13]). The orbit shrinks until the mass of the secondary equals that of the primary, and then it widens again until the mass transfer stops. At this point the secondary has become the most massive star of the two, rejuvenated when it has adapted itself to its new mass. The primary has become a helium star, identical or very similar to a Wolf-Rayet star. In case A, the primary is still burning hydrogen after the phase of mass transfer. Its hydrogen-envelope mass has been much reduced and it keeps filling its Roche lobe and slowly transfers mass to the original secondary, which now is the more massive of the two. When it has transferred its entire H-rich envelope, the system may become a Wolf-Rayet binary.

Table 1 lists the masses of Wolf-Rayet plus O-star binaries recently compiled by Cherepashchuk. In most systems the WR star is less massive than its O-star companion, which is expected because of stellar-wind mass loss and mass transfer. On evolutionary grounds, the WN phase (referring to its spectral appearance) should be followed by WC and WO; the average mass of the WN stars in this sample is 20.1 $M_\odot$, compared to 12.6 $M_\odot$ for the WC stars (the average O-star companion mass is 31.5 and 29.4 $M_\odot$, respectively). There is no evidence for a bimodal WR mass distribution, not in the total sample, nor amongst the WC stars. This suggests that not only the mass of the progenitor, but also other factors (such as rotation and magnetic fields) determine the mass of the compact remnant (cf. Ergma & Van den Heuvel [6]).

At some point the primary star will go into supernova; the mass and momentum loss results in a kick velocity of the system (~ 50 km s^{-1}, Nelemans et al. [10], Van den Heuvel et al. [14]). The system remains bound because less than half of its total mass is lost from the system (Boersma [2], Blaauw

Table 1. The masses, spectral types, and orbital parameters of Wolf-Rayet plus O-star binaries, ranked according to mass of the WR star. In two systems the WR star is more massive than the O-star companion. The distribution of the WR-star masses is continuous, there is no evidence for a bimodal distribution.

WR+O system	M(WR) ($M_\odot$)	M(O) ($M_\odot$)	Sp. Type	P_{orb} (days)	e	i (degrees)
HD 92740	55.3	20.6	WN7+06.5-8.5	80.34	0.56	70
HDE 331884	48	57	WN6+05V	6.34	0	70
HD 193928	40	30	WN6+0	21.64	0.02	68
HD 193793	27:	50:	WC7+04V	2900	0.84	44:
γ^2 Vel	21	39	WC8+09I	78.5002	0.40	70
CX Cep	20	28.3	WN5+05V	2.1267	0	74
HD 94305	19	40	WC6+06-8V	18.82	0	70:
HD 90657	17	36	WN4+05	8.255	0	57
CQ Cep	17	20	WN6-7+09II-Ib	1.641246	0	78
HD 186943	16	35	WN4+09V	9.5550	0	56
AB8	14	52	WO4+04V	16.644	0.19	41
CV Ser	14	29	WC8+08-9V-III	29.707	0	67
HD 211853	14:	26:	WN6+O+?	6.6884	0	78
HD 190918	13	29	WN5+09.5I	112.8	0.48	15
B22	12:	35:	WC6+05-6V-III:	14.926	0.17	71:
HD 97152	11	19	WC7+07V	7.886	0	43
V 444 Cyg	9.3	27.9	WN5+O6	4.212424	0	78.7
HD 94546	9	26	WN4+07	4.9	0	74
AB6	8	47	WN3+07	6.681	0	68
HD 5980	8:	27:	WN4+07I	19.266	0.49	89
HD 63099	7	11	WC5+07	14.305	0	64
HD 137603	>5	>27	WC8+BOIa	26.9	0	<20
B32	5	30	WC4+06V-III:	1.91674	0	29
HD 152270	5	14	WC7+O5	8.893	0	45

[1]). When the secondary evolves into a supergiant, the accretion of the dense stellar wind (and later Roche-lobe overflow) onto the compact object results in the production of X-rays and we observe the system as a high-mass X-ray binary (HMXB). After this relatively short phase (~10,000 years) spiral-in of the compact star and the subsequent supernova of the secondary might result in a binary consisting of two compact objects or in the disruption of the system.

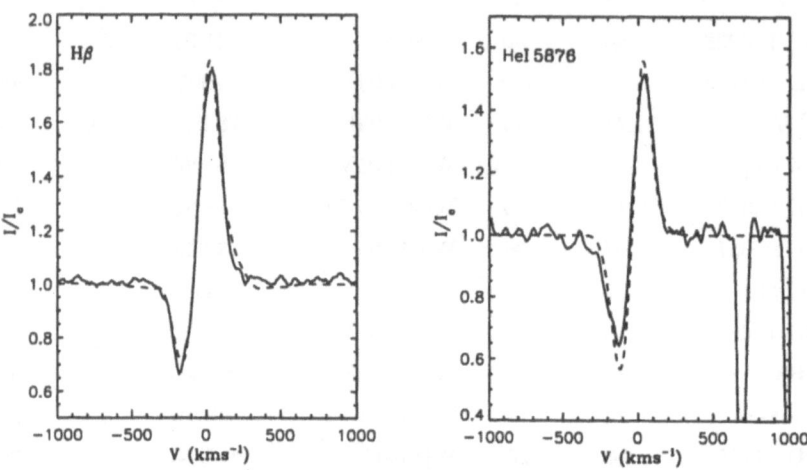

Fig. 1. *Left:* Model fit to the Hβ line of Wray 977, the B hypergiant companion of GX301-2 (from Kaper & Najarro [8]). In normal B supergiants the Hβ line is not in emission, but appears in emission for hypergiants and LBVs. It is well fit by Najarro's atmospheric model for extreme supergiants; *Right:* Model fit to the He I line at 5876 Å.

2   Wray 977: the Most Massive HMXB Hosting an X-Ray Pulsar

Naively one would predict that the X-ray binaries hosting a black hole originate from the most massive binaries. The black hole is the remnant of the initially most massive star, and a phase of mass transfer substantially increased the mass of the secondary so that the system remained bound after the supernova explosion forming the black hole. However, there are only a few black-hole candidates with an OB-star companion (Cyg X-1, LMC X-1, LMC X-3), most of the firm black-hole candidates are found in soft X-ray transients which have low-mass companions. The latter systems must have a different evolutionary origin (see contributions of Charles, Lasota, and Kalogera); here we focus on the evolution of massive binaries.

A disadvantage of black-hole binaries is that only for one component the orbit can be derived, yielding only the so-called mass function, but not the masses of the individual stars. In case of an X-ray pulsar companion the masses of both stars can be measured, given an estimate of the system's inclination. Assuming case B binary evolution (which may not always be a correct assumption, see Wellstein & Langer [15]) one can state that, for an initial mass ratio of 0.5, the progenitor of the neutron star must initially have been more massive than the present mass of the secondary. Under these assumptions, the most massive HMXB hosting an X-ray pulsar sets the empirical lower mass limit for black-hole formation in a massive binary.

Wray 977 (GX301-2) is the most massive HMXB with an X-ray pulsar. Based on a new spectral classification of Wray 977 as a B hypergiant, Kaper et al. ([7]) determined a mass of ~ 40 $M_\odot$ for Wray 977, which has recently been confirmed by atmospheric modelling of the hypergiant spectrum (Fig. 1, Kaper & Najarro [8]). With case B evolution, and assuming an initial mass ratio of about 0.5, the progenitor of the neutron star was initially more massive than 40 $M_\odot$. However, if one drops this condition on the mass ratio, also a less massive progenitor is possible. Wellstein & Langer ([15]) show that this system can also evolve according to case A from a $26 + 25$ $M_\odot$ binary.

3   4U1700-37: a Non-Pulsating Neutron Star

HD153919 is the O6 Iaf+ counterpart of 4U1700-37, a non-pulsating X-ray source. The position of HD153919 in the HRD for single stars indicates a main sequence mass exceeding $50 - 60$ $M_\odot$. Conti ([5]), however, already noted that the OB supergiants in HMXBs might well be undermassive (and overluminous), so that its spectral type is not representative for its (initial) mass. Brown et al. ([3]) argue that 4U1700-37 is the only example of a well-studied HMXB that does not pulse, and could well be a low-mass black hole. Recent BeppoSAX observations of 4U1700-37 show that its X-ray spectrum (Fig. 2) is well characterized by the standard accreting pulsar model (Reynolds et al. [11]. There is some evidence for the presence of a broad cyclotron feature at ~ 37 keV; if real, this would prove that the X-ray source is a neutron star.

Since the X-ray source is not pulsating, only the mass function can be derived, so that this system cannot be used to determine a lower mass limit for black-hole formation. If the X-ray source is a 1.4 $M_\odot$ neutron star, the mass of HD153919 is about 35 $M_\odot$. The short orbital period of 3.41 days makes it unlikely that this system followed a similar case A evolutionary scenario as proposed by Wellstein & Langer ([15]) for Wray 977. Therefore, it is probable that the initial mass of the progenitor of 4U1700-37 was 35 $M_\odot$ or more, which would set the lower mass limit for black-hole formation in a massive binary to 35 $M_\odot$.

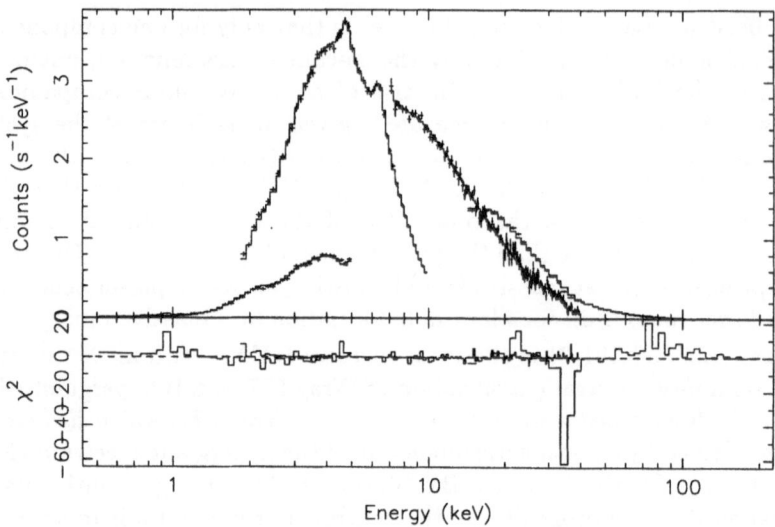

Fig. 2. The different model components to the X-ray spectrum of 4U1700-37 suggests the presence of cyclotron absorption feature at an energy of ~ 37 keV (Reynolds et al. [11]), best visible in the contribution to χ^2 between model (without cyclotron feature) and observed spectrum. The compact object most likely is a neutron star rather than a low-mass black hole.

References

1. Blaauw, A. 1961, Bull. Astron. Inst. Netherlands 15, 265
2. Boersma, J. 1961, Bull. Astron. Inst. Netherlands 15, 291
3. Brown, G.E., Weingartner, J.C., Wijers, R.A.M.J. 1996, ApJ 463, 297
4. Charles, P.A. 1998, in "Theory of black-hole accretion disks", Eds. Abramowica, Bjornsson, Pringle, Cambridge Univ. Press, p. 1
5. Conti, P.S. 1978, A&A 63, 225
6. Ergma, E., Van den Heuvel, E.P.J. 1998, A&A 331, L29
7. Kaper, L. et al. 1995, A&A 300, 446
8. Kaper, L., Najarro, P. 2000, in preparation
9. Maeder, A. 1992, A&A 264, 105
10. Nelemans, G., Tauris, T.M., van den Heuvel, E.P.J. 1999, A&A 352, L87, and these proceedings
11. Reynolds, A.P., Owens, A., Kaper, L., et al. 1999, A&A 349, 873
12. Timmes, F.X., Woosley, S.E., Weaver, T.A. 1996, ApJ 457, 834
13. Van den Heuvel, E.P.J. 1994, in "Interacting Binaries", Saas-Fee Advanced Course 22, Springer-Verlag
14. Van den Heuvel, E.P.J., Portegies Zwart, S.F., Bhattacharya, D., Kaper, L. 2000, A&A in press
15. Wellstein, S., Langer, N. 1999, A&A 350, 148, see also these proceedings
16. Woosley, S.E., Langer, N., Weaver, T.A. 1995, ApJ 448, 315

Constraints on the Initial Mass Limit for Black-Hole Formation from the Massive X-Ray Binary Wray 977

Stephan Wellstein and Norbert Langer

Universität Potsdam, Institut für Physik, Astrophysik, Am Neuen Palais 10,
D-14469 Potsdam

Abstract. The progenitor evolution of the massive X-ray binary Wray 977 is investigated using new models of massive close binary evolution. These models yield constraints on the mass limit for neutron-star/black-hole formation in single stars, M_{BH}. We argue for quasi-conservative evolution in this system, and we find $M_{BH} > 13...21\,M_\odot$ from the existence of a neutron star in Wray 977, the uncertainty being due to uncertainties in the treatment of convection. Our results revise earlier published much larger values of M_{BH} derived from Wray 977. Then, on the basis of a grid of 37 evolutionary models for massive close binaries with various initial masses, mass ratios and periods, we show that in binaries the critical initial mass limit for neutron star/black hole formation is very different from the corresponding value in single stars. This implies that neutron-star and black-hole mass functions obtained for single stars cannot per se be compared to the masses of compact objects in binary systems.

1 Introduction

The evolution of stars is considerably complicated by the presence of a close companion (for a recent review and references see Vanbeveren 1998ab). But close binaries provide unique possibilities to test and to constrain uncertainties inherent in stellar evolution theories in general. For example Ergma & van den Heuvel (1998) showed that massive X-ray binaries can yield constraints for the initial mass limit for black hole formation M_{BH}. In principle, a neutron-star binary contains information on the initial mass of the neutron-star progenitor, and thus provides a lower limit on M_{BH}. The most massive X-ray binary containing a neutron star — Wray 977/GX 301-2 — is an ideal object for such a study, because it yields the largest neutron-star progenitor mass. This system contains an X-ray pulsar (GX 301-2) and the B supergiant Wray 977 (BP Cru). The most resent determination of the stellar parameters of Wray 977 has been performed by Kaper & Najarro (2000). They found log T_{eff} [K] = 4.23...4.30, a radius of R = 60...70 $R_\odot$ and a lower mass limit from the absence of eclipses and from the mass function (cf. Kaper et al. 1995) of 39...40 $M_\odot$. An independent lower mass limit of $\approx 40\,M_\odot$ for Wray 977 has been derived spectroscopically from the velocity amplitude by Kaper &

Najarro (2000). Ergma & van den Heuvel (1998) concluded that the neutron-star progenitor had a larger initial mass than the present mass of Wray 977, which results in a very massive neutron-star progenitor. Here we present binary models from the zero-age main sequence until beyond the death of the primary component which show the possibility of much smaller initial masses of the progenitor of GX 301-2. The computational methods and the complete grid of models can be found in Wellstein & Langer (1999).

2   A Progenitor Model for Wray 977/GX 301-2

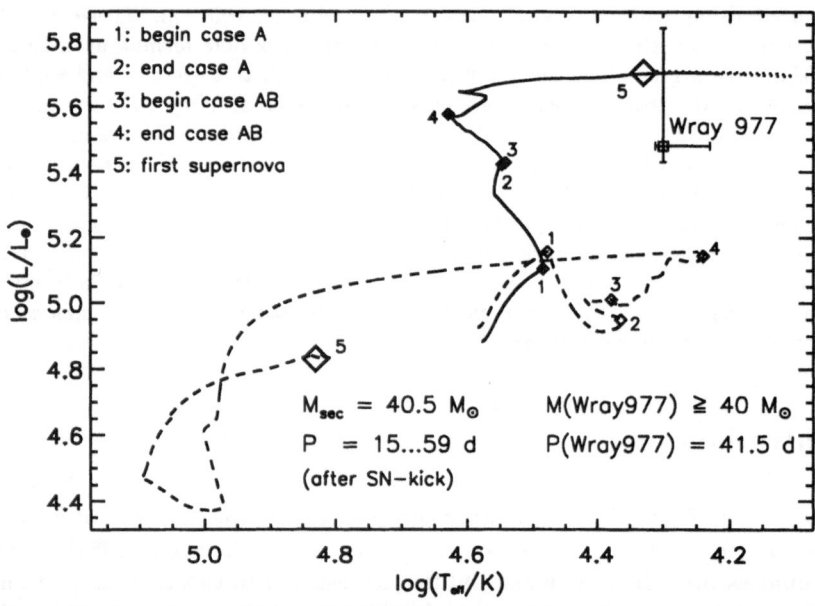

Fig. 1. Evolutionary tracks in the HR diagram of the primary component (initial mass 26 $M_\odot$, dashed line) and of the secondary component (initial mass 25 $M_\odot$, solid line) of our progenitor binary model for the massive X-ray binary Wray 977 (initial period 3.5 days). The diamonds mark the time of the supernova explosion of the primary on both tracks. The solid line of the secondary's track end when it fills its Roche lobe and mass transfer onto the neutron star would commence; the dotted continuation of the solid line shows its continued evolution as if it were in isolation. The square marks the observed position of the B supergiant Wray 977 and its error bars. In the lower part of the figure we show a comparison of mass and period of our model with empirical values (see Wellstein & Langer 1999).

Our binary model for Wray 977 consists of a conservative Case A evolution (mass transfer during central hydrogen burning) starting with a mass ratio near 1 — this system yields the lowest achievable progenitor mass for the neutron star — and a mass close to the present system mass of $\approx 42\,M_\odot$. I.e., our model system initially contains a 26 $M_\odot$ and a 25 $M_\odot$ star with a period of 3.5 days. The parameters of this model system after the supernova explosion of the primary agrees well with observations (cf. Fig. 1). Even the relative large period of Wray 977 is in agreement with a conservative model. We conclude that the progenitor of the neutron star GX 301-2 must have been born with at least 26 $M_\odot$.

3 M_{BH}: From Binaries to Single Stars

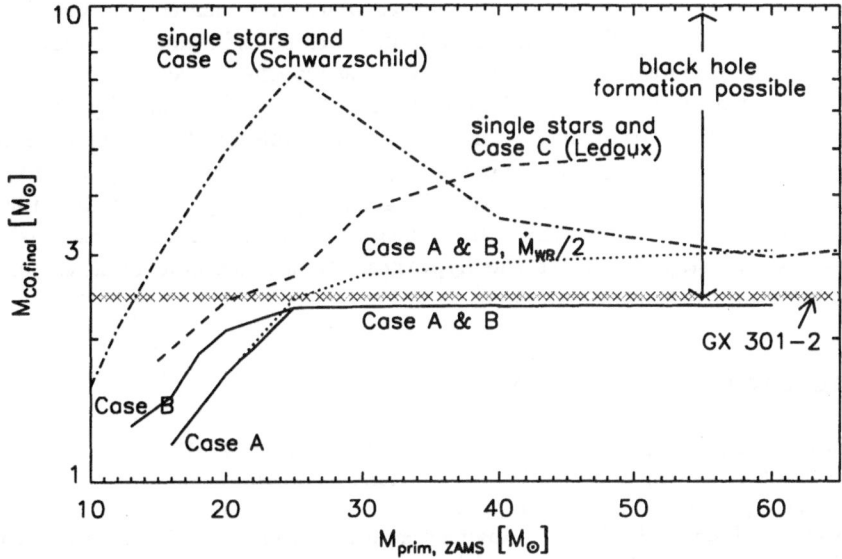

Fig. 2. Final CO-core masses of the primaries of close binaries as function of the initial mass (solid line, from Wellstein and Langer 1999). CO-core masses derived with a Wolf-Rayet-wind mass loss rate dived by a factor 2 are shown by the dotted line. The shaded horizontal line marks the CO-core mass of our progenitor of the neutron star GX 301-2 (cf. sect. 2). The dash-dotted line marks CO-core masses in single stars as derived by Maeder (1992), the dashed line as derived by Langer & Henkel (1995).

Since the mass loss imposed on the primary components of close binary systems is much larger than the mass loss of a single star of comparable initial mass, their final core masses can be significantly smaller, as shown in Fig. 2 for the final CO-core masses. Assuming that the final iron core mass depends monotonously on the final CO-core mass, this implies that, for example, a

26 $M_\odot$ star in a Case A binary may form a neutron star (cf. Sect. 2), while perhaps a 26 $M_\odot$ single star may form a black hole. In binaries, the initial mass limit for NS/BH formation is larger than 26 $M_\odot$ since the neutron star in Wray 977/GX 301-2 had an initial mass of at least 26 $M_\odot$. Therefore, we can derive a lower initial mass limit for black-hole formation in single stars of 13...21 $M_\odot$ from Fig. 2, the uncertainty being due to the uncertainty in the treatment of convection.

The fact that even the most massive primaries computed with the standard WR-wind mass-loss rates form no larger CO-cores than the progenitor of Wray 977 leads to the question of whether Case A/B binaries can form black holes at all. Even assuming a WR-wind mass loss rate multiplicated by a factor 1/2 the CO-cores become only a little bit larger (cf. dotted line in Fig. 2), and they do not provide the required mass (their maximum mass is approximately 4 $M_\odot$) to form the observed black-hole masses (e.g. Cyg X-1, LMC X-1, LMC X-3). We conclude that the black holes in massive X-ray binaries are likely to have formed via Case C mass transfer, as recently proposed by Brown et al. (1999). As can be seen in Fig. 2, only the Case C primaries, which are supposed to form CO-cores with approximately the same masses as single stars because they lose their hydrogen envelope very late, appear to be capable to provide enough mass. Also the generally short periods of all known black-hole X-ray binaries, which are a clear indication of a non-conservative mass transfer stage, support the hypothesis that black-hole binaries form predominantly in Case C systems.

References

1. Brown G.E., Lee C.-H., Bethe H.A., 1999, New Astron. 4, 313
2. Ergma E., van den Heuvel E.P.J., 1998, A&A 331, L29
3. Kaper L., Najarro F., 2000, preprint
4. Kaper L., Lamers H.J.G.L.M., Ruymaekers E., van den Heuvel E.P.J., Zuidervijk E.J. 1995, A&A 300, 446
5. Langer N., Henkel C., 1995, Space Sci. Rev. 74, 343
6. Maeder 1992, A&A 264, 105
7. Vanbeveren D., Van Rensbergen W., De Loore C., 1998a, *The Brightest Binaries*, Kluwer Academic Publishers, Dordrecht
8. Vanbeveren D., De Loore C., Van Rensbergen W., 1998b, A&A Rev. 9, 63
9. Wellstein S., Langer N., 1999, A&A 350, 148

Formation of Black-Hole X-Ray Binaries with Low-Mass Donors

Vassiliki Kalogera

Harvard-Smithsonian Center for Astrophysics, Cambridge MA 02138, USA

Abstract. The characteristics of black-hole X-ray binaries can be used to obtain information about their evolutionary history and the process of black-hole formation. In this paper I focus on systems with donor masses lower than the inferred black-hole masses. Current models for the evolution of hydrogen-rich, massive stars and of helium stars losing mass in a wind cannot explain the current sample of black-hole mass measurements. Assuming that the radial evolution of mass-losing massive stars is at least qualitatively accurate, I show that the properties of the BH companions lead to constraints on the masses of black-hole progenitors (at most twice the black-hole mass) and on the strength of winds in helium stars (fractional amount of mass lost smaller than about 50%). Constraints on common-envelope evolution are also derived.

1 Introduction

Radial velocity measurements of the non-degenerate donors in X-ray binaries combined with information about donor spectra and optical light curves allow us to measure the masses of accreting compact objects [1]. At present, measured masses for nine X-ray transients exceed the optimum maximum neutron-star mass [2] and the binaries are thought to harbor black holes (BH). Studies of the properties of such systems can shed light on their evolutionary history and the process of BH formation.

Black-hole X-ray transients correspond to low-mass X-ray binaries with neutron stars (NS) in that mass transfer is driven by Roche-lobe overflow and that the donors are less massive than the BH. This critical (maximum) mass ratio of about unity allows the donor to transfer mass stably to the compact object. However, for the majority of the BH X-ray transients (including six BH candidate systems based on their spectral properties [3]) the donors are less massive than $\sim 1\,M_\odot$, much less massive than the typical BH masses. Only two systems, J1655-40 and 4U1543-47, have donors of $\sim 2.3\,M_\odot$ and $\sim 2-3\,M_\odot$, respectively. This apparent paucity of intermediate-mass donors (more massive than $\sim 2\,M_\odot$) cannot be explained by mass-transfer stability considerations alone.

In what follows, we address the issue of donor masses in BH binaries by studying a larger set of constraints imposed on their progenitors. We find that there is a strong dichotomy between the formation of systems with low- and intermediate-mass donors. Formation of both at the appropriate relative fraction requires little mass loss at BH formation, weak helium-star winds, and possibly energy sources other than the orbit for common-envelope (CE) ejection. This study and the results obtained are described in more detail in [4].

2 Evolutionary Constraints

We consider BH X-ray binary formation from primordial binaries with extreme mass ratios evolving through a CE phase, similar to the formation channel for NS low-mass X-ray binaries, e.g., [5]. The primary must be massive enough so that its helium core exposed at the end of the CE phase collapses into a BH. The X-ray phase is initiated when the donor fills its Roche lobe because of orbital shrinkage through magnetic braking (for low-mass donors) or of radial expansion through nuclear evolution on the main sequence (for intermediate-mass donors).

Black-hole binary progenitors evolve through this path provided that the following constraints are satisfied:

- The orbit is small enough that the primary fills its Roche lobe and the binary enters a CE phase.
- At the end of the CE phase the orbit is wide enough so that both the helium-rich primary and its companion fit within their Roche lobes. The constraint for the companion turns out to be stricter.
- The system remains bound after the collapse of the helium star. In the absence of kicks imparted to the BH, this sets an upper limit on the mass of the BH progenitor.
- After the collapse, the orbit must be small enough so that mass transfer from the donor starts before it leaves the main sequence and within 10^{10} yr.
- Mass transfer from the donor proceeds stably and at sub-Eddington rates. This sets an upper limit to the donor mass on the zero-age main sequence and to the orbital size for more evolved donors.

3 Donor Masses in Black-Hole X-Ray Binaries

For a specific value of the BH mass, the above constraints translate into limits on the properties, circularized post-collapsed orbital sizes (A) and donor masses (M_d), of BH binaries with Roche-lobe filling donors. The relative positions of these limits on the $A - M_d$ plane and the resulting allowed M_d ranges are exactly determined by three well-constrained model parameters:

- The amount of mass loss from the binary during BH formation, characterized by the ratio $M_{He,f}/M_{BH}$, where $M_{He,f}$ is the mass of the helium-rich BH progenitor at the time of the collapse. For the post-collapse system to remain bound it must be $1 \leq M_{He,f}/M_{BH} \leq 3$.

- The amount of mass lost in the helium-star wind between the end of the CE phase and the BH formation, characterized by the ratio $M_{He,f}/M_{He}$, where M_{He} is the initial helium-star mass (at the end of the CE phase). This ratio must lie in the range $0-1$.
- The CE efficiency, α_{CE}, defined as the ratio of the CE binding energy to the orbital energy released during the spiral-in of the companion. Although the absolute normalization of α_{CE} is not well determined [4], values higher than unity imply the existence of energy sources other than the orbit (ionization or nuclear burning energy).

Note that the last two constraints depend *only* on the BH mass, while α_{CE} affects only the upper limit on A (the first constraint in § 2). For different values of these three parameters, the positions of the limits on the $A - M_d$ plane change and three different outcomes with respect to the donor masses are possible: BH binaries can be formed with (i) *only* low-mass; (ii) *only* intermediate-mass; (iii) *both* low- and intermediate-mass donors.

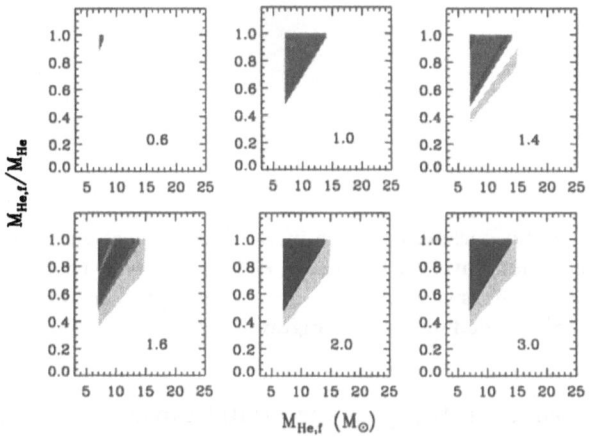

Fig. 1. Limits on the parameter space of the final (pre-collapse) helium-star mass, $M_{He,f}$, and the ratio, $M_{He,f}/M_{He}$, for six values of α_{CE} =0.6, 1.0, 1.4, 1.6, 2.0, 3.0, and for a $7\,M_\odot$ BH. Conditions in the unshaded areas do not allow the formation of BH binaries with main-sequence Roche-lobe filling donors; conditions in the light-gray, dark-gray, and black areas allow the formation of systems with only low-mass, only intermediate-mass, and both types of donors, respectively.

The donor types as a function of the three parameters, $M_{He,f}$, $M_{He,f}/M_{He}$, and α_{CE}, are shown in Fig. 1, for a $7\,M_\odot$ BH. For α_{CE} smaller than ~ 0.5, the orbital contraction is so large that the donor stars cannot fit in the post-CE orbits, and hence no BH X-ray binaries are formed. As α_{CE} increases, CE ejection without the need of strong orbital contraction becomes possible for the more massive of the donors, while formation of binaries with low-mass donors occurs only if $\alpha_{CE} > 1.5$. The results become independent of α_{CE} for values in excess of ~ 2, when the upper limit for CE evolution (first of the constraints) lies at high enough values of A that it never interferes with the other limits.

The dependence of these results on the two mass-loss parameters (wind and collapse) are determined by their association with orbital expansion. For strong helium-star wind mass loss ($M_{He,f}/M_{He} < 0.35$), the progenitor orbits expand so much that donors less massive than the BH can never fill their Roche lobes on the main sequence. Both low- and intermediate-mass donors are formed only if less than 50% of the initial helium-star mass is lost in the wind. Mass loss at BH formation is limited to BH progenitors less massive than about twice the BH mass so that post-collapse systems with low-mass donors remain bound.

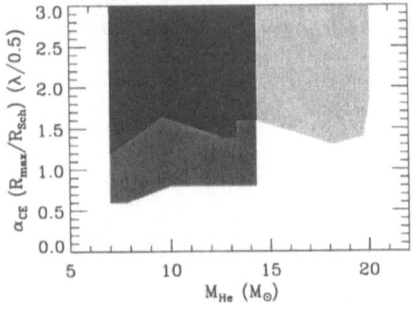

Fig. 2. Limits on the parameter space of the initial (post-CE) helium-star mass, M_{He}, and the common-envelope efficiency, α_{CE}, properly normalized (by the maximum stellar radii of massive stars [6] and the central-concentration parameter, λ), for a $7\,M_\odot$ BH. Shade coding is as in Figure 1.

The dependence on $M_{He,f}$ of the orbital expansion during helium-star wind mass loss and BH formation is such that the ratio of circularized post-collapse over post-CE orbital separations becomes independent of $M_{He,f}$. This means that, for a specific BH mass, the position of the limits on the $A - M_d$ plane depend only on the initial helium-star mass and the CE efficiency. Indeed, in Fig. 1, the change of donor types occurs along straight lines in the $M_{He,f}/M_{He}-M_{He,f}$ plane, or else along lines of constant M_{He}. This simplifying property allows us to combine the panels in Fig. 1 into one plot (Fig. 2). It is evident that formation of $7\,M_\odot$ BH X-ray binaries with both low- and intermediate-mass donors (as required by the observed sample) constrains the common-envelope efficiency to relatively high values and the initial helium-star progenitors at most twice as massive as the BH (corresponding to initial primaries in the range 25-45 $M_\odot$).

Additional constraints can be obtained by examining the relative numbers of systems with low- and intermediate-mass donors formed for the parameters in the black-shaded areas in Figs. 1 and 2. The lifetimes for the two different types are determined by the process that drives mass transfer. The magnetic-braking timescale, for low-mass donors is comparable to the nuclear evolution timescale of intermediate-mass stars [4]. The number ratio then becomes equal to the ratio of birth rates. The latter can be calculated using the derived limits on A and M_d and assumed distributions of mass ratios and orbital separations of primordial binaries. The results indicate that even when low-mass companions in primordial binaries are strongly favored, BH binaries with intermediate-mass donors are much more easily formed because of the larger range of orbital separations allowed to their progenitors. Models predict a paucity of intermediate-mass donors only for rather high α_{CE} values (> 3) or for moderate α_{CE} values ($1.5 - 2$) and BH progenitors either slightly more massive or twice as massive as the BH.

4 Discussion

We find that the models for BH formation are consistent with the properties of the observed sample if (i) wind mass-loss from helium stars is limited so that they lose less than half of their initial mass (ii) helium stars that form black holes are at most twice more massive than the black holes, and (iii) CE efficiencies are relatively high and, depending on the exact radii of massive stars and their density profiles, significant contributions from energy sources other than the orbit may be required. Note that amounts of mass lost in helium-star winds and in BH formation are actually anti-correlated. If one is close to the maximum allowed then the other must be minimal (see Fig. 1). These results are quite robust and do not depend on the assumed BH mass nor on the properties of primordial binaries.

The present study allows us to place constraints on the extent of wind mass loss from helium stars as they evolve towards collapse, primarily because helium cores are exposed at the end of the CE phase *prior* to core helium exhaustion. Current models of helium-star evolution through core helium burning [7] predict amounts of mass lost in the wind significantly larger than the maxima allowed for BH X-ray transient formation ($< 50\%$). In fact, the final helium-star masses in these models are $\sim 4 \, M_\odot$, far too small to explain the BH mass measurements. Therefore, if the CE phase is initiated early in the core helium burning phase of the primary, then helium-star winds must be much weaker than thought until now. It is noteworthy that more recent empirical estimates of wind mass loss rates [8] show a downward trend.

The strength of helium-star winds becomes irrelevant to the process of BH X-ray binary formation, if the CE phase is initiated late in the evolution of the massive primary, i.e., after core helium exhaustion. In this case the helium star is exposed only through carbon burning and later evolutionary phases. The total duration of these phases is so short that the wind mass loss is insignificant and the helium-star mass remains essentially constant [7].

Current models of massive star evolution [6], though, permit CE evolution at such late stages only for primary masses lower than $\sim 25\,M_\odot$ and for an extremely narrow range of orbital separations [9]. For more massive stars, there is not enough radial expansion (in fact the radius decreases) to counterbalance the orbital expansion due to wind mass loss from the hydrogen-rich primary during core helium burning, and bring the primary to Roche-loe overflow. Therefore, relying on CE evolution occurring only at late stages cannot account for $\sim 10\,M_\odot$ BH and it is possible only for a tiny fraction of primordial binaries leading to uncomfortably low birth rates for BH X-ray binaries [10]. All these problems can be circumvented only if the radial evolution predicted by the current models of mass-losing massive star evolution is incorrect both qualitatively and quantitatively and stars more massive than $\sim 25\,M_\odot$ expand significantly after core helium exhaustion. Such a modification of the massive-star models has been assumed by Wellstein & Langer [11] and no reduction of helium-star wind mass loss rates is then required. Given the uncertainties in models of massive star evolution such a modification cannot be excluded at present. In any case, it becomes clear that the existence of X-ray transients with $\sim 10\,M_\odot$ BH requires that either the hydrogen-rich massive star models or the strength of helium-star winds be modified.

Acknowledgments. I would like to thank N. Langer and S. Wellstein for useful discussions. Support by the Smithsonian Astrophysical Observatory through a Harvard-Smithsonian Center for Astrophysics Postdoctoral Fellowship is also acknowledged.

References

1. Charles, P.A. 1998, in Theory of Black Hole Accretion Disks, eds. M.A. Abramowicz, G. Bjornsson & J.E. Pringle, (Cambridge: Cambridge University Press), 1
2. Kalogera, V. & Baym, G. 1996, ApJ, 470, L61
3. Chen, W., Shrader, C.R., & Livio, M. 1997, ApJ, 491, 312
4. Kalogera, V. 1999, ApJ, 521, 723
5. van den Heuvel, E.P.J. 1983, in Accretion-Driven Stellar X-Ray Sources, ed. W. H. G. Lewin, & E. P. J. van den Heuvel (Cambridge: Cambridge University Press), 303
6. Schaller, G., Schaerer, D., Meynet, G., & Maeder, A. 1992, A&A Suppl. Series, 269, 331
7. Woosley, S.E., Langer, N., & Weaver, T.A. 1995, ApJ, 448, 315
8. Hamann, W.-R. & Koesterke, L. 1998, A&A, 335, 1003
9. Kalogera, V. & Webbink, R.F. 1998, ApJ, 493, 351
10. Portegies–Zwart, S.P., Verbunt, F., & Ergma, E. 1997, A&A, 321, 207
11. Wellstein, S. & Langer, N. 1999, A&A, 350, 148

The History of Cygnus X–2

Ulrich Kolb[1], Melvyn Davies[2], Andrew King[2], and Hans Ritter[3]

[1] Dept. of Physics & Astronomy, The Open University, Walton Hall,
  Milton Keynes MK7 6AA, UK
[2] Dept. of Physics & Astronomy, University of Leicester, Leicester, LE1 7RH, UK
[3] Max–Planck–Institut für Astrophysik, D–85740 Garching, Germany

Abstract. We present a detailed evolutionary sequence for Cygnus X–2, and calculate possible dynamical orbits of the system in a realistic Galactic potential.

1   The Evolutionary State

Cygnus X–2 is a long–period neutron–star X–ray binary with a low–mass donor star. As the donor's spectral type is too early for a Hayashi line star (Casares et al. 1998), King & Ritter (2000; KR) and Podsiadlowski & Rappaport (2000; PR) suggested that Cygnus X–2 could be a descendent of an intermediate–mass X–ray binary (IMXB). Such systems undergo a rapid mass transfer phase, previously regarded as fatal, with transfer rates exceeding the Eddington value by several orders of magnitude.

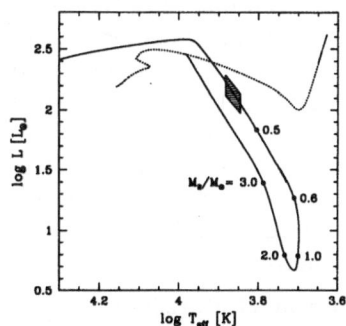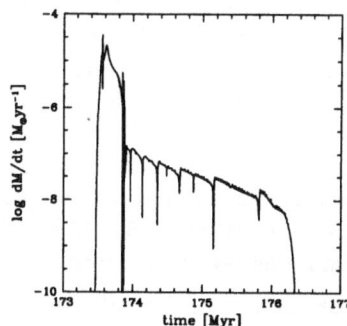

Fig. 1. *Left:* HR diagram showing the evolutionary track of a Cygnus X–2 like IMXB system (solid), and a track of a $3.5 M_\odot$ single star (dotted), calculated without core overshooting. Labels along the track indicate the donor mass. The shaded error box marks the observed location of Cygnus X–2. *Right:* Mass transfer rate $\dot{M}$ vs time (elapsed since the beginning of core hydrogen burning).

To check this we calculated detailed binary evolution sequences, using Mazzitelli's stellar code (Mazzitelli 1989). We find that an early case B mass transfer starting with a $3.5 M_\odot$ donor star is the most likely evolutionary solution for Cygnus X–2. This makes the currently observed state rather short–lived, of order 3 Myr (see Fig. 1). We note that in PR's preferred solution for Cygnus X–2 the donor is still on the main sequence when mass transfer begins. This is possible only if convective core overshooting is very large — larger than what is usually assumed. IMXBs like Cygnus X–2 are progenitor systems of millisecond pulsar binaries with high–mass white dwarfs but short to moderate orbital periods of 10-100 days (e.g. Taam et al. 2000, Tauris et al. 2000).

2   The Dynamical State

Cygnus X–2 is in an unusual dynamical state. The system has a measured line-of-sight velocity of $v = -208.6$ km/s (Casares et al. 1998) and is 2.28 kpc above the Galactic plane, at a distance of 11.6 kpc from the sun.

We integrated the equations of motion in a realistic Galactic potential to investigate possible trajectories of Cygnus X–2. We found that there are viable trajectories, compatible with the observed line-of-sight velocity, that cut the Galactic plane within the solar circle 170 - 250 Myr ago, consistent with the age of the system as given by the evolutionary calculations. If the system had previously been in a circular orbit in the Galactic plane, the required kick velocities imparted to the neutron star at birth are only moderate, ≤ 200 km/s.

By Monte Carlo simulation we produced a large number of systems at various initial Galactocentric radii, assumed to be initially on circular orbits, with kick velocities drawn from the standard distributions (e.g. Hansen & Phinney 1997), and followed their trajectories in the Galactic potential. Assuming that the formation rate of binaries scales as the surface density of stars we computed the relative number of Cygnus X–2–like systems which would be produced by integrating over the entire disc within the solar circle. We found that $\sim 7\%$ of all binaries will produce Cygnus X–2–like binaries on trajectories that would place them at a radius similar to Cygnus X–2 today. A further $\sim 5 - 15\%$ will be further out than Cygnus X–2, whilst the remainder will be located closer to the Galactic centre.

More details of our calculations are presented in Kolb et al. (2000).

References

1. Casares J., Charles P.A., Kuulkers E. 1998, ApJ, 493, 39
2. Hansen B.M.S., Phinney E.S., 1997, MNRAS, 291, 569
3. King A.R., Ritter H. 1999, MNRAS, 309, 253
4. Kolb U., Davies M., King A.R., Ritter H. 2000, MNRAS, submitted
5. Podsiadlowski P., Rappaport S.A. 2000, ApJ, in press (astro-ph/9906045)
6. Taam R.E., King A.R., Ritter H. 2000, ApJ, submitted
7. Tauris T.M., van den Heuvel E.P.J., Savonije G.J. 2000, ApJ, in press (astro-ph/0001013)

Evidence of a Supernova Origin for the Black Hole in GRO J1655-40 (Nova Scorpii 1994)

Garik Israelian

Instituto de Astrofísica de Canarias, E-38200 La Laguna, Tenerife, Spain

Abstract. We report the discovery of a strong overabundance of several α-elements in the atmosphere of the star orbiting around the massive black hole in the binary system GRO J1655-40 (Nova Scorpii 1994). A high-resolution spectrum of this star obtained with the Keck I 10m telescope reveals the presence of extra strong absorption lines of O, Mg, Si, and S. Our analysis using spectral synthesis techniques and a suitable model atmosphere indicates that these elements are 6 to 10 times more abundant in the star than in the Sun. The overall spectrum of the companion resembles that of a normal subgiant star with a solar iron abundance. We interpret the enhanced abundances of α-elements as a direct result of the nucleosynthesis in a supernova/hypernova explosion event associated with the progenitor of the black hole. This is to our knowledge up to now the most direct evidence for a link between a supernova explosion and the formation of a black hole.

1   Introduction

Massive stars ($M \geq 10\,M_\odot$) are expected to end their lives either with a supernova explosion leading to the formation of a compact remnant (neutron star or black hole) or alternatively with a direct collapse into a black hole without an explosion (Maeder 1992, Brown et al. 1999). The existence of pulsars in supernova remnants can be considered as evidence for a link between a supernova explosion and the formation of a neutron star. No such compelling evidence is available for the formation of black holes as a result of the explosive deaths of very massive stars. In fact, there are claims that such stars may end their lives without ejecting matter in a supernova explosion in order to avoid overproduction of oxygen and other α-elements in the Galaxy (Wheeler et al. 1989). However, this point cannot be considered as an argument against explosive deaths of very massive stars since it has been demonstrated recently that the [O/Fe] ratio increases for [Fe/H]$\leq$ 1.5 and there is no *plateau* at [O/Fe]=0.5 (Israelian et al. 1998). This result has been confirmed by Boesgaard et al. (1999) on the basis of Keck high resolution and high-S/N spectra. The continuous increase of [O/Fe] for [Fe/H]$\leq$ -2.5 can be due to, for example, the existence of very massive (up to $500 M_\odot$) population III objects in a pregalactic epoch (Woosley & Weaver 1982). These superstars were probably responsible for the observed large primordial C and N abundances in several extremely metal-poor dwarfs (Norris et al. 1997). These observations strongly suggest that stars with masses larger than 35–40 $M_\odot$ do explode as supernovae.

The α-elements are expected to be significantly overproduced with respect to iron group elements in the thermonuclear reactions that take place during the Type II supernova phase (Arnett 1996, Woosley and Weaver 1995, Thielemann et al. 1996). Low-mass X-ray binaries (LMXBs) are ideal astrophysical sites to investigate the link between the explosive end of massive stars and the formation of black holes and neutron stars. The chemical composition of the atmosphere of the secondary star can provide a unique opportunity to investigate the nucleosynthesis of elements in any supernova event associated with the precursor of the compact object. The matter ejected during the explosion will pollute the environment and may contaminate the atmosphere of the low mass companion. A detailed abundance analysis could then show evidence for any nucleosynthesis products from the supernova, which in turn can yield an important constraint on the mass of the progenitor. Here we present results on the chemical composition of the secondary of GRO J1655-40, an eclipsing LMXB containing a black hole with a mass estimated in the range 4.1–7.9$M_\odot$ (Orosz and Bailyn 1997, van der Hooft et al. 1999, Shahbaz et al. 1999). The mass of the optical companion lies in the range 1.6–3.1 $M_\odot$ and agrees well with the mass obtained from a comparison of the luminosity and effective temperature of the star with evolutionary models (Orosz and Bailyn 1997, van der Hooft et al. 1999).

2   Observations

We obtained a high resolution spectrum of the secondary star in GRO J1655-40, in a quiescence phase, on May 24 1998 with the Keck I 10 m telescope and the High Resolution Echelle Spectrograph (Vogt 1994). The used configuration gives a resolving power of R=31,000 (0.2 Å at 670.0 nm) and a wavelength range of 6430 to 8750 Å. The S/N per pixel (0.1 Å) in the continuum was 35. The spectral type of the secondary is constrained to the range F3-F8IV/III (Orosz and Bailyn 1997, Shahbaz et al. 1999) and the rotational velocity found from the Keck spectrum is 93±3 km s^{-1}, in good agreement with previous studies.

3   Analysis

We shall not discuss the details of the analysis since this information can be found in the paper of Israelian et al. (1999).

Fig. 1 shows the comparison between the synthetic and observed line profiles for different oxygen abundances ([O/H]=log (O/H)$_*$–log (O/H)$_\odot$). In the upper panel we present our results for the O I triplet at 7771-5 Å and in the bottom panel the O I triplet at 8446 Å. It follows from the upper panel in the figure that we can only reproduce the observed profile if we adopt

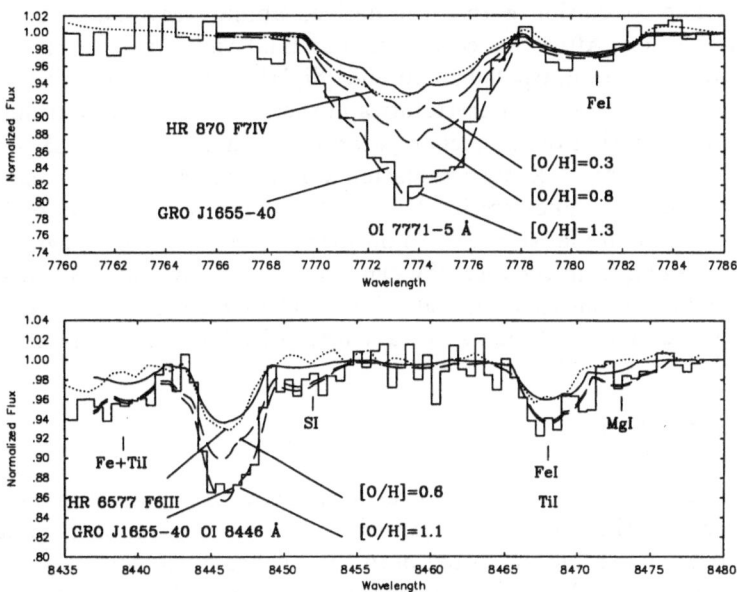

Fig. 1. Spectrum of the secondary star in GRO J1655-40 (histogram) and synthetic spectra computed in non-LTE for a range of oxygen abundances. High-resolution spectra of two stars whose spectral types provide the best classification of the secondary star (Shahbaz et al. 1999) are shown (solid lines) for comparison (HR 870 F7 IV, upper panel) obtained with the Hamilton echelle spectrograph at the 3 m Lick Telescope; the spectrum of HR 6577 F6 III (lower panel) was obtained from the literature (Andrillat et al. 1995).

an oxygen abundance a factor 20 higher than solar while a solar oxygen abundance reproduces well the template spectrum as expected. In the same figures we can see how the Fe I 7780 Å line has a a remarkably similar strength to that in the template spectrum one and can be fit using a solar abundance. Iron lines do not show any significant difference in our spectrum as compared to the templates. From them we infer that the iron abundance is solar and, more generally, that the atmospheric structure of the star is very similar to that of normal F7/6 subgiant/giant stars. The comparison of the equivalent widths measured in GRO J1655-40 with high-quality spectra of many other F-type standards (Andrillat et al. 1995) provides additional confirmation that α-elements have stronger absorption lines in this otherwise normal F-type star (Israelian et al. 1999).

A large overabundance of α-elements may have an impact on the atmospheric structure and consequently α-element enhanced model atmospheres

may be more suitable for the chemical analysis. We generated such models
for fixed $[\alpha/H]=1.0$ (the α's are O, Mg, Si, S, Ti and Ca) and repeated our
analysis. The conclusion is that the derived abundances had to be reduced
by $\sim$0.1–0.2 dex. After this correction our final values are $[O/H]=1.0\pm0.3$,
$[S/H]=0.75\pm0.2$, $[Mg/H]=0.9\pm0.40$ and $[Si/H]=0.9\pm0.3$. The given 1-σ errors take into account the uncertainties in the location of the continuum and
in the determination of the stellar parameters ($\Delta T_{eff}=250$ K, $\Delta \log g=0.3$
dex). No signature of X-ray irradiation was found in our quiescence spectrum: the hydrogen lines are in absorption, Fe and some Fe-group elements
like Ni, Cr show similar strengths in normal F5–F8 stars and no evidence was
found for veiling from the disc. We can think of no physical mechanism that
selectively acts on the spectral lines of the α-elements while the Fe-group, the
hydrogen lines and the broad spectral energy distribution remain unaffected.

4   Discussion

Our findings of enhanced abundances of α-elements in the secondary of the
system demand that matter synthesized in the inner regions of the progenitor
was ejected, mixed and deposited onto the surface of the low mass companion. This rules out the direct collapse, without explosion, of a massive progenitor as the explanation for the formation of the massive black hole in the
GRO J1655-40 system. Current scenarios for stellar collapse and supernova
explosion (Woosley and Weaver 1995) can lead to the formation of massive
black holes through an intermediate stage of short duration where a neutron
star is initially formed, and later, on a timescale of a few seconds to several
hours, accretes matter that due to hydrodynamical interactions in the ejecta
is pushed back on to it and forms the black hole. When applied to our system
this scenario also helps to explain the measured kick velocity (Brandt et al.
1995) of 114 ± 19 km s^{-1}, which is rather high compared to other LMXBs
containing black holes.

It is well known that in a spherically symmetrical supernova explosion the
remnant binary will not be disrupted if the mass of the secondary is larger
than the difference between the mass of the helium core of the progenitor
and twice the mass of the black hole (Brandt et al. 1995). From this simple
consideration we infer that the maximum mass of the helium core which would
allow preservation of the binary is in the range 10-16 $M_\odot$ in our case. Current
models for the evolution of a progenitor of 25-40 $M_\odot$ predict the formation
of helium cores below the requested maximum mass for preservation of the
binary (Nomoto et al. 1997).

Models of explosive nucleosynthesis for a 40 $M_\odot$ case show that the layers
above those going into a 4 $M_\odot$ black hole will be ejected, containing O, S, Si
and Mg in proportions similar to those seen in the companion star. However,
if the mass cut was located at more than 5 $M_\odot$ then standard SN progenitor
yields cannot explain the observed abundances. If such a massive progen-

itor had a large angular momentum and a magnetic field possibly associated with the spiralling-in of the companion star, the collapse of the massive Fe core can lead to the formation of even more massive black holes and induce a hypernova explosion (Paczyński 1998) with nucleosynthesis yields from a $12-15\,M_{CO}$ core progenitor (Nomoto et al. 1999) consistent with our observations. This suggests that GRO J1655-40 is the first Galactic hypernova identified less than one million years after its tremendous explosion.

5   Acknowledgements

This article summarizes the work carried out in collaboration with R. Rebolo, G. Basri, J. Casares and E. Martín.

References

1. Andrillat, Y., Jaschek, C. & Jaschek, M. (1995) A&AS **112**, 475
2. Arnett, W. D. (1996), *Nucleosynthesis And Supernovae*, Princeton Univ. Press
3. Boesgaard, A. M., Kimg, J. R., Deliyannis, C. P., & Vogt, S. S. (1999), AJ **117**, 492
4. Brandt, N., Podsiadlowski, Ph. & Sigursson, S. (1995), MNRAS **277**, L35
5. Brown, G. E., Lee, C-H. & Bethe, H. A. (1999), New Astronomy, in press
6. van der Hooft, F., Heemskerk, M. H. M., Alberts, F. & van Paradijs, J. (1998), A&A **329**, 538
7. Israelian, G., Rebolo, R., Basri, G., Casares, J. and Martin E. (1999), Nature **401**, 142
8. Israelian, G., García López, R., & Rebolo, R. (1998), ApJ **507**, 805
9. Maeder, A. (1992), A&A **264**, 105
10. Nomoto, K., Nakamura, T., Iwamoto, K., Umeda, H. & Mazzali, A. (1999), In *Nuclei In The Cosmos V*, Eds. Prantzos, N. & Harissopulus, S., Editions Frontieres, p. 252
11. Nomoto, K., *et al.*, (1997), Nuclear Physics **A616**, 79
12. Norris, J., Ryan, S., Beers, T. (1997), ApJ **488**, 350
13. Orosz, J. A. & Bailyn, C. D. (1997), ApJ **477**, 876
14. Paczyński, B. (1998), ApJ **494**, L45
15. Shahbaz, T., van der Hooft, F., Casares, J., Charles, P. A. & van Paradijs, J. (1999), MNRAS **306**, 89
16. Thielemann, F-K., Nomoto, K. & Hashimoto, M-A. (1996), ApJ **460**, 408
17. Vogt, S., *et al.* (1994), in *Proc. S.P.I.E.*, **2198**, 362
18. Wheeler, J., Sneden, C., and Lambert, D. (1989) ARA&A **27**, 279
19. Woosley, S. E. & Weaver, T. A. (1995), ApJS **101**, 181
20. Woosley, S. E. & Weaver, T. A. (1982), in *Supernovae*, ed.Rees, M. and Stoneham, R., Dordrecht, Reidel. p. 79

Constraints on Mass Ejection in Black Hole Formation Derived from Black Hole X-Ray Binaries

G. Nelemans, T.M. Tauris, and E.P.J. van den Heuvel

Astronomical Institute, "Anton Pannekoek", University of Amsterdam, Kruislaan 403, NL-1098 SJ Amsterdam, The Netherlands

Abstract. Both the recently observed high runaway velocities of Cyg X-1 ($\sim$ 50 km s^{-1}) and X-ray Nova Sco 1994 ($\geq$ 100 km s^{-1}) and the relatively low radial velocities of the black hole X-ray binaries with low mass donor stars, can be explained by symmetric mass ejection in the supernovae (SNe) which formed the black holes in these systems. We find that at least 2.6 $M_\odot$ and 4.1 $M_\odot$ must have been ejected in the formation of Cyg X-1 and Nova Sco, respectively. A possible kick at the formation of the black hole is not needed to explain their space velocities.

1 Introduction

There is increasing evidence that in events in which black holes are formed, mass is ejected[1][2]. Also, White & van Paradijs[3] conclude from the velocity dispersion of black hole X-ray binaries with low mass companions that an extra velocity of 30 – 40 km s^{-1} is given to these systems in the formation of the black hole. This requires substantial mass ejection in the formation of a black hole if no asymmetric kicks are involved.

Recent determinations of the space velocity of Cyg X-1[4] and the radial velocity of Nova Sco[5] demonstrate that these black holes binaries have significantly higher runaway velocities.

2 Runaway Velocities from Symmetric SNe

We investigate the velocities of black hole X-ray binaries assuming symmetric mass ejection[6], which gives the system a space velocity, due to the well know recoil effect[7]. After the mass ejection, the orbit is eccentric. However before the bright X-ray phase, in which mass from the companion is accreted by the black hole, the orbit is re-circularized. The orbital period after the re-circularization is given by[6]

$$v_{\text{sys}} = 213 \left(\frac{\Delta M}{M_\odot} \right) \left(\frac{m}{M_\odot} \right) \left(\frac{P_{\text{re-circ}}}{\text{day}} \right)^{-\frac{1}{3}} \left(\frac{M_{\text{BH}} + m}{M_\odot} \right)^{-\frac{5}{3}} \text{ km s}^{-1} \quad (1)$$

Before applying this equation, a possible evolution before and during the X-ray phase must be considered. We apply this equation to the cases of Nova Sco 1994 and Cygnus X-1 (Figure 1).

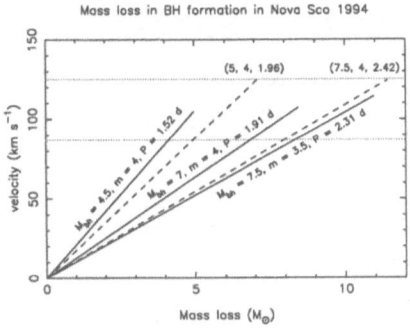
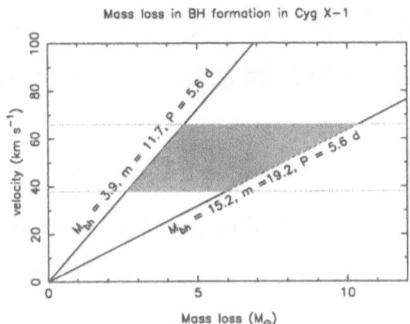

Fig. 1. Limits on the amount of mass ejected in the SN explosion that is required to explain the measured velocities[6]. Left: Nova Sco 1994 with three possibilities of the stellar masses and orbital period at the onset of the X-ray phase (solid lines) and the two possibilities in the case 0.5 M$_\odot$ has been lost since the onset of the mass transfer (dashed lines). Right: Cygnus X-1 with two different solutions for the stellar masses.

3   Results and Conclusion

Applying (1) we find that at least 48% and 28% of the black hole progenitor mass must have been lost, for the case of Nova Sco and Cygnus X-1. An assumed mass loss fraction of 35%, gives velocities for the remaining black hole X-ray binaries between 15 and 35 km/s; in the right range to explain their velocity dispersion[3]. This removes the need to invoke a large kick for Nova Sco (and Cyg X-1) and at the same time a small or no kick for the remaining systems[8], which seems highly unlikely to us.

The fact that black holes in X-ray binaries have lost significant amounts of mass in their formation is in clear disagreement with the results from some stellar evolution models that all Wolf-Rayet stars have a mass < 3.5M$_\odot$ at the moment they explode in a supernova[9], and prevents double black holes to form in a close orbits at all.

References

1. Iwamoto, K. et al., 1998, Nature 395, 672
2. Israelian, G., et al., 1999, Nature 401, 142
3. White, N. E. and van Paradijs, J., 1996, ApJ 473, L25
4. Kaper, L., Camerón, A., and Barziv, O., 1999, in K. A. van der Hucht, G. Koenigsberger, and R. J. Eenens (eds.), Wolf-Rayet phenomena in massive stars and starburst galaxies, IAU Symp. 193, p. 316
5. Bailyn, C., Orosz, J., McClintock, J., and Remillard, R., 1995, Nature 378, 157
6. Nelemans, G., Tauris, T. M. and van den Heuvel, E. P. J., 1999, A&A 352, L87
7. Blaauw, A., 1961, BAN 15, 165
8. Brandt, W. N., Podsiadlowski, P., and Sigurdsson, S., 1995, MNRAS 277, L35
9. Woosley, S., Langer, N., and Weaver, T., 1995, ApJ 448, 315

A Search for the Presence of Bow Shocks around High-Mass X-Ray Binaries

Fredrik Huthoff and Lex Kaper

Astronomical Institute "Anton Pannekoek", University of Amsterdam, Kruislaan 403, 1098 SJ Amsterdam, The Netherlands

Abstract. The measured radial velocities, and the recently derived tangential velocities based on *Hipparcos* observations, have demonstrated that several high-mass X-ray binaries (HMXBs) with OB-supergiant companion travel through interstellar space with velocities on the order of 50 km s^{-1}. On theoretical grounds it is expected that all HMXBs are runaway systems, since the loss of mass from the system during the supernova explosion of the compact star's progenitor results in a kick to the remaining (bound) binary system.

When a massive star is running through space with a supersonic velocity, the interaction of its stellar wind with the interstellar medium will result in the formation of a bow shock. The detection of a wind bow shock around a high-mass X-ray binary would be an indication that the system is moving with a high space velocity, and from its shape the direction of motion could be derived. Especially for systems at larger distances, for which no *Hipparcos* measurements are available, the detection of a bow shock would be a valuable diagnostic.

We searched for the presence of bow shocks around high-mass X-ray binaries using infrared observations. Up to now, we have only one firm detection out of 12 candidates, while for (apparently single) OB runaway stars the detection rate is about 30 % [5]. A new hypothesis to explain this difference is put forward.

1   Why Do Only Few HMXBs Show a Bow Shock?

Table 1 shows the sample of HMXBs investigated for the presence of a wind bow shock. For none of the Be/X-ray binaries we detected a bow shock, but this might be due to their relatively weak winds. Only for one of the HMXBs with OB-supergiant companion a wind bow shock is observed [2]. Although we are obviously dealing with small number statistics, it looks as if HMXBs do not as often show a bowshock as (supposedly) single OB-runaways [5].

We propose that the observed difference in the detection rate of wind bow shocks is related to a difference in kinematical age of the runaways. Massive stars ejected by dynamical interactions should have escaped the OB association at a relatively early stage, when the cluster is still dense and thus the probability for close encounters higher. The OB stars in HMXBs went through a stage of binary evolution that resulted in the supernova of the initially most massive companion. Therefore, the kinematical age of HMXBs is likely to be smaller on average than that of OB runaways ejected by dynamical interactions. As a consequence, HMXBs have a higher probability to be still enclosed in the hot and rarefied regions (e.g. superbubbles) surrounding OB associations. These superbubbles are thought to be a result of the constant heating by the intense radiation fields of the enclosed OB stars, by their stellar winds, and by the occasional supernovae. The temperatures in these regions can be of the order of 10^6 K implying sound velocities of the order ~100 km/s. Under such circumstances a space velocity of 50 km/s would not be sufficient to create a bow shock.

Table 1. The sample of HMXBs inspected for the presence of a wind bow shock in the IRAS database. The spatial resolution is enhanced using the HIRAS program [1]. Only Vela X-1 is clearly associated with a bowshock [2]. The distances (r) are based on the spectral type and reddening of the star [3],[4]. Columns 3, 4 and 5 show the distance above the galactic plane (z), the velocity with respect to the local standard of rest (corrected for peculiar solar motion and galactic rotation), and the spectral type of the massive component, respectively.

| Object | r | z | V_{space} | Sp. Type |
(X-ray counterpart)	(kpc)	(pc)	(km/s)	
0114+650	3.8	170	31	B0.5b
1223-624	5.0	-3	-2^r	B1.5 Ia
1700-377	1.7	64	71	O6.5 Iaf
1907+097	4.0	37		B0.5 Ib
QV Nor	5.5	207	-81^r	B0 Iab
Cen X-3	8.0	40	16^r	O6.5 II-III
Cyg X-1	2.5	135	41	O9.5 Iab
Vela X-1	1.8	123	46	B0.5 Ib
V615 Cas	2.0	41	38	B0e
V725 Tau	2.0	-90	71	O9.7 III-IVe
X Per	0.8	-230	35	O9e
γ Cas	0.2	-13	15	B0.5 III-IVe

^{r} Radial velocity component only.

References

1. Bontekoe, Tj.R., Koper, E., & Kester, D.J.M., 1994, A&A, 284, 1037
2. Kaper, L., et al. 1997, ApJ, 475, L37
3. Moffat, A.F.J., et al. 1997, A&A, 331, 949
4. Steele, I.A., Negueruela, I., Coe, M.J., & Roche, P., 1998, MNRAS, 297, L5
5. Van Buren, D., Noriega-Crespo, A., & Dgani, R. 1995, AJ, 110, 2914

Gamma-Ray Bursts
in Relation to Black-Hole Formation

J.A. van Paradijs[1,2,3]

[1] Died 2 November 1999; talk presented by E.P.J. van den Heuvel
[2] Astronomical Institute "Anton Pannekoek", University of Amsterdam,
   Kruislaan 403, 1098 SJ Amsterdam, The Netherlands
[3] Department of Physics, University of Alabama, Huntsville, AL 35899, USA

Abstract. The relatively nearby (40 Mpc) peculiar Type Ic supernova 1998bw was found in the error box of the Gamma Ray Burst GRB980425, and its explosion time coincided within less than one day with the time of the burst. The probability that this was a chance coincidence is less than 10^{-6}, which strongly suggests that the GRB and the supernova are related. The spectrum and lightcurve of this supernova indicate that the exploding star was a carbon-oxygen star of about 10 $M_\odot$, and that the collapsing core had a mass of at least 3 $M_\odot$ (Iwamoto et al. 1998), too large for forming a stable neutron star. This indicates that here most probably a black hole was formed. The large radio emission of the supernova, with a synchrotron spectrum, indicates that here mass ejection took place with mildly relativistic velocities ($v > 0.9c$), presumably in the form of a jet, which likely was responsible for the GRB. Although this was intrinsically a very weak GRB, all this seems to provide support for the "collapsar"- or "hypernova"-model which links GRBs with the core collapse of a very massive star to a black hole, as was suggested by Woosley (1993) and Paczynski (1998), respectively. The recent discovery that the optical afterglow lightcurves of two "normal" GRBs at cosmological distances also show evidence for underlying supernova lightcurves similar to that of SN1998bw, provides further strong support for such a link. It thus appears that here for the first time we have been witnessing a new type of supernovae, which are the birth events of black holes as the final products of massive star evolution. Further studies will have to reveal what fraction of all GRBs can be connected with such events.

1   Introduction

Theoretical considerations indicate that the cores of stars that started out more massive than about 8 $M_\odot$ will collapse to either a neutron star or a black hole. Black-hole formation is expected only for the most massive stars, but the lower limiting (initial) mass for this to occur is poorly known from theory. Evidence from observations of the black hole X-ray binaries suggests this limit to be in the range 20 - 30 $M_\odot$ (Ergma and van den Heuvel 1998). The relation between supernovae and the formation of neutron stars is well established, but until very recently (see Section 3) we knew nothing about a possible relation between supernovae and the formation of black holes. We know more than a dozen radio pulsars inside young supernova remnants (e.g.

Camilo 1999). Still, it is a long-standing puzzle why so many young supernova remnants do not seem to show the slightest evidence for the presence of a young radio pulsar, neither directly nor indirectly in the form of a relativistic pulsar wind nebula, a so-called "plerion".

Recently, part of this lack of supernova remnant – neutron star associations was filled in with the discovery by Kouveliotou et al. (1998) and Woods et al. (1999a,b) that several soft gamma-ray repeaters (SGRs) coincide with young supernova remnants and are slow X-ray pulsars (not radio pulsars!) that are rapidly spinning down. This indicates the presence of a superstrong magnetic field in these neutron stars, with a dipole surface strength of order 10^{15} Gauss. These "magnetars" owe the bulk of their energy emission in the form of X-rays and soft gamma-ray outbursts to the decay of their magnetic fields. Their existence had been theoretically predicted several years earlier by Duncan and Thompson (1992) and Thompson and Duncan (1995, 1996). They appear to be related to the anomalous X-ray pulsars (AXPs) which also have relatively long pulse periods, in the range 5 to 10 seconds and also show relatively rapid spin-down. The AXPs are also often found associated with young supernova remnants (cf. Mereghetti and Stella 1995, and van Paradijs et al. 1995, for a review of their properties). Together these two new classes of slowly spinning neutron stars with very strong magnetic fields constitute at least some 10 to 20 per cent of all newborn neutron stars (cf. Kouveliotou et al. 1998; Gaensler 1999), thus filling in part of the "missing" SNR-neutron star associations. Furthermore, some 30 per cent of all supernovae in galaxies like our own are of Type Ia, which are not expected to leave any compact remnant (cf. Nomoto et al. 1997).

Taking these, plus the remnants that contain magnetars and AXPs into account, one is still left with several tens of per cents of young SNRs in our galaxy that must be due to the explosions of massive stars, but did not leave a trace of a compact object. Could these supernovae have been the formation events of black holes? Or, more generally speaking, does black-hole formation give rise to a supernova event, including substantial mass ejection? I will argue from recent evidence from gamma-ray burst afterglows that this is indeed the case. In section 2 I first briefly review the observed characteristics of gamma-ray bursts (GRBs) and their optical and X-ray afterglows. In a striking twist, the wealth on new data on GRBs revives an old idea of Colgate (1968,1974) that links GRBs to supernovae. In Section 3 I describe in detail the characteristics of SN 1998bw, the first-discovered GRB-SN association (Galama et al. 1998), and the arguments indicating that here a black hole was formed. Also the two other GRB-SN associations that were recently discovered are described here, and in Section 4 the implications of these findings for the understanding of the GRB phenomenon and black hole formation are discussed.

2   Gamma-Ray Bursts

Gamma-ray Bursts were discovered in 1967 with the military Vela satellites (Klebesadel et al. 1973). They are energetic astronomical events that last from tens of milliseconds to tens of minutes, with energies generally up to 1 MeV, but occasionally even up to tens of GeV (for a review see Fishman and Meegan 1995 and Piran 1999). During the peaks of the brightest events, their energy flux received on Earth is similar to that of the brightest stars visible to the naked eye. A great step forward in the study of the bursts was the launch in 1991 of the *Compton Gamma-ray Observatory* (CGRO), with its *Burst And Transient Source Experiment* (BATSE, cf. Meegan et al. 1992). The BATSE observations showed that the bursts are uniformly spread out over the sky, and in fact exclude the possibility that they occur in our galaxy as had been previously thought by most astrophysicists.

A wealth of new data is providing clues to the origin of these events. Distances to GRBs were measured for the first time in 1997 after the detection of long-lived X-ray, optical and radio afterglows of GRBs (Costa et al. 1997, van Paradijs et al.1997, Frail et al. 1997). The ability to measure these afterglows was the direct consequence of the rapid availability of accurate arc-minute positions, measured with the wide field hard X-ray cameras on the Italian-Dutch satellite *Beppo*SAX (Heise et al. 1998). So far, more than a dozen optical afterglows have been observed. Many of them were found in faint host galaxies, and their redshifts were measured (e.g. Metzger et al. 1997). They appear to range between 0.4 and 3.4, with an average of about 1.0 which unambiguously established that their distance scale is cosmological, i.e. that most of them originate in the distant and early universe. GRBs are by far the most luminous photon emitters in the known universe. Assuming isotropic emission, their peak luminosities are up to several 10^{52} erg s^{-1} (that is: 10^{19} solar luminosities). Their optical afterglows are also extraordinarily luminous. For instance, at its redshift 0.835, the peak magnitude of the optical afterglow of GRB970508 corresponds to an optical luminosity about two orders of magnitude higher than that of a type Ia supernova, one of the brightest extra-solar system objects in the sky. Even brighter was GRB990123, the only GRB for which optical emission was detected simultaneously with the gamma-ray emission; for a brief moment this optical emission became as bright as 10 million type Ia supernovae (Akerlof et al. 1999).

Figure 1 shows the R-band lightcurve of this event which, after its initially very bright phase exhibited the characteristic power-law decay (in this case with exponent -1.12 ± 0.03), which is typical for all GRB afterglows. The ability to determine the location and the distance of a GRB and observe its burst and decay in detail at a variety of different wavelengths has allowed the testing of GRB models. An early model, developed long before the present observations, proposed that GRBs were caused by a shock wave emerging from the photosphere of a supernova (Colgate 1968, 1974). However, this model was soon discarded as the GRB discovery paper already

R–band lightcurve of GRB 990123

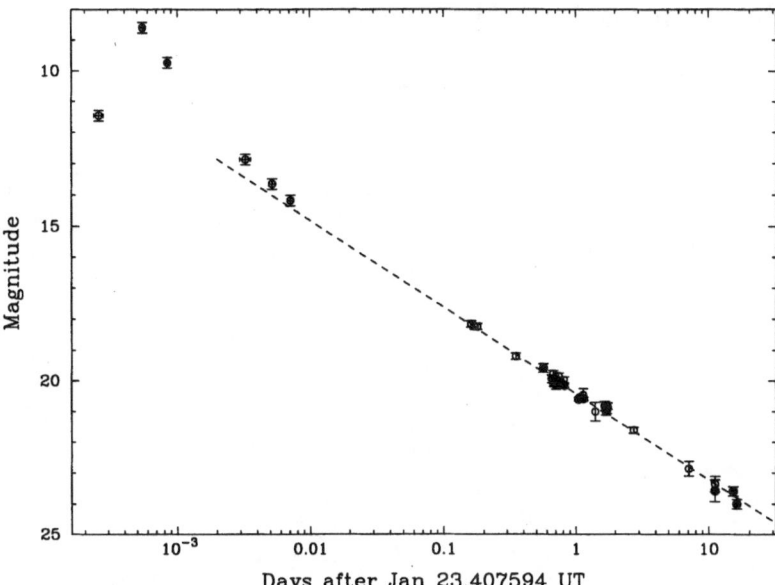

Fig. 1. R-band lightcurve of the afterglow of GRB990123 with data from a variety of observatories (after Galama, T.J. et al. 1999a). The dashed line indicates a power-law fit to the lightcurve for $t > 0.1$ days, with exponent -1.12 ± 0.03. The first 6 points are ROTSE data (Akerlof et al. 1999) of which the first two are simultaneous with the gamma-ray burst itself.

showed that there was no known correlation of GRBs with observed super-novae. It has been revived, in different form, in the early nineties by Woosley (1993), and now, after detailed studies of several GRBs in the late 1990s have revealed more information, theorists again propose a relation between GRBs and supernovae (cf. Paczynski 1998; Woosley 1998). There are at present two main models for GRBs. In the first, two compact objects (such as a neutron star and a black hole) coalesce (Lattimer and Schramm 1974; Eichler et al. 1989; Mochkovitch et al. 1993). In the second, the core of a presumably very massive star collapses, leaving behind a black hole surrounded by a solar mass accretion disk of dense degenerate matter (Woosley 1993; Paczynski 1998). In this second, so-called "collapsar" or "hypernova" model, the GRB is produced by a relativistic jet that pierces through the exploding star along its rotation axis. In addition to the fast GRB, a much slower (10^4 km s^{-1}) outflow is produced, giving rise to the thermal emission from an expanding photosphere, that is, a supernova. If this model is correct one would thus expect a (relatively weak, because of the large distance) supernova to be apparent in the afterglow lightcurve of a GRB.

3   Observational Evidence for a Relation Between GRBs and Supernovae

3.1   GRB980425 and Supernova 1998bw

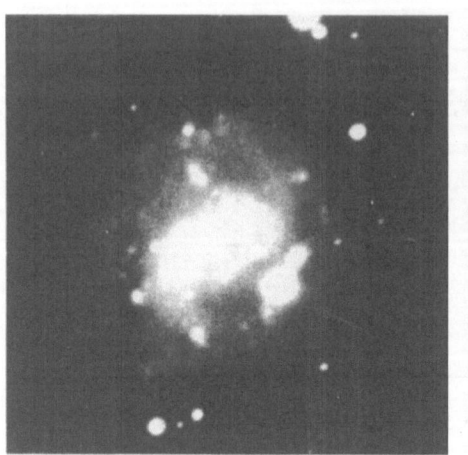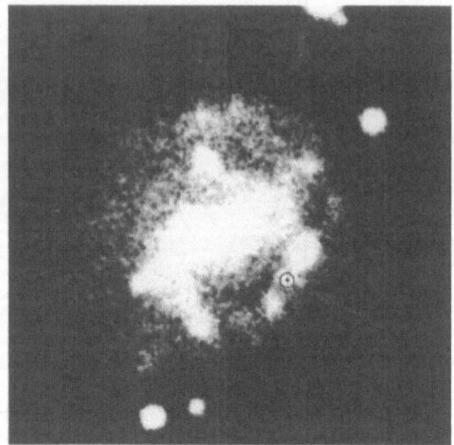

Fig. 2. The galaxy ESO 184-G82 several years before the supernova 1998 bw (right) and within a day of the gamma-ray burst GRB980425 (left), which occured at or near SN 1998bw (arrow). The circle in the right panel indicates the future position of SN 1998bw; after Galama et al. (1998)

The first direct evidence for a connection between GRBs and supernovae came from observations of GRB980425, which occurred on 25 April 1998 (see Figure 2). One of the the wide field cameras of *Beppo*SAX determined the error box in which this GRB must have occurred. This error box contained a supernova, SN1998bw, located in a spiral arm of the galaxy ESO 184-G82 (Galama et al. 1998). This galaxy lies at a distance of only about 40 Mpc from the Milky Way, cosmologically speaking very nearby. As its intensity profile measured at the Earth looked very "normal" for a GRB, i.e. similar to that of the bursts at high redshift, the energy budget of this GRB must have been very small, some 10^5 times smaller than that of normal bursts. The probability per year of finding a supernova as bright as SN1998bw that would explode inside the error box of GRB 980425 is less than 10^{-4}. However, it also appears that the time of the supernova explosion itself coincided within one day with that of the GRB, as the lightcurves of the supernova in different energy bands, depicted in Figure 3, show. The probability for this combined coincidence in position and time is less than 3×10^{-7}. This implies that a physical association between the two is likely. Some researchers have rejected the association because of the presence of a faint X-ray source in the GRB error box, (which does not coincide with the supernova) which could be the

GRB afterglow. But a recent analysis of this X-ray source by Pian et al. (1999) shows that it is unlikely to be a GRB afterglow.

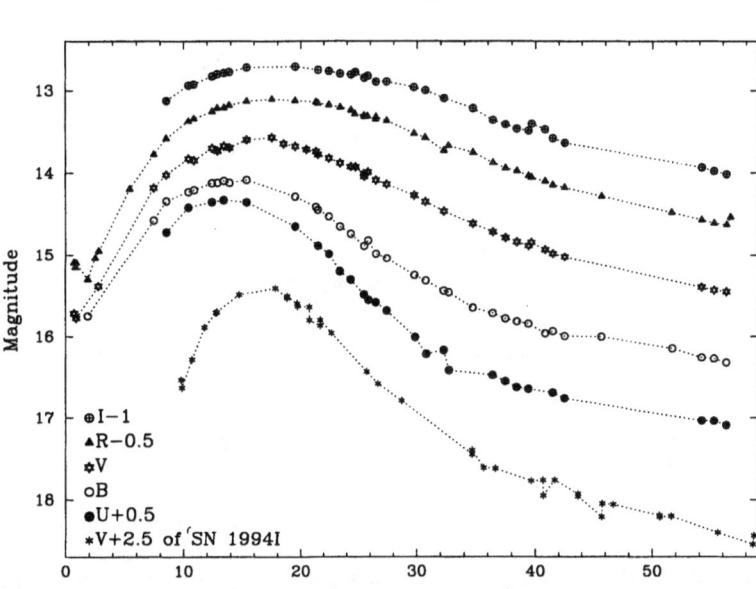

SN 1998 bw light curves

Fig. 3. UBVRI lightcurves of SN 1998bw, corrected for galactic foreground extinction, A_v=0.20 (after Galama et al. 1998). Time is in days since April 25.91 UT. As indicated in the figure, for clarity lightcurves R and I are shifted by 0.5 and 1.0 magnitude, respectively, relative to V, and U is shifted by 0.5 magnitude relative to B.

Independent of its association with the GRB, SN 1998bw was a most remarkable event. It showed prompt radio emission (which is rare for supernovae) and is the most luminous supernova to date at radio wavelengths. The spectral properties of its radio emission are characteristic for a synchrotron source and indicated the presence of a mildly relativistic outflow at about 90% the speed of light (Kulkarni et al. 1998). It has been classified as a type Ic supernova; that is, hydrogen as well as helium are absent in its spectrum. Such supernovae are rare (less than 7 per cent of all supernovae) and are believed to arise from stars that have lost their hydrogen- and helium-rich envelope in a strong stellar wind or as a result of mass transfer in a binary system.

Figure 4 shows its spectrum as compared with that of a typical type Ic supernova (Galama et al. 1998). It is clear that even among the type Ic supernovae this one is quite peculiar. An analysis of the optical lightcurve and spectra of SN 1998bw by Iwamoto et al. (1998) indicates that the exploding

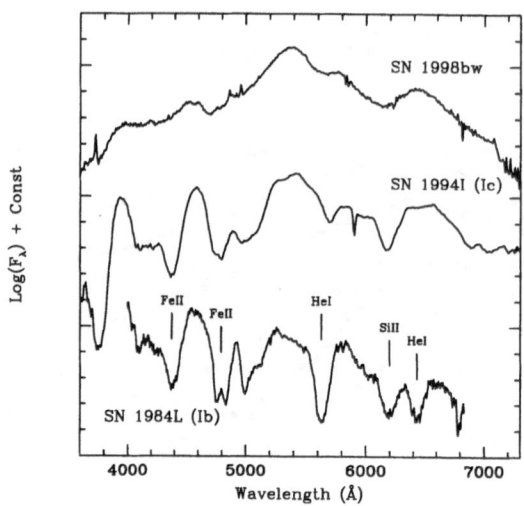

Fig. 4. Representative spectra near maximum light of SN 1998bw, SN 1994I (Type Ic, ESO Supernova archive, courtesy M. Turatto) and SN 1984L (Type Ib, Wheeler and Leireault 1985). Hydrogen lines, characteristic of type II supernovae, and Si II characteristic of type Ia supernovae, are absent in the spectrum of SN 1998bw. The strong HeI 5876 line which characterizes type Ib supernovae is weak in SN 1994I and absent in SN 1998 bw. The overall shape of the spectrum of SN 1998bw is similar to that of a type Ic supernova, although its spectral features are less pronounced. The difference is strongest in the 3500-5000 Å region, where the Ca II an Fe II lines are much weaker than in SN 1994I. In this respect, SN 1998bw appears to represent an extreme case in the old class of type Ic supernova.

star was a CO star of about 10 solar masses, whose collapsing core had a mass of at least some 3 solar masses. As the latter is too large for having left a stable neutron star (cf. Nauenberg and Chapline 1973, and Srinivasan, this volume), this core collapse most probably left a black hole. A subsequent analysis by Woosley et al. (1999) has completely confirmed these conjectures. From the great intrinsic brightness of the supernova the amount of ^{56}Ni produced in the event has been estimated to be about 0.75 solar masses, an order of magnitude higher than that of typical Ic supernovae, with a total explosion energy of several 10^{52} ergs, an order of magnitude larger than usual (Iwamoto et al. 1998; Woosley et al. 1999). In a recent study Bloom et al. (1999) have shown that the decay properties of GRB980425 and SN 1998bw can be fitted well with a model similar to the collapsar model, i.e.: involving the production of a relativistic jet when the core of the CO-star collapsed to a black hole surrounded by a temporary high-density accretion disk of substantial mass.

The jet, which partially pierced the envelope of the collapsing star along the rotation axis, is here responsible for the mildly relativistic mass ejection that produced the radio emission and the remainder of the envelope is ejected with typical supernova ejection velocities (a few times 10^4 km s^{-1}) and produced the observed supernova event.

As to the formation and characteristics of the 10 solar mass CO-star that produced the supernova: this star has all the characteristics typical of a so-called WC-star: a carbon-type Wolf-Rayet star. Such stars are expected to be the end-products of the evolution of the most massive single stars, with initial masses in excess of some 40 to 50 solar masses (see for example Maeder and Conti 1994 and van der Hucht 2000 for reviews). On the main sequence such stars have spectral types earlier than about O5, and throughout hydrogen- and helium burning they exhibit very strong mass loss by stellar wind, at a rate of order 10^{-5} solar masses per year. This makes that by the end of their evolution they will have lost all of their hydrogen- and helium-envelopes, and will be bare CO-stars. Such CO-stars can also be produced in close binaries by mass transfer. To produce a CO-star of some 10 solar masses in this way, again requires an initially quite massive star, probably with a mass upwards of 35 to 40 solar masses (cf. van Beveren and van Rensbergen 1994). Since at least half of the known WC-stars are in close binaries, the probability that SN 1998bw took place in a binary system is at least 50%.

3.2   Two More GRBs with Evidence for Underlying Supernovae: GRB980326 and GRB970228

Evidence for a second GRB coinciding with a supernova was found in the case of GRB980326 (Bloom et al. 1999). By the time spectra could be taken of the optical afterglow of this burst it had become too faint to determine its redshift. However, its lightcurve is well-determined (Figure 5). Several weeks after the burst, its afterglow brightened by a factor 60 compared with the extrapolation of the earlier power-law decay. Afterwards, the flux decay continued, showing that the flattening of the lightcurve does not reflect the steady light of an underlying host galaxy. Bloom et al. (1999) were able to successfully model the lightcurve of GRB980326 by combining a power-law afterglow - as is usual for the optical decay of GRB afterglows (see, e.g. Figure 1) - with the lightcurve of SN 1998bw, redshifted by about 1. The steep power-law afterglow decay of GRB980326 can be explained by a blast wave propagating in the wind of a massive star. As explained above for the case of SN 1998bw, such a wind is expected to have been produced by the progenitor of the exploding CO core.

A similar connection between GRBs and supernovae was recently suggested by Reichart (1999) for the optical lightcurve of GRB970228, the first GRB to be optically identified (van Paradijs et al. 1997). This suggestion was confirmed by Galama et al. (1999b) who produced the fits shown in Figure 6. In this case the redshift is known (0.69), and the only free parameter in the fit

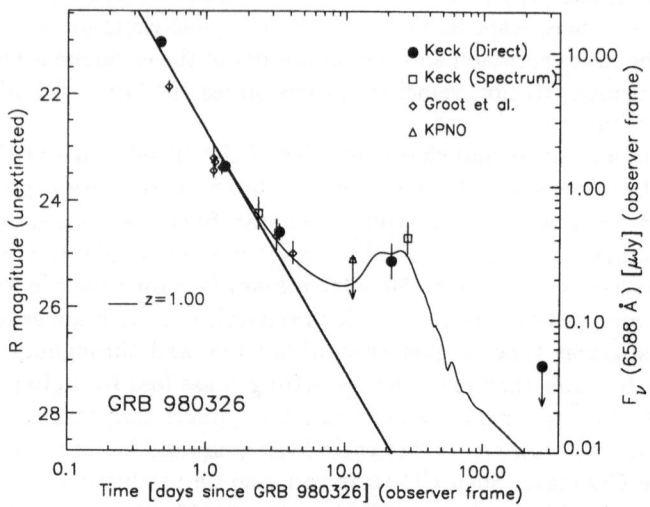

Fig. 5. R-band light emission of GRB980326 as measured by various sources, superimposed on a relatively faint supernova light curve (dashed) at a redshift of about 1, which fits the data well (Bloom et al. 1999).

light curves of GRB 970228

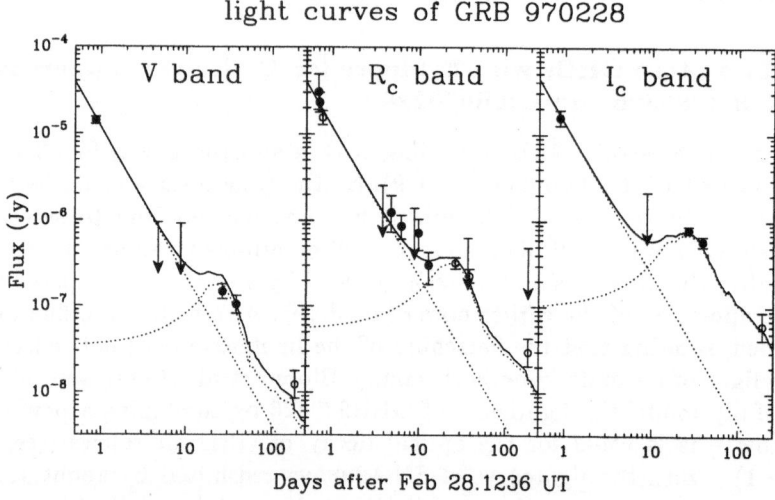

Fig. 6. The V-, R_c- and I_c-band lightcurves of GRB970228 (flux versus time). The dotted curves indicate power-law decays with $\alpha = -1.73$, and redshifted SN 1998bw lightcurves. The thick line is the resulting sum of SN and power-law decay light curves (after Galama et al. 1999b; see also Reichart 1999).

to the lightcurve is the exponent of the power-law component. Chevalier and Li (1999) have shown that after subtraction of the supernova contribution, the lightcurve of GRB970228 is also steep and best explained by an explosion in a massive stellar wind, lending further support to the association between GRBs, massive stars and supernovae.

4   Discussion and Conclusions

Gamma-ray Bursts are rare events. Now that we know their average redshift to be of order 1, one can compare their frequency of occurrence of a few per day with the number of galaxies at such redshifts. This results in a rate of order one per galaxy per 10^6 years. On the other hand the rate of type Ic supernovae is of order one per galaxy (like ours) per 10^3 years, and assuming all stars more massive than about 30 to 40 solar masses to produce black holes, one expects with a Salpeter IMF that the rate of black-hole formation is about an order of magnitude smaller than the rate of neutron star formation, i.e. at least of order one per 10^3 years. This fits well with the formation rate derived from the incidence of black-hole X-ray binaries with low-mass donor stars (see van den Heuvel 1992, 1994). Thus the GRB rate is of order 10^3 times smaller than the SN Ic rate and the black-hole formation rate. So, if GRBs are connected with SN Ic and black-hole formation, which seems likely from the last section, then we apparently *observe* only one out of 10^3 SN Ic/black-hole formation events as a GRB. This is understandable in terms of the collapsar model, because of the strong beaming expected in this model. If the opening half angle of the beams is of order 5 degrees, one will observe only one out of 10^3 black-hole formation events as a GRB.

The exciting consequence of these recent studies is that they open the perspective that a large fraction of the GRBs may originate from the core collapse of very massive stars. The supernova connection thus provides a direct link between GRBs and crucial events in the evolution of massive stars and galaxies. A word of caution is in place, however, since there is evidence that the GRBs fall in at least two distinct categories (Fishman and Meegan 1995): relatively "long" ones, which have relatively soft spectra, and relatively "shorter" ones, with systematically harder spectra. Their rates of incidence are of the same order. The *Beppo*SAX bursts, of which we know the afterglows, exclusively belong to the "long-soft" category. About the other category we know nothing. They may be the ones that might be connected with the other GRB model, i.e. the merger of two compact objects (cf. Piran 1999), which, by the way, will also lead to the formation of a black hole. For the "long-soft" category, the framework for describing GRBs in terms of a connection with supernovae and black-hole formation promises a rapid increase in our understanding of this violent phenomenon. It is to be expected that HETE-II, which will promptly provide accurate GRB positions at a rate of once per one to two weeks, will also rapidly increase our knowledge about the other category of bursts, and thus hopefully, also about the "other" type of black-hole formation.

References

1. Akerlof, C.A., et al. 1999, Nature 398, 400
2. Bloom, J.S., et al. 1999, Nature 401, 453
3. Camilo, F. 1999, in: "Pulsar Timing, General Relativity and the Internal Structure of Neutron Stars"; Eds. Z. Arzoumanian, F. Van der Hooft, and E.P.J. van den Heuvel; published by KNAW
4. Chevalier, R., Li, Z. 1999, ApJ 520, 271
5. Colgate, S.A. 1968, Can. J. Phys. 46, 5476
6. Colgate, S.A. 1974, ApJ 187, 333
7. Costa, E., et al. 1997, Nature 387, 783
8. Duncan, R., Thompson, C. 1992, ApJ 392, L29
9. Eichler, D., et al. 1989, Nature 340, 126
10. Ergma, E., van den Heuvel, E.P.J. 1998, Astron. Astrophys. 331, L29
11. Fishman, G.J., Meegan, C.A. 1995, Ann. Rev. Astron. Astrophys. 35, 415
12. Frail, D., et al. 1997, Nature 389, 261
13. Gaensler, B. 1999, Proc. IAU Colloq. Pulsars (Bonn 1999), in press
14. Galama, T., et al. 1998, Nature 395, 672
15. Galama, T., et al. 1999a, Nature 398, 394
16. Galama, T., et al. 1999b, ApJ in press, preprint astro-ph/9907264
17. Heise, J., et al. 1998, in: "Gamma Ray Bursts" (part 1, ed. C.A. Meegan, R.D. Preece and T.M. Koshut), American Inst. of Physics, Conf. Proc. 428, 397
18. Iwamoto, K., et al. 1998, Nature 395, 672
19. Klebesadel, R.W., et al. 1973, ApJ 182, L85
20. Kouveliotou, C., Dieters, S., Strohmayer, T., van Paradijs, J., Fishman, G.J., Meegan, C.A., Hurley, K., Kommers, J., Smith, I., Frail, D. and Murakami, T. 1998, Nature 393, 235
21. Kulkarni, S.R., et al. 1998, Nature 395, 663
22. Lattimer, J.M., Schramm, D.N. 1974, ApJ 192, L145
23. Maeder, A., Conti, P.S. 1994, Ann. Rev. Astron. Astrophys. 32, 227
24. Meegan, C., et al. 1992, Nature 355, 143
25. Mereghetti, S., Stella, L. 1995, ApJ 442, L17
26. Metzger, M., et al. 1997, Nature 387, 878
27. Mochkovitch, R., et al. 1993, Nature 361, 236
28. Nauenberg, M., Chapline, G. 1973, ApJ 179, 277
29. Nomoto, K., Iwamoto, K., Kishimoto, N. 1997, Science 276, 1378
30. Paczynski, B. 1998, ApJ 454, L45
31. Pian, E., et al. 1999, private communication
32. Piran, T. 1999, Phys. Rep. 314, 575
33. Reichart, D. 1999, ApJ 521, L111
34. Thompson, C., Duncan, R. 1995, MNRAS 275, 255
35. Thompson, C., Duncan, R. 1996, ApJ 473, 322
36. Vanbeveren, D., Van Rensbergen, W. 1994, in Proceedings of a meeting held at the Brussels University (VUN) from 3 to 6 August, 1993, Dordrecht: Kluwer; Eds. Vanbeveren, D., Van Rensbergen, W. and De Loore, C.

37. Van den Heuvel, E.P.J. 1992, in: "Proc. Internat. Space Year" Conf. (Satellite Symp. 3), ESA ISY-3, 29-36
38. Van den Heuvel, E.P.J. 1994, in: "Interacting Binaries" (by S.N. Shore, M. Livio and E.P.J. van den Heuvel), Springer Verlag, Heidelberg, 263-474
39. Van der Hucht, K.A. 2000, New Astronomy Rev., in press
40. Van Paradijs, J., Taam, R.E., van den Heuvel, E.P.J. 1995, Astron. Astrophys. 299, L41
41. Van Paradijs, J., et al. 1997, Nature 386, 686
42. Woods, P.M., Kouveliotou, C., van Paradijs, J., Hurley, K., Kippen, R.M., Finger, M.H., Briggs, M.S., Dieters, S., Fishman, G.J. 1999, ApJ 519, L139
43. Woods, P.M., Kouveliotou, C., van Paradijs, J., Finger, M.H., Thompson, C. 1999, ApJ 518, L103
44. Woosley, S.E. 1993, ApJ 405, 273
45. Woosley, S.E., et al. 1999, ApJ 516, 788

Black Hole Formation, Gamma-Ray Bursts, and Stellar Winds

Chris L. Fryer

UC Santa Cruz, Santa Cruz, CA 95060, USA

Abstract. Many of the black hole accretion disk gamma-ray burst (BHAD GRB) models require the formation of black holes in short-period binaries. Although many uncertainties in the evolution of close binaries make it difficult to study the characteristics of these GRB models, one of the most nefarious is our lack of understanding of stellar winds, both in giant/supergiant stars and in Wolf-Rayet stars. Here, we review the effects of winds on both the rates and redshift distribution of GRBs. These winds affect GRB models and X-ray binaries alike, and we can use observations of X-ray binaries to constrain mass-loss rates from winds.

1   Massive Star Winds and Binaries

Short-period binaries (which are the progenitors of the observed black hole binaries as well as many GRB models) all go through a mass transfer phase in which the primary star (which collapses down into a black hole) loses much of its hydrogen. If most of the hydrogen layer is lost early in the star's life, it will continue to lose mass via a Wolf-Rayet wind. Understanding this mass loss is crucial to understanding black hole formation, because it dictates whether the massive star collapses to form a black hole or a neutron star.

Wellstein & Langer [1],[2] have recently calculated the evolution of binaries with mass loss from Wolf-Rayet winds. They find that massive stars which lose their hydrogen envelope prior to the end of helium core burning lose so much mass in a Wolf-Rayet wind that, regardless of their initial mass, they end their lives with masses less than $\sim 3 M_\odot$. These strong winds constrain black hole binary formation to those systems which do not undergo mass transfer until after helium core burning. The mass loss rates of single stars (or stars which lose their hydrogen envelope after helium core burning) further limits the maximum black hole mass to $12 M_\odot$. Combined with our current understanding of stellar collapse (Fryer [3]), such strong winds exclude the formation of prompt collapsars. Here we discuss the effects these winds have on GRB redshift distributions and the constraints black hole binaries might have on these winds.

2   Winds and the Redshift of GRBs

Black holes can form in the collapse of massive stars either 1) in weak supernova explosions where fallback pushes the proto-neutron star beyond

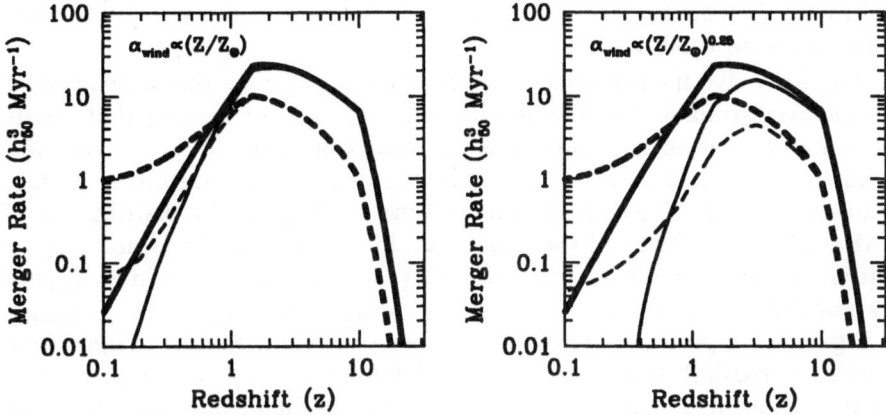

Fig. 1. Rates of prompt collapsars (solid lines give the rate/300) and NS/BH (dashed lines) mergers as a function of redshift assuming no mass loss (thick lines) and a mass loss estimated by stellar models (thin lines). We use the metallicity vs. redshift distribution of Pei, Fall, & Hauser [8], a star formation history which rises from the present day out to redshift of 1.5, is flat out to redshift of 10 and then decreases exponentially (this is the same as the $A = 3., z_p = 1.5, B = 0.$ model of Fryer et al. [7] with a cut-off at 10 instead of 5). We assume the mass loss depends on metallicity with the following form: $\alpha_w(Z) \propto Z^p$ with p=1 and p=0.25.

the maximum neutron-star mass or 2) in the direct collapse of the massive star core (Woosley & Weaver [4]). Core-collapse simulations of massive stars (Fryer [3]) found that, without winds, stars more massive than $40M_\odot$ collapse directly into black holes. It is these systems that form the prompt collapsars which have been studied most in the literature (MacFadyen & Woosley [5]). However, the single-star winds of Langer & Henkel [6] predict that stars more massive than $40M_\odot$ will lose so much mass that their cores will resemble a zero wind star of much lower mass. These stars will not go through a prompt collapse, but instead will form fallback black holes, or more likely, neutron stars. Thus, winds drastically reduce the rates of collapsars and NS/BH binary mergers.

Do these results exclude the production of prompt collapsars? No! Even if the winds of massive stars are this strong for solar metallicity stars, they are likely to decrease dramatically in strength at higher redshift (and lower metallicity). In the population synthesis study by Fryer, Woosley, & Hartmann [7], mass-loss from winds is parameterized by assuming the mass loss for a given star $\Delta M = \alpha_w \times \Delta M_{StarModels}$ where $\Delta M_{StarModels}$ is the mass loss as a function of initial star mass given by stellar models with winds (e.g. Langer & Henkel [6]) and α_w is a free parameter ranging from 0 to 1. They found that even though the rate of BHAD GRBs can be quite low for $\alpha_w = 1$, there is a sharp rise in the rates of these GRB progenitors for α_w's below 1 and by $\alpha_w = 0.5$, the rates are within a factor of 2 of the maximum rates achieved for no wind models.

We can revive the collapsar model in two ways: assume that current wind models overestimate the mass loss by a factor of 2, or assume that winds depend upon metallicity. The former has been considered by Fryer, Woosley, & Hartmann [7] and we do not address it here. The latter has not been fully studied, but it has interesting implications on the redshift distribution of GRBs. The dependence of the total mass loss on metallicity is not easy to determine because both the mass loss $rate$ ($\dot{M}_{\mathrm{Winds}}$) and the lifetime (t_{Star}) of the star depend on metallicity: $\alpha_w(Z) = \dot{M}_{\mathrm{Winds}}(Z) \times t_{\mathrm{Star}}(Z)/\Delta M_{\mathrm{StarModels}}$. In Figure 1, we plot the rates of prompt collapsars and NS/BH mergers as a function of redshift assuming $\alpha_w(Z) \propto Z^p$ for $p = 1$ and $p = 0.25$. $p = 1$ best fits the stellar models of Langer & Henkel [6] who found that the total mass loss decreased by roughly a factor of 10 when they decreased the metallicity by a factor of 10. Note that the mean redshift of collapsars, NS/BH mergers increases from 2.26,1.53 respectively for a zero mass loss case to 2.34,1.87 for p=1 and 3.69,2.58 for p=0.25. The bulk of the gamma-ray bursts are formed between redshifts of 1 and 10.

3   Constraints on Winds from Black Hole Binaries

When we include the effect of Wolf-Rayet winds in binaries, it is not only difficult to form black holes from prompt collapse, but it is difficult to form any black holes whatsoever (Wellstein & Langer [1])! Because stars which lose their hydrogen envelope in binary mass transfer before helium exhaustion in the core will lose most of their remaining mass in a strong Wolf-Rayet wind, these stars will form neutron stars rather than black holes. Only those massive stars which avoid a mass transfer phase until after helium core exhaustion will still form black holes. With current estimates of stellar radii (e.g. Kalogera & Webbink [9]), this would decrease the number of black holes formed in binaries by over an order of magnitude!

Either 1) the black hole binary formation rate is very low (<1% that of neutron stars), 2) the current estimates of Wolf-Rayet winds (Wellstein & Langer [1]) are too high, or 3) the radii of stars must be wrong. If the radii of stars prior to helium core burning are smaller than we find in current stellar models, a larger fraction of binaries would avoid mass transfer until after helium core exhaustion, producing more black holes. In Figure 2, we test the dependence of the formation of GRB progenitors on stellar radii. Lowering the radii of stars before helium ignition by a factor of 10 yields rates in all of the GRB progenitors which are within a factor of 2 of the rates produced assuming no Wolf-Rayet winds and it is likely that we can explain all X-ray binaries without requiring weaker winds. However, winds limit the maximum black hole mass to $\sim 10 M_\odot$ which is barely consistent with the observed black hole binaries. In the end, observations of massive black holes in binaries will probably force weaker stellar winds, but they do not require it yet.

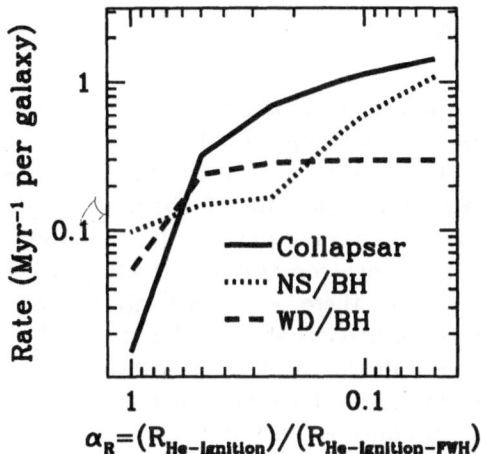

Fig. 2. Rates of GRB progenitors as a function of stellar radius prior to helium ignition. The collapsar rate is scaled down by a factor of 1000 and assumes no mass loss for single stars.

References

1. Wellstein, S., & Langer, N 1999, submitted to A&A, astro-ph/9904256
2. Wellstein, S., & Langer, N 2000, this issue
3. Fryer, C.L. 1999, ApJ, 522, 413
4. Woosley, S. E., & Weaver, T. A. 1995, ApJS, 101, 181
5. MacFadyen, A., & Woosley, S.E. 1999, accepted by ApJ, astro-ph/9810274
6. Langer, N., & Henkel, C., 1995, Space Science Reviews, 74, 343
7. Fryer, C.L., Woosley, S.E., & Hartmann, D., accepted by ApJ, astro-ph/9904122
8. Pei, Y.C., Fall, S.M., & Hauser, M.G. 1999, ApJ, 522, 604
9. Kalogera, V., & Webbink, R. 1998, ApJ, 493, 351

Fallback in Supernovae and Black Hole Formation

Alexander Heger, Andrew I. MacFadyen, and Stan E. Woosley

Department of Astronomy and Astrophysics, University of California, Santa Cruz, CA 95064, U.S.A.

When the iron core of a (non-rotating) massive star collapses to a neutron star there are three possible outcomes. First, if a neutrino powered explosion of sufficient intensity develops, almost all the material outside the neutron star is ejected and a typical supernova of about 10^{51} erg produced. If the neutrino powered explosion does not develop, then a black hole forms in the time it takes for the proto-neutron star to accrete a few tenths of a solar mass – a second or so. But there is a third intermediate possibility that we explored in some detail here. A shock is be launched initially, but lacks adequate energy to eject all the overlying matter. Then, after a time that is highly sensitive to the initial shock energy, mass falls back to the origin, accretes, and possibly produce a black hole. In SN 1987A, an explosion of roughly $1.2 \cdot 10^{51}$ erg in a 20 $M_\odot$ blue supergiant, it was previously estimated that roughly $0.1\,M_\odot$ re-imploded in this way (Chevalier 1989).

Here, fallback as a function of shock energy was explored in a rotating presupernova star of 25 $M_\odot$ (Heger et al. 2000a,b). A shock of variable energy was initiated artificially using a piston (Woosley & Weaver 1995) and the subsequent hydrodynamical evolution of the star was followed using two one-dimensional codes - an implicit Lagrangian hydrodynamics code with a reflecting inner boundary condition (KEPLER, Weaver et al. 1978) and an explicit, Eulerian code with absorbing inner boundary (PROMETHEUS). The results of these two calculations should bracket the actual fall back behavior. We found a variable range of accreted masses and time-scales. For shock energies of 0.3, 0.7, 1.2, and $1.5 \cdot 10^{51}$ erg, 1.8, 1.4, 0.5, and 0.3 $M_\odot$ (half of the total fallback) fell back in approximately 2, 7, 5, and 3 min. This resulted in accretion rates of about 10^{-4} to $10^{-2}\,M_\odot\,s^{-1}$ (Fig. 1). For the lower energies explored (energies $\lesssim 10^{51}$ erg), an appreciable fraction of the helium core, imploded and a black hole surely formed. Still, a bright, somewhat sub-luminous optical supernova would have resulted. At late times, our numerically derived accretion rates verify the predicted scaling of Chevalier (1989), an energy dependent constant times $t^{-5/3}$. Delayed black hole formation does not alter the neutrino signal, e.g., as observed from SN 1987A, but if the material falling back has sufficient angular momentum, an accretion disk may form and launch a jet (MacFadyen et al. 1999, 2000; Woosley et al. 2000). This jet *may increase the power* of the explosion, affect its symmetry, or even lead to a X-ray/gamma-ray burst as it penetrates through the stellar surface. The

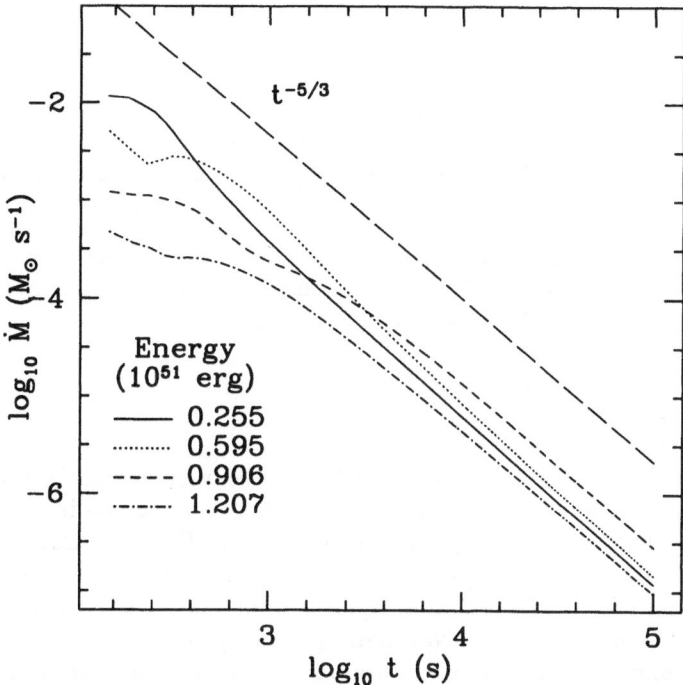

Fig. 1. *Mass accretion rate* onto the central object as a function of time after core collapse for five different (final kinetic) explosion energies (at infinity). A typical explosion energy, as, e.g., observed for SN 1987A, is $1.2 \cdot 10^{51}$ erg. The dashed blue line gives the scaling law predicted by Chevalier (1989) times an energy dependent constant.

issue of fallback is also important for the nucleosynthesis yields of supernovae and, in particular, on the amount of iron and other heavy elements ejected.

References

1. Chevalier, R.A. (1989), ApJ, **346**, 847
2. Heger, A., Langer, N., Woosley, S.E. (2000a), ApJ, **528**, in press, astro-ph/9904132
3. Heger, A., Woosley, S.E., Langer, N. (2000b), ApJ, in preparation
4. MacFadyen, A., Woosley, S.E., Heger, A. (2000), ApJ, submitted, astro-ph/9910034
5. MacFadyen, A., Woosley, S.E., Heger, A. (1999), in these proceedings
6. Woosley, S.E., MacFadyen, A., Heger, A. (1999), in: Supernovae and Gamma-Ray Bursts, eds. M. Livio, K. Sahu, & N. Panagia (Cambridge: Cambridge University Press), astro-ph/9909034
7. Weaver, T.A., Zimmerman, G.B., Woosley, S.E. (1978), ApJ, **225**, 1021
8. Woosley, S.E., Weaver, T.A. (1995), ApJS, **101**, 181.

Black-Hole Formation and Gamma-Ray Bursts

Remo Ruffini

I.C.R.A.–International Center for Relativistic Astrophysics
and Physics Department, University of Rome "La Sapienza", I-00185 Rome, Italy

Abstract. Recent work on the dyadosphere of a black hole is reviewed with special emphasis on the explanation of gamma-ray bursts. A change of paradigm in the observations of black holes is presented.

1 Introduction

An "effective potential" technique had been used very successfully by Carl Størmer in the 1930s in studying the trajectories of cosmic rays in the Earth's magnetic field (Størmer 1934). In the fall of 1967 Brandon Carter visited Princeton and presented his remarkable mathematical work leading to the separability of the Hamilton-Jacobi equations for the trajectories of charged particles in the field of a Kerr-Newmann geometry (Carter 1968). This visit had a profound impact on our small group working with John Wheeler on the physics of gravitational collapse. Indeed it was Johnny who had the idea to use the Størmer "effective potential" technique in order to obtain physical consequences from the set of first order differential equations obtained by Carter. I still remember the $2m \times 2m$ grid plot of the effective potential for particles around a Kerr metric I prepared which finally appeared in print (Ruffini and Wheeler 1971; Rees, Ruffini and Wheeler 1973, 1974; see Fig. 1).

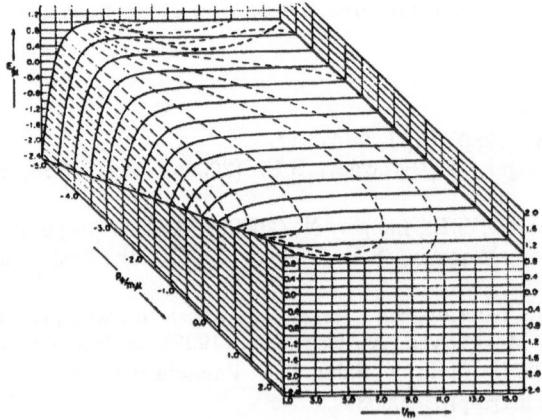

Fig. 1. "Effective potential" around a Kerr black hole, see Ruffini and Wheeler (1971).

From this work came the celebrated result of the maximum binding energy of $1 - \frac{1}{\sqrt{3}} \sim 42\%$ for corotating orbits and $1 - \frac{5}{3\sqrt{3}} \sim 3.78\%$ for counter-rotating orbits in the Kerr geometry. We were very pleased to be associated with Brandon Carter in a "gold medal" award for this work presented by Yevgeny Lifshitz: in the last edition of volume 2 of the Landau and Lifshitz series (*The Classical Theory of Fields*), both Brandon's work and my own work with Wheeler were proposed as named exercises for bright students! During this meeting it was also gratifying to hear in the talks of Rashid Sunyaev and others that these results have become the object of direct observations in X-ray sources.

The "uniqueness theorem" stating that black holes can only be characterized by their mass-energy E, charge Q and angular momentum L had been advanced in our article "Introducing the Black Hole" (Ruffini and Wheeler 1971) with its very unconventional figure in which TV sets, bread, flowers and other objects lose their characteristic features and merge in the process of gravitational collapse into the three fundamental parameters of a black hole. That picture became the object of a great deal of lighthearted discussion in the physics community. A proof of this uniqueness theorem, satisfactory for most cases of astrophysical interest, has been obtained after twenty five years of meticulous mathematical work (see e.g., Regge and Wheeler 1957, Zerilli 1970, 1974, Teukolsky 1973, C.H. Lee 1976, 1981, Chandrasekhar 1976, 1983). However the proof still presents some outstanding difficulties in its most general form. Possibly some progress will be reached in the near future with the help of computer algebraic manipulation techniques to overcome the extremely difficult mathematical calculations (see e.g., Cruciani 1999).

The "maximum mass of a neutron star" was the subject of the thesis of Clifford Rhoades, my second graduate student at Princeton. A criterium was found there to overcome fundamental unknowns about the behaviour of matter at supranuclear densities by establishing an absolute upper limit to the neutron star mass based only on general relativity, causality and the behaviour of matter at nuclear and subnuclear densities. This work, presented at the 1972 Les Houches summer School (B. and C. de Witt 1973), only appeared after a prolonged debate (see the reception and publication dates!; Rhoades and Ruffini 1974).

- The "black hole uniqueness theorem", implying the axial symmetry of the configuration and the absence of regular pulsations from black holes,
- the "effective potential technique", determining the efficiency of the energy emission in the accretion process, and
- the "upper limit on the maximum mass of a neutron star" discriminating between an unmagnetized neutron star and a black hole

were the three essential components in establishing the paradigm for the identification of the first black hole in Cygnus X-1 (Leach and Ruffini 1973). These results were also presented in a widely attended session chaired by John

Wheeler at the 1972 Texas Symposium in New York, extensively reported by the New York Times. The New York Academy of Sciences which hosted the symposium had just awarded me their prestigious Cressy Morrison Award for my work on neutron stars and black holes. Much to their dismay I never wrote the paper for the proceedings since it coincided with the one submitted for publication (Leach and Ruffini 1973).

The definition of the paradigm did not come easily but slowly matured after innumerable discussions, mainly on the phone, both with Riccardo Giacconi and Herb Gursky. I still remember an irate professor of the Physics Department at Princeton pointing publicly to my outrageous phone bill of $274 for one month, at the time considered scandalous, due to my frequent calls to the Smithsonian, and a much more relaxed and sympathetic attitude about this situation by the department chairman, Murph Goldberger. This work was finally summarized in two books: one with Herbert Gursky (Gursky and Ruffini 1975), following the 1973 AAAS Annual Meeting in San Francisco, and the second with Riccardo Giacconi (Giacconi and Ruffini 1975) following the 1975 LXV Enrico Fermi Summer School (see also the proceedings of the 1973 Solvay Conference).

The effective potential technique, see Figure (2), was also essential in order to explore a suggestion, presented by Penrose at the first meeting of the European Physical Society in Florence in 1969, that rotational energy could be extracted from black holes.

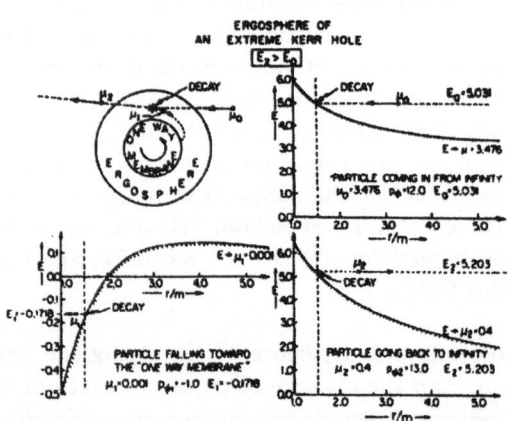

Fig. 2. (Reproduced from Ruffini and Wheeler with their kind permission.) Decay of a particle of rest-plus-kinetic energy E_0 into a particle which is captured by the black hole with positive energy as judged locally, but negative energy E_1 as judged from infinity, together with a particle of rest-plus-kinetic energy $E_2 > E_0$ which escapes to infinity. The cross-hatched curves give the effective potential (gravitational plus centrifugal) defined by the solution E of Eq.(2) for constant values of p_ϕ and μ (figure and caption reproduced from Christodoulou 1970).

The first specific example of such an energy extraction process by a gedanken experiment was given in Ruffini and Wheeler (1970), see Figure (2), and later by Floyd and Penrose (1971). The reason for showing this figure here is not just to recall the first explicit computation and the introduction of the "ergosphere", but to emphasize how contrived and difficult such a mechanism can be: it can only work for very special parameters and is in general associated with a reduction of the rest mass of the particle involved in the process. To slow down the rotation of a black hole and to increase its horizon by accretion of counter-rotating particles is almost trivial, but to extract the rotational energy from a black hole by a reversible transformation in the sense of Christodoulou and Ruffini (1971), namely to slow down the black hole *and* keep its surface area constant, is extremely difficult, as also clearly pointed out by the example in Figure (2).

The above gedanken experiments, extended as well to electromagnetic interactions, became very relevant not for their direct astrophysical significance but because they gave the tool for testing the physics and identifying the general mass-energy formula for black holes (Christodoulou and Ruffini 1971):

$$E^2 = M^2 c^4 = \left(M_{\mathrm{ir}} c^2 + \frac{Q^2}{2\rho_+} \right)^2 + \frac{L^2 c^2}{\rho_+^2}, \tag{1}$$

$$S = 4\pi \rho_+^2 = 4\pi (r_+^2 + \frac{L^2}{c^2 M^2}) = 16\pi \left(\frac{G^2}{c^4} \right) M_{\mathrm{ir}}^2, \tag{2}$$

with

$$\frac{1}{\rho_+^4} \left(\frac{G^2}{c^8} \right) (Q^4 + 4L^2 c^2) \leq 1, \tag{3}$$

where M_{ir} is the irreducible mass, r_+ is the horizon radius, ρ_+ is the quasi-spheroidal cylindrical coordinate of the horizon evaluated at the equatorial plane, S is the horizon surface area, and extreme black holes satisfy the equality in Eq. (3). The crucial point is that transformations at constant surface area of the black hole, namely reversible transformations, can release an energy up to 29% of the mass-energy of an extremal rotating black hole and up to 50% of the mass-energy of an extremely magnetized and charged black hole. Since my Les Houches lectures "On the energetics of black holes" (B.C. De Witt 1973), one of my main research goals has been to identify an astrophysical setting where the extractable mass-energy of the black hole could manifest itself: I give reasons below why I think that gamma ray bursts (GRBs) are outstanding candidates for observing this extraction process.

2   The Dyadosphere

At the time of the AAAS meeting in San Francisco (Gursky and Ruffini 1975), we had heard about the observations of the military Vela satellites which had just been unclassified and we asked Ian B. Strong to report for the first time in a public meeting on gamma-ray bursts (Strong 1975). Since those days thousands of publications have appeared on the subject, most irrelevant. One of the reasons for this is that the basic energetic requirements for GRBs have become clear only recently. The observations of the Compton satellite, through thousands of GRB observations, clearly pointed to the isotropic distribution of these sources in the sky. However, it was only with the very unexpected and fortuitous observations of the Beppo-SAX satellite that the existence of a long lasting afterglow of these sources was identified: this has led to the determination of a much more accurate position for these sources in the sky, which permitted for the first time their optical and radio identification, which in turn has led to the determination of their cosmological distances and to their paramount energetic requirements (see e.g., Frontera and Piro 1999 and references therein). The very fortunate interaction and resonance between X-ray, optical and radio astronomy which in the seventies allowed the maturing of the physics and astrophysics of neutron stars and black holes (see e.g. Giacconi and Ruffini 1977) promises to be active again today in unravelling the physics and astrophysics of the gamma ray burst sources.

In 1975, following the work on the energetics of black holes (Christodoulou and Ruffini 1971), we pointed out (Damour and Ruffini, 1975) the existence of the vacuum polarization process à la Heisenberg-Euler-Schwinger (Heisenberg and Euler 1931, Schwinger 1951) around black holes endowed with electromagnetic structure (EMBHs). Such a process can only occur for EMBHs of mass smaller then $7.2 \cdot 10^6 M_\odot$. The basic energetics implications were contained in Table 1 of Damour and Ruffini (1975), where it was also shown that this process is almost reversible in the sense introduced by Christodoulou and Ruffini (1971) and that it extracts the mass energy of an EMBH very efficiently. We also pointed out that this vacuum polarization process around an EMBH offered a natural mechanism for explaining GRBs.

The recent optical observations of GRBs (see e.g. Kulkarni et al. 1998), pointing clearly to their cosmological origin and their enormous energy requirements, have convinced us to return to our earlier work in defining more accurately the region of pair creation around an EMBH. This has led to the new concept of the dyadosphere of an EMBH (named after the Greek word for pair) and to the concept of a plasma-electromagnetic (PEM) pulse and its evolution which can generate signals with the characteristic features of a GRB. I am proposing and giving reasons to support the claim that with gamma-ray bursts, we are witnessing for the first time the moment of gravitational collapse to a black hole in real time. Even more importantly, the tremendous energies involved in the energetics of these sources clearly point to the necessity for and give the opportunity to use the extractable energy of black holes as an energy source for these objects as in Eqs. (1)–(3) above.

Various models have been proposed in order to tap the rotational energy of black holes by processes of relativistic magnetohydrodynamics (see e.g., Ruffini and Wilson 1975; Damour et al. 1978). It should be expected, however, that these processes are relevant over the long time scales characteristic of accretion processes. In the present case of gamma-ray bursts a sudden mechanism appears to be at work on time scales shorter than a second for depositing the entire energy in the fireball at the moment of the triggering process of the burst, similar to the vacuum polarization process introduced in Damour and Ruffini (1975). The fundamental new points we have found re-examining this work can be simply summarized, see Preparata, Ruffini and Xue (1998a,b) for details:

- The vacuum polarization process can occur in an extended region around the black hole called the dyadosphere, extending from the horizon radius r_+ out to the dyadosphere radius r_{ds}. Only black holes with a mass larger than the upper limit of a neutron star and up to a maximum mass of $7.2 \cdot 10^6 M_\odot$ can have a dyadosphere.
- The efficiency of transforming the mass-energy of a black hole into particle-antiparticle pairs outside the horizon can approach 100%, for black holes in the above mass range.
- The pair created are mainly positron-electron pairs and their number is much larger than the quantity Q/e one would have naively expected on the grounds of qualitative considerations. It is actually given by $N_{\text{pairs}} \sim \frac{Q}{e} \frac{r_{ds}}{\hbar/mc}$, where m and e are respectively the electron mass and charge. The energy of the pairs and consequently the emission of the associated electromagnetic radiation peaks in the gamma X-ray region, as a function of the black hole mass.

Let us now recall the main results on the dyadosphere obtained in Preparata, Ruffini and Xue (1998a,b). Although the general considerations presented by Damour and Ruffini (1975) did refer to a Kerr-Newmann field with axial symmetry about the rotation axis, for simplicity, we have considered the case of a nonrotating Reissner-Nordstrom EMBH to illustrate the basic gravitational-electrodynamical process. The dyadosphere then lies between the radius

$$r_{ds} = \left(\frac{\hbar}{mc}\right)^{\frac{1}{2}} \left(\frac{GM}{c^2}\right)^{\frac{1}{2}} \left(\frac{m_p}{m}\right)^{\frac{1}{2}} \left(\frac{e}{q_p}\right)^{\frac{1}{2}} \left(\frac{Q}{\sqrt{GM}}\right)^{\frac{1}{2}}. \tag{4}$$

and the horizon radius

$$r_+ = \frac{GM}{c^2}\left[1 + \sqrt{1 - \frac{Q^2}{GM^2}}\right]. \tag{5}$$

The number density of pairs created in the dyadosphere is

$$N_{e^+e^-} \simeq \frac{Q - Q_c}{e}\left[1 + \frac{(r_{ds} - r_+)}{\frac{\hbar}{mc}}\right], \tag{6}$$

where $Q_c = 4\pi r_+^2 \frac{m^2 c^3}{\hbar e}$. The total energy of pairs, converted from the static electric energy, deposited within the dyadosphere is then

$$E_{e^+e^-}^{\text{tot}} = \frac{1}{2}\frac{Q^2}{r_+}(1 - \frac{r_+}{r_{\text{ds}}})(1 - \left(\frac{r_+}{r_{\text{ds}}}\right)^2) .\tag{7}$$

3 The PEM Pulse

The analysis of the radially resolved evolution of the energy deposited within the e^+e^--pair and photon plasma fluid created in the dyadosphere of an EMBH is much more complex then we had initially anticipated. Explaining our first attempt to Jim Wilson led him to pronounce the Salomonic sentence "Remo, your bomb will not kill anyone!" Some basic ingredients well known to Livermore scientists were missing. We decided to join forces and propose a new collaboration with the Livermore group renewing the successful collaboration with Jim of 1974 (Ruffini and Wilson 1975). We proceeded in parallel: in Rome with simple almost analytic models to then be validated by the Livermore codes (Wilson, Salmonson and Mathews 1997,1998).

In Wilson (1975,1977), a black-hole charge of the order 10% was formed. Thus we assumed a Reissner-Nordstrom black hole with charge $Q = 0.1 Q_{max}$, where $Q_{max} = \sqrt{G}M$. For the evolution we assumed the relativistic hydrodynamic equations, for details see ref. [23]. We assumed the plasma fluid of e^+e^--pairs, photons and baryons to be a simple perfect fluid in the curved spacetime. The baryon-number and energy-momentum conservation laws are

$$(n_B U^\mu)_{;\mu} = (n_B U^t)_{,t} + \frac{1}{r^2}(r^2 n_B U^r)_{,r} = 0 ,\tag{8}$$

$$(T_\mu^\sigma)_{;\sigma} = 0,\tag{9}$$

and the rate equation:

$$(n_{e^\pm} U^\mu)_{;\mu} = \overline{\sigma v}[n_{e^-}(T)n_{e^+}(T) - n_{e^-}n_{e^+}] ,\tag{10}$$

where U^μ is the four-velocity of the plasma fluid, n_B the proper baryon-number density, $n_{e^\pm}$ are the proper densities of electrons and positrons($e^\pm$), σ is the mean pair annihilation-creation cross-section, v is the thermal velocity of $e^\pm$, and $n_{e^\pm}(T)$ are the proper number-densities of $e^\pm$ at an appropriate equilibrium temperature T. The calculations are continued until the plasma fluid expands, cools and the e^+e^- pairs recombine and the system becomes optically thin.

The results of the Livermore computer code are compared and contrasted with three almost analytical models: (i) spherical model: the radial component of the four-velocity is of the form $U(r) = U\frac{r}{\mathcal{R}}$, where U is the four-velocity at the surface ($\mathcal{R}$) of the plasma, similar to a portion of a Friedmann model; (ii) slab 1: $U(r) = U_r = $ const., an expanding slab with constant width

$\mathcal{D} = R_0$ in the coordinate frame in which the plasma is moving; (iii) slab 2: an expanding slab with constant width $R_2 - R_1 = R_0$ in the comoving frame of the plasma.

We compute the relativistic Lorentz factor γ of the expanding e^+e^- pair and photon plasma.

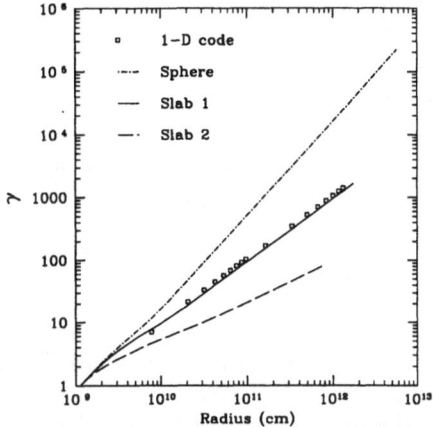

Fig. 3. Lorentz γ as a function of radius. Three models for the expansion pattern of the e^+e^- pair plasma are compared with the results of the one dimensional hydrodynamic code for a $1000M_\odot$ black hole with charge $Q = 0.1Q_{max}$. The 1-D code has an expansion pattern that strongly resembles that of a shell with constant coordinate thickness.

In Figure 3 we see a comparison of the Lorentz factor of the expanding fluid as a function of radius for all the models. We can see that the one-dimensional code (only a few significant points are plotted) matches the expansion pattern of a shell of constant coordinate thickness.

In analogy with the notorious electromagnetic radiation EM pulse of some explosive events, we called this relativistic counterpart of an expanding pair electromagnetic radiation shell a PEM pulse.

In Figure 4 we plot correspondingly the time t_{90} over which 90% of the emission is received from a PEM pulse reaching transparency, as a function of the black hole mass, details given in reference [23]. These theoretical predictions can be compared and contrasted with the observations.

4   Conclusions

It is well known that pulsars are powered by the rotational energy of neutron stars, which provided evidence for the existence of neutron stars. Binary X-ray sources extract their energy from the deep relativistic potential well of

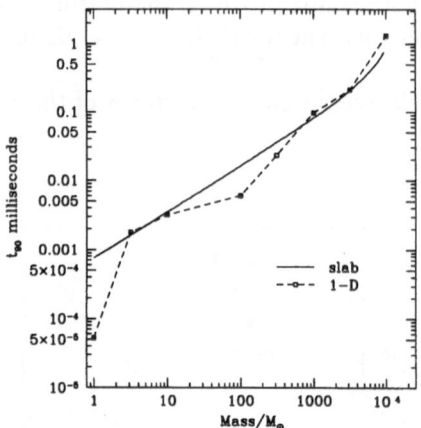

Fig. 4. The duration of the emission at decoupling is represented by t_{90} plotted over a range of black hole masses.

neutron stars and black holes, and provided evidence for the presence of black holes in our galaxy with Cygnus X-1. We propose that the gamma-ray bursts get their energy from the mass-energy of black holes.

The vacuum polarization process we consider can occur in two very distinct regimes: in the collapse of systems leading to black holes of a few solar masses, and in the collapse of very large black holes in the range 10^3 to $10^6 M_{\odot}$. While the mechanism of formation for the systems of the first type is well understood, further work is left to be done in understanding the astrophysical settings leading to the collapse of very large EMBHs. Such 10^3 to $10^6 M_{\odot}$ black holes should be considered as "seed black holes" leading by subsequent process of accretion to active galactic nuclei and quasars.

By refining the theoretical models we should be able to retrace the basic parameters of EMBHs from the timing and energy spectrum of GRBs.

Further work is directed toward:

- studying the interaction of the PEM pulse with the baryonic matter of the remnant;
- generalizing our treatment to the rotating case leading to the breakdown of spherical symmetry;
- analyzing the process of formation of the dyadosphere during the process of gravitational collapse itself.

References

1. Carter, B. (1968), *Phys. Rev.*, **174**, 1559
2. Christodoulou, D. (1970), *Phys. Rev. Lett.*, **25**, 1596
3. Christodoulou, D., Ruffini, R. (1971), *Phys. Rev. D*, **4**, 3552

4. Chandrasekhar, S., (1983), *The Mathematical Theory of Black Holes*, Clarendon Press, Oxford, see also Chandrasekhar S., (1976), *Proc. R. Soc. Lond.* **A 349**, 571

5. Cruciani, G.L. (1999), in *Proceedings of the Third Icra Network Workshop*, Cherubini C. and Ruffini R. Ed., *Nuovo Cim. B* in press

6. Damour, T., Ruffini, R., (1975) *Phys. Rev. Lett.* **35**, 463

7. Damour, T., Hanni, R.S., Ruffini, R., Wilson, J. (1978), *Phys. Rev.* **D17**, 1518

8. de Witt, B., de Witt, C. Ed., *Black Holes*, Gordon and Breach 1973

9. Floyd, R.M., Penrose, R. (1971), Nature **229**, 177, submitted 16 December 1970

10. Frontera, F., Piro, L. (1999) *GRB in the Afterglow Era*, *A&A Suppl. Ser.***138**

11. Giacconi, R., Ruffini, R., Ed. and coauthors (1978), *Physics and Astrophysics of Neutron Stars and Black Holes*, North Holland Amsterdam

12. Heisenberg, W., Euler, H. (1931), *Zeits. Phys.*, **69**, 742

13. Gursky, H., Ruffini, R., Ed. and coauthors (1975), *"Neutron Stars, Black Holes and Binary X-ray Sources"* , D. Reidel Dordrecht

14. Kulkarni, S. R. *et al.* (1998), *Nature*, **395**, 663, see also Frontera and Piro

15. Lee, C.H., (1976), *J. Math. Phys* **17**, 1226, and (1981), *Prog. Theor. Phys.* **66**, 180

16. Leach, R.W., Ruffini, R. (1973), *Astrophys. J.* **180**, L15-L18

17. Preparata, G., Ruffini, R., Xue, S.-S. 1998a, submitted to *Phys. Rev. Lett.* and 1998b, *A&A* **338**, L87.

18. Rees, M., Ruffini, R., Wheeler, J.A. (1974) *"Black Holes, Gravitational Waves and Cosmology"*, Gordon and Breach New York (also in Russian MIR 1973)

19. Regge, T., Wheeler, J.A., (1957), *Phys. Rev.* **108**, 1063

20. Rhoades, C., Ruffini, R. (1974) *On the maximum mass of neutron stars*, *Phys. Rev.letters* **32**, 324

21. Ruffini, R., Wheeler, J.A. (1971), *Introducing the Black Hole* Physics Today, **24**, (1), 30

22. Ruffini, R., Wheeler, J.A., "Relativistic Cosmology from Space Platforms" in *Proceedings of the Conference on Space Physics*, Hardy V. and Moore H., Eds., E.S.R.O. Paris, (1971). The preparation of this report took more than one year and the authors were unwilling to publish parts of it before the final publication. In order to avoid delays, the results of the energy extraction process from a Kerr black hole, as well as the definition of the "ergosphere", were inserted as Fig.2 in the Christodoulou 1970 paper, published 30 November 1970.

23. Ruffini, R., Salmonson, J.D., Wilson, J.R., Xue, S.-S. (1998), *A&A Suppl. Ser.* **138**, 511-512 and (1999) *A&A* **350**, 334-343

24. Ruffini, R. & Wilson, J.R. (1975), *Phys. Rev.* **D12**, 2959

25. Schwinger, J. (1951), *Phys. Rev. D*, 82, 664

26. Størmer, C. (1934) Astrophysica Norvegica 1, 1

27. Teukolsky, S.A. (1973), *Astrophys. J.* **185**, 635

28. Wilson, J.R. (1975), *Annals of the New York Academy of Sciences*, **262**, 123

29. Wilson, J.R. (1977), in *Proc. of the First Marcel Grossmann Meeting on General Relativity*, ed. Ruffini R. (North-Holland Pub. Amsterdam) 1977, p. 393

30. Wilson, J.R., Salmonson, J.D., Mathews, G.J. (1997), in *Gamma-Ray Bursts: 4th Huntsville Symposium*, ed. C.A. Meegan, R.D. Preece, T.M. Koshut (A.I.P.)

31. Wilson, J.R., Salmonson, J.D., Mathews, G.J. (1998), in *2nd Oak Ridge Symposium on Atomic and Nuclear Astrophysics* (IOP Publishing Ltd)

32. Zerilli, F.J., (1970), *Phys. Rev. D* **2**, 2141 and (1974), *Phys. Rev. D* **9**, 860

The Final Fate of Coalescing Binary Neutron Stars: Collapse to a Black Hole?

Frederic A. Rasio

Department of Physics, Massachusetts Institute of Technology,
Cambridge, MA 02139, USA

Abstract. Coalescing compact binaries with neutron star (NS) or black hole (BH) components are important sources of gravitational waves for the laser-interferometer detectors currently under construction, and may also be sources of gamma-ray bursts at cosmological distances. This paper focuses on the final hydrodynamic coalescence and merger of NS–NS binaries, and addresses the question of whether black-hole formation is the inevitable final fate of these systems.

1 Introduction

Many theoretical models of gamma-ray bursts (GRBs) rely on coalescing compact binaries (NS–NS or BH–NS) to provide the energy of GRBs at cosmological distances (e.g., Eichler et al. 1989; Narayan, Paczyński, & Piran 1992; Mészáros & Rees 1992). The close spatial association of some GRB afterglows with faint galaxies at high redshifts is not inconsistent with a compact binary origin, in spite of the large recoil velocities acquired by compact binaries at birth (Bloom, Sigursson, & Pols 1999). Currently the most popular models all assume that the coalescence leads to the formation of a rapidly rotating Kerr BH surrounded by a torus of debris. Energy can then be extracted either from the rotation of the BH or from the material in the torus so that, with sufficient beaming, the gamma-ray fluxes observed from even the most distant GRBs can be explained (Mészáros, Rees, & Wijers 1999). However, it is important to understand the hydrodynamic processes taking place during the final coalescence before making assumptions about its outcome. In particular, as will be argued in §3, it is not clear that the coalescence of two $1.4\,M_\odot$ NS will form an object that must collapse to a BH on a dynamical time, and it is not certain either that matter will be ejected during the merger and form an outer torus around the central object.

Coalescing compact binaries are also the most promising sources of gravitational waves for detection by the large laser interferometers currently under construction, such as LIGO (Abramovici et al. 1992) and VIRGO (Bradaschia et al. 1990). In addition to providing a major new confirmation of Einstein's theory of general relativity (GR), including the first direct proof of the existence of black holes (Flanagan & Hughes 1998; Lipunov, Postnov, & Prokhorov 1997), the detection of gravitational waves from coalescing binaries at cosmological distances could provide accurate independent measurements of the Hubble constant and mean density of the Universe (Schutz 1986; Chernoff & Finn 1993; Marković 1993). Expected rates of NS–NS binary coalescence in the Universe, as well as expected event rates in laser interferometers, have now been calculated by many groups. Although there is some disparity between various published results, the estimated rates are generally encouraging (see Kalogera 2000 for a recent review). Many calculations of gravitational wave emission from coalescing binaries have focused on the waveforms emitted during the last few thousand orbits, as the frequency sweeps upward from $\sim 10\,\mathrm{Hz}$ to $\sim 300\,\mathrm{Hz}$. The waveforms in this frequency range, where the sensitivity of ground-based interferometers is highest, can be calculated very accurately by performing high-order post-Newtonian (PN) expansions of the equations of motion for two *point masses* (see, e.g., Owen & Sathyaprakash 1999 and references therein). However, at the end of the inspiral, when the binary separation becomes comparable to the stellar radii (and the frequency is $\gtrsim 1\,\mathrm{kHz}$), hydrodynamics becomes important and the character of the waveforms must change. Special purpose narrow-band detectors that can sweep up frequency in real time will be used to try to catch the last ~ 10 cycles of the gravitational waves during the final coalescence (Meers 1988; Strain & Meers 1991). These "dual recycling" techniques are being tested right now on the German-British interferometer GEO 600 (Danzmann 1998). In this terminal phase of the coalescence, when the two stars merge together into a single object, the waveforms contain information not just about the effects of GR, but also about the interior structure of a NS and the nuclear equation of state (EOS) at high density. Extracting this information from observed waveforms, however, requires detailed theoretical knowledge about all relevant hydrodynamic processes. If the NS merger is followed by the formation of a BH, the corresponding gravitational radiation waveforms will also provide direct information on the dynamics of rotating core collapse and the BH "ringdown" (see, e.g., Flanagan & Hughes 1998).

2   Hydrodynamics of Binary Coalescence

The final hydrodynamic merger of two NS is driven by a combination of relativistic and fluid effects. Even in Newtonian gravity, an innermost stable circular orbit (ISCO) is imposed by *global hydrodynamic instabilities*, which can drive a close binary system to rapid coalescence once the tidal interaction between the two stars becomes sufficiently strong. The existence of these global instabilities for close binary equilibrium configurations containing a compressible fluid, and their particular importance for binary NS systems, were demonstrated for the first time by Rasio & Shapiro (1992, 1994, 1995; hereafter RS1–3) using numerical hydrodynamic calculations. These instabilities can also be studied using analytic methods. The classical analytic work for close binaries containing an incompressible fluid (e.g., Chandrasekhar

1969) was extended to compressible fluids in the work of Lai, Rasio, & Shapiro (1993a,b, 1994a,b,c, hereafter LRS1–5). This analytic study confirmed the existence of dynamical instabilities for sufficiently close binaries. Although these simplified analytic studies can give much physical insight into difficult questions of global fluid instabilities, fully numerical calculations remain essential for establishing the stability limits of close binaries accurately and for following the nonlinear evolution of unstable systems all the way to complete coalescence.

A number of different groups have now performed such calculations, using a variety of numerical methods and focusing on different aspects of the problem. Nakamura and collaborators (see Nakamura & Oohara 1998 and references therein) were the first to perform 3D hydrodynamic calculations of binary NS coalescence, using a traditional Eulerian finite-difference code. Instead, RS used the Lagrangian method SPH (Smoothed Particle Hydrodynamics). They focused on determining the ISCO for initial binary models in strict hydrostatic equilibrium and calculating the emission of gravitational waves from the coalescence of unstable binaries. Many of the results of RS were later independently confirmed by New & Tohline (1997) and Swesty, Wang, & Calder (1999), who used completely different numerical methods but also focused on stability questions, and by Zhuge, Centrella, & McMillan (1994, 1996), who also used SPH. Zhuge et al. (1996) also explored in detail the dependence of the gravitational wave signals on the initial NS spins. Davies et al. (1994) and Ruffert et al. (1996, 1997) have incorporated a treatment of the nuclear physics in their hydrodynamic calculations (done using SPH and PPM codes, respectively), motivated by cosmological models of GRBs. All these calculations were performed in *Newtonian gravity*, with some of the more recent studies adding an approximate treatment of energy and angular momentum dissipation through the gravitational radiation reaction (e.g., Janka et al. 1999; Rosswog et al. 1999), or even a full treatment of PN gravity to lowest order (Ayal et al. 2000; Faber & Rasio 2000).

All recent hydrodynamic calculations agree on the basic qualitative picture that emerges for the final coalescence. As the ISCO is approached, the secular orbital decay driven by gravitational wave emission is dramatically accelerated (see also LRS2, LRS3). The two stars then plunge rapidly toward each other, and merge together into a single object in just a few rotation periods. In the corotating frame of the binary, the relative radial velocity of the two stars always remains very subsonic, so that the evolution is nearly adiabatic. This is in sharp contrast to the case of a head-on collision between two stars on a free-fall, radial orbit, where shock heating is very important for the dynamics (RS1; Shapiro 1998). Here the stars are constantly being held back by a (slowly receding) centrifugal barrier, and the merging, although dynamical, is much more gentle. After typically $1 - 2$ orbital periods following first contact, the innermost cores of the two stars have merged and the system resembles a single, very elongated ellipsoid. At this point a secondary instability occurs: *mass shedding* sets in rather abruptly. Material (typically $\sim 10\%$ of the total mass) is ejected through the outer Lagrange points of the effective potential and spirals out rapidly. In the final stage, the inner spiral arms widen and merge together, forming a nearly axisymmetric torus around the inner, maximally rotating dense core.

In GR, strong-field gravity between the masses in a binary system is alone sufficient to drive a close circular orbit unstable. In close NS binaries, GR effects combine nonlinearly with Newtonian tidal effects so that the ISCO is encountered at larger binary separation and lower orbital frequency than predicted by Newtonian hydrodynamics alone, or GR alone for two point masses. The combined effects of relativity and hydrodynamics on the stability of close compact binaries have only very recently begun to be studied, using both analytic approximations (basically, PN generalizations of LRS; see, e.g., Lai & Wiseman 1997; Lombardi, Rasio, & Shapiro 1997; Shibata & Taniguchi 1997), as well as numerical calculations in 3D incorporating simplified treatments of relativistic effects (e.g., Baumgarte et al. 1998; Marronetti, Mathews & Wilson 1998; Wang, Swesty, & Calder 1998). Several groups have been working on a fully general relativistic calculation of the final coalescence, combining the techniques of numerical relativity and numerical hydrodynamics in 3D (Baumgarte, Hughes, & Shapiro 1999; Landry & Teukolsky 1999; Seidel 1998; Shibata & Uryu 1999). However this work is still in its infancy, and only very preliminary results of test calculations have been reported so far.

3   Black-Hole Formation

The final fate of a NS–NS merger depends crucially on the NS EOS, and on the extraction of angular momentum from the system during the final merger. For a stiff NS EOS, it is by no means certain that the core of the final merged configuration will collapse on a dynamical timescale to form a BH. One reason is that the Kerr parameter J/M^2 of the core may exceed unity for extremely stiff EOS (Baumgarte et al. 1998), although Newtonian and PN hydrodynamic calculations suggest that this is never the case (see, e.g., Faber & Rasio 2000). More importantly, the rapidly rotating core may in fact be dynamically stable. Take the obvious example of a system containing two identical $1.35\,M_\odot$ NS. The total baryonic mass of the system for a stiff NS EOS is then about $3\,M_\odot$. Almost independent of the spins of the NS, all hydrodynamic calculations suggest that about 10% of this mass will be ejected into the outer torus, leaving at the center a *maximally rotating* object with baryonic mass $\simeq 2.7\,M_\odot$ (Any hydrodynamic merger process that leads to mass shedding will produce a maximally rotating object since the system will have ejected just enough mass and angular momentum to reach its new, stable quasi-equilibrium state). Most stiff NS EOS (including the recent "AU" and "UU" EOS of Wiringa et al. 1988) allow stable, maximally rotating NS with baryonic masses exceeding $3\,M_\odot$ (Cook, Shapiro, & Teukolsky 1994), i.e., well above the mass of the final merger core. Differential rotation (not taken into account in the calculations of Cook et al. 1994) can further increase this maximum stable mass very significantly (see Baumgarte, Shapiro, & Shibata 2000). However, for slowly rotating stars, the same EOS give maximum stable baryonic masses in the range $2.5 - 3\,M_\odot$. Thus the final fate of the merger depends critically on its rotational profile and total angular momentum.

Note that other processes, such as electromagnetic radiation, neutrino emission, and the development of various secular instabilities (e.g., r-modes), which may also lead to angular momentum losses, take place on timescales much longer than the dynamical timescale (see, e.g., Baumgarte & Shapiro 1998, who show that neutrino emission is probably negligible). These processes are therefore decoupled from the hydrodynamics of the coalescence. Unfortunately their study is plagued by many fundamental uncertainties in the microphysics.

The question of the final fate of the merger also depends crucially on the evolution of the fluid vorticity during the final coalescence. Close NS binaries are likely to be *nonsynchronized*. Indeed, the tidal synchronization time is almost certainly much longer than the orbital decay time (Kochanek 1992; Bildsten & Cutler 1992). For NS binaries that are far from synchronized, the final coalescence involves some new, complex hydrodynamic processes (Rasio & Shapiro 1999). Consider for example the case of an irrotational system (containing two nonspinning stars at large separation; see LRS3). Because the two stars appear to be counter-spinning in the corotating frame of the binary, a *vortex sheet* (where the tangential velocity jumps discontinuously by $\Delta v \sim 0.1\,c$) appears when the stellar surfaces come into contact. Such a vortex sheet is Kelvin-Helmholtz unstable on all wavelengths and the hydrodynamics is therefore extremely difficult to model accurately given the limited spatial resolution of 3D calculations. The breaking of the vortex sheet generates some turbulent viscosity so that the final configuration may no longer be irrotational. In numerical simulations, however, vorticity is quickly generated through spurious shear viscosity, and the merger remnant is observed to evolve rapidly (in just a few rotation periods) toward uniform rotation.

The final fate of the merger will be affected drastically by these processes. In particular, the shear flow inside the merging stars (which supports a highly triaxial shape; see Rasio & Shapiro 1999) may in reality persist long enough to allow a large fraction of the total angular momentum in the system to be radiated away in gravitational waves. In this case the final merged core may resemble a Dedekind ellipsoid, i.e., it will have a triaxial shape supported entirely by internal fluid motions, but with a stationary shape in the inertial frame (so that it no longer radiates gravitational waves). This state will be reached on the gravitational radiation reaction timescale, which is no more than a few tens of rotation periods. On the (possibly much longer) *viscous timescale*, the core will then evolve to a uniform, slowly rotating state and will likely collapse to a BH. In contrast, in all 3D numerical simulations performed to date, the shear is quickly dissipated, so that gravitational radiation never gets a chance to extract more than a small fraction ($\sim 10\%$) of the angular momentum, and the final core appears to be a uniform, maximally rotating object exactly as in calculations starting from synchronized binaries. However this behavior is most likely an artefact of the large spurious shear viscosity present in the 3D simulations.

In addition to their obvious significance for gravitational wave emission, these issues are also of great importance for models of GRBs that depend on energy extraction from a torus of material around the central BH. Indeed, if a large fraction of the total angular momentum is removed by the gravitational waves, rotationally-induced mass shedding may not occur at all during the merger, leaving a BH with no surrounding matter and no way of extracting energy from the system.

Acknowledgements

This work was supported by NSF Grant AST-9618116, NASA ATP Grant NAG5-8460, and by an Alfred P. Sloan Research Fellowship. The computational work was also supported by the National Computational Science Alliance and utilized the SGI/Cray Origin2000 supercomputer at NCSA.

References

1. Abramovici, M., et al. 1992, Science, 256, 325
2. Ayal, S., Piran, T., Oechslin, R., Davies, M.B., & Rosswog, S. 2000, ApJ, submitted
3. Baumgarte, T.W., Cook, G.B., Scheel, M.A., Shapiro, S.L., & Teukolsky, S.A. 1998, PRD 57, 7299
4. Baumgarte, T.W., Hughes, S.A., & Shapiro, S.L. 1999, PRD, 60, 087501
5. Baumgarte, T.W., & Shapiro, S.L. 1998, ApJ, 504, 431
6. Baumgarte, T.W., Shapiro, S.L., & Shibata, M. 2000, ApJL, 528, L29
7. Bildsten, L., & Cutler, C. 1992, ApJ, 400, 175
8. Bloom, J.S., Sigurdsson, S., & Pols, O.R. 1999, MNRAS, 305, 763
9. Bradaschia, C., et al. 1990, Nucl. Instr. Methods A, 289, 518
10. Chandrasekhar, S. 1969, Ellipsoidal Figures of Equilibrium (New Haven: Yale University Press); Revised Dover edition 1987
11. Chernoff, D.F., & Finn, L.S. 1993, ApJL, 411, L5
12. Cook, G. B., Shapiro, S. L., & Teukolsky, S. L. 1994, ApJ, 424, 823
13. Danzmann, K. 1998, in Relativistic Astrophysics, eds. H. Riffert et al. (Proc. of 162nd W.E. Heraeus Seminar, Wiesbaden: Vieweg Verlag), 48
14. Davies, M.B., Benz, W., Piran, T., & Thielemann, F.K. 1994, ApJ, 431, 742
15. Eichler, D., Livio, M., Piran, T., & Schramm, D.N. 1989, Nature, 340, 126
16. Faber, J., & Rasio, F.A. 2000, PRD, submitted
17. Flanagan, E.E., & Hughes, S.A. 1998, PRD, 57, 4566

18. Kalogera, V. 2000, to appear in the Proceedings of the 3rd Amaldi Conference on Gravitational Waves, ed. S. Meshkov [astro-ph/9911532]
19. Kochanek, C.S. 1992, ApJ, 398, 234
20. Janka, H., Eberl, T., Ruffert, M., & Fryer, C.L. 1999, ApJL, in press
21. Lai, D., Rasio, F.A., & Shapiro, S.L. 1993a, ApJS, 88, 205 [LRS1]
22. Lai, D., Rasio, F. A., & Shapiro, S.L. 1993b, ApJ, 406, L63 [LRS2]
23. Lai, D., Rasio, F. A., & Shapiro, S.L. 1994a, ApJ, 420, 811 [LRS3]
24. Lai, D., Rasio, F. A., & Shapiro, S.L. 1994b, ApJ, 423, 344 [LRS4]
25. Lai, D., Rasio, F. A., & Shapiro, S.L. 1994c, ApJ, 437 742 [LRS5]
26. Lai, D., & Wiseman, A.G. 1997, PRD, 54, 3958
27. Landry, W., & Teukolsky, S.A. 1999, PRD, submitted
28. Lipunov, V.M., Postnov, K.A., & Prokhorov, M.E. 1997, AstL, 23, 492
29. Lombardi, J.C., Rasio, F.A., & Shapiro, S.L. 1997, PRD, 56, 3416
30. Marković, D. 1993, PRD, 48, 4738
31. Marronetti, P., Mathews, G.J., & Wilson, J.R. 1998, PRD, 58, 107503
32. Meers, B.J. 1988, PRD, 38, 2317
33. Mészáros, P., & Rees, M.J. 1992, ApJ, 397, 570
34. Mészáros, P., Rees, M.J., & Wijers, R.A.M.J. 1999, NewA, 4, 303
35. Nakamura, T., & Oohara, K. 1998, preprint [gr-qc/9812054]
36. Narayan, R., Paczyński, B., & Piran, T. 1992, ApJ, 395, L83
37. New, K.C.B., & Tohline, J.E. 1997, ApJ, 490, 311
38. Owen, B.J., & Sathyaprakash, B.S. 1999, PRD, 60, 022002
39. Rasio, F.A., & Shapiro, S.L. 1992, ApJ, 401, 226 [RS1]
40. Rasio, F.A., & Shapiro, S.L. 1994, ApJ, 432, 242 [RS2]
41. Rasio, F.A., & Shapiro, S.L. 1995, ApJ, 438, 887 [RS3]
42. Rasio, F.A., & Shapiro, S.L. 1999, CQG, 16, 1
43. Rosswog, S., Liebendoerfer, M., Thielemann, F.-K., Davies, M.B., Benz, W., & Piran, T. 1999, A&A, in press
44. Ruffert, M., Janka, H.-T., & Schäfer, G. 1996, A&A, 311, 532
45. Ruffert, M., Janka, H.-T., Takahashi, K., & Schäfer, G. 1997, A&A, 319, 122
46. Ruffert, M., Rampp, M., & Janka, H.-T. 1997, A&A, 321, 991
47. Schutz, B.F. 1986, Nature, 323, 310
48. Seidel, E. 1998, in Relativistic Astrophysics, eds. H. Riffert et al. (Proc. of 162nd W.E. Heraeus Seminar, Wiesbaden: Vieweg Verlag), 229
49. Shapiro, S.L. 1998, PRD, 58, 103002
50. Shibata, M. & Taniguchi, K. 1997, PRD 56, 811
51. Shibata, M., & Uryu, K. 1999, PRD, in press
52. Strain, K.A., & Meers, B.J. 1991, PRL, 66, 1391
53. Swesty, F.D., Wang, E.Y.M., & Calder, A.C. 1999, preprint [astro-ph/9911192]
54. Wang, E.Y.M., Swesty, F.D., & Calder, A.C. 1998, in Proceedings of the Second Oak Ridge Symposium on Atomic and Nuclear Astrophysics, [astro-ph/9806022]
55. Wiringa, R.B., Fiks, V., & Fabrocini, A. 1988, PRC, 38, 1010
56. Wiseman, A.G. 1993, PRD, 48, 4757
57. Zhuge, X., Centrella, J.M., & McMillan, S.L.W. 1994, PRD, 50, 6247
58. Zhuge, X., Centrella, J.M., & McMillan, S.L.W. 1996, PRD, 54, 7261

Supermassive Black Holes: Their Formation, and Their Prospects as Probes of Relativistic Gravity

Martin J. Rees

Institute of Astronomy
Madingley Road, Cambridge, CB3 0HA, UK

Abstract. The existence of supermassive collapsed objects in the cores of most galaxies poses still-unanswered questions. First, how did they form, and how does their mass depend on the properties of the host galaxy? Second, can observations probe the metric in the strong-field domain, testing whether it indeed agrees with the Kerr geometry predicted by general relativity (and, if so, what the spin is)?

1 Introduction

Compact dark objects, with deep gravitational potential wells, seem to lurk in most galactic centres; but current evidence cannot 'diagnose' the metric in the innermost region where Newtonian approximations break down. Several other speakers have described the status of the observations, as well as some aspects of theoretical models. This written text addresses two issues. How did supermassive holes form? And do they have Schwarzschild/Kerr metrics, thereby offering real prospects of testing our theories of strong-field gravity.

It has long been suspected that supermassive holes are implicated in the power output from active galactic nuclei (AGNs), and in the production of relativistic jets that energise strong radio sources. But the demography of these massive holes has been clarified by studies of relatively nearby galaxies: the centres of most of these display either no activity or a rather low level, but most seem to harbour dark central masses. Recent observational progress brings into sharper focus the question of how and when supermassive black holes formed, and how this process relates to galaxy formation.

There are now two spectacularly convincing cases of massive collapsed objects in nearby galaxies. The first, in the peculiar spiral NGC 4258, has been revealed by amazingly precise mapping of gas motions via the 1.3 cm maser-emission line of H_2O. [1,2]. The spectral resolution of this microwave line is high enough to pin down the velocities with accuracy of 1 km s^{-1}. The Very Long Baseline Array achieves an angular resolution better than 0.5 milliarc seconds (100 times sharper than the HST, as well as far finer spectral resolution of velocities!). These observations have revealed, right in NGC 4258's core, a disc with rotational speeds following an exact Keplerian law around a compact dark mass. The inner edge of the observed disc is orbiting at 1080 km s^{-1}. It would be impossible to circumscribe, within its radius, a stable and long-lived star cluster with the inferred mass of $3.6 \times 10^7 \, M_\odot$.

The second utterly convincing candidate lies in our own Galactic Centre. Most nearby large galaxies seem to harbour massive central holes, so our own would seem underendowed if it did not have one too. Some have advanced this view for many years (eg ref [3]). Also, an unusual radio source has long been known to exist right at the dynamical centre of our Galaxy, which can be interpreted in terms of accretion onto a massive hole [4-6]. Direct evidence used to be ambiguous because intervening gas and dust in the plane of the Milky Way prevents us from getting a clear optical view of the central stars, as we can in, for instance, M 31. A great deal was known about gas motions, from radio and infrared measurements, but these were hard to interpret because gas does not move ballistically like stars, but can be influenced by pressure gradients, stellar winds, and other non-gravitational influences.

The situation was transformed by remarkable observations of stars in the near infrared band, where obscuration by intervening material is less of an obstacle. These are presented by Eckhart and by Ghez at this meeting. The speeds scale as $r^{-1/2}$ with distance from the centre, consistent with a hole of mass $2.5 \times 10^6 \, M_\odot$.

As other speakers will discuss, there is a crude proportionality between the hole's mass and that of the central bulge or spheroid in the stellar distribution (which is of course the dominant part of an elliptical galaxy, but only a subsidiary component of a disc system like M31 or our own Galaxy). But how did the holes form?

2   AGN Demography and Black-Hole Formation

Many of the faint smudges visible in the Hubble Deep Field [7] are galaxies with redshifts of order 3, being viewed at (or even before) the era when their spheroids formed. Physical conditions in the central potential wells of young and gas-rich galaxies should be propitious for black-hole formation, and such processes are presumably connected with high-z quasars. It now seems clear that most galaxies that existed at $z = 3$ would have participated subsequently in a series of mergers; giant present-day elliptical galaxies are the outcome of such mergers. Any black holes already present would tend to spiral inwards, and coalesce (unless a third body fell in before the merger was complete, in which case a Newtonian slingshot could eject all three: a binary in one direction; the third, via recoil, in the opposite direction).

The issues are then:

(a) how much does a black hole grow (and how much electromagnetic energy does it radiate) at each stage? and

(b) how far up the 'merger tree' did the first massive holes form? A single big galaxy can be traced back to the stage when it was in dozens of smaller components with individual internal velocity dispersions as low as 20 km/sec.

Did central black holes form even in these small and weakly bound systems? If so, they could have coalesced, in a hierarchical fashion, during subsequent mergers.

Perhaps black holes form with the same efficiency in small galaxies (with shallow potential wells), or maybe their formation had to await the buildup of substantial galaxies with deeper potential wells (i.e. with V above some threshold). This issue is important for the detectability of high-z miniquasars; it also determines whether the ionizing UV background at high redshifts has a nonthermal component that is able to ionize He as well as H.

The actual formation mechanism is still uncertain. More than 20 years ago, I presented a 'flow diagram' [8] which carried the message that it seemed likely – indeed almost inevitable - that large masses would collapse in galactic centres: there was indeed a variety of possible routes. We have now got used to the idea that black holes exist within most galaxies, but it is rather depressing that we still cannot decide which formation route is most likely.

One possibility is that the gas in a 'proto-spheroid' does not all break up into stellar-mass condensations, but that a supermassive star forms, which then collapses. As the gas evolved (through loss of energy and angular momentum) to higher densities and more violent internal dissipation, radiation pressure would prevent fragmentation, and puff it up into a single superstar [9,10]. Ordinary star formation may be suppressed even at less extreme densities – i.e. before the gas has become a single superstar – by other effects. For example, a magnetic field, even if not dynamically important overall, could inhibit fragmentation, especially because the free-electron concentration is unlikely to fall low enough to permit ambipolar diffusion, whereby the magnetic flux can escape from protostars in present-day molecular clouds.

Once a large mass of gas started to behave like a single superstar, it would continue to contract and deflate. Some mass would inevitably be shed, carrying away angular momentum, but the remainder would undergo complete gravitational collapse. This could be a substantial fraction – for example, if 10 percent of the mass had to be shed in order to allow contraction by a factor of 2, about 20 percent could form a black hole [10,11].

The mass of the hole would depend on that of its host galaxy, though not necessarily via an exact proportionality: the angular momentum of the protogalaxy and the depth of its central potential well are relevant factors too. Firmer and more quantitative conclusions will have to await elaborate numerical simulations. But it certainly seems in no way implausible that massive black holes form directly from gas (some, albeit, already processed through stars), perhaps after a transient phase as a supermassive object.

However, we cannot exclude the alternative 'scenario' where a massive star builds up within a dense central cluster of ordinary stars. The most detailed calculations were done by Quinlan and Shapiro ([12] and other references cited therein). These authors showed that stellar coalescence, followed by the segregation of the resultant high-mass stars towards the centre, could trigger runaway evolution without (as earlier and cruder work had suggested) requiring implausible initial starting points. It would be well worthwhile extending these simulations to a wider range of initial conditions, and also to follow the build-up from stellar masses to a supermassive object.

It is worth noting, incidentally, that whereas activity in low-z galaxies may be correlated with some unusual disturbance due to a tidal encounter or merger, this may not be the right way to envisage the more common high-z quasars, since almost all high-z galaxies are 'disturbed', in the sense that they are nearly always experiencing a merger or disturbance that is sufficient to perturb axisymmetry or to trigger a large inflow of gas.

The most massive black holes would have gained mass through a succession of mergers, as well as through accretion of gas at each stage. Haehnelt and Kauffmann ([13], and these proceedings) have modelled this in the context of semi-analytic schemes for galaxy evolution, and have achieved a good fit with the luminosity function and z-dependence of quasars.

3   Do the Candidate Holes Obey the Kerr Metric?

3.1   Probing Near the Hole

The observed molecular disc in NGC 4258 lies a long way out: at around 10^5 gravitational radii. We can exclude all conventional alternatives (dense star clusters, etc); however, the measurements tell us nothing about the central region where gravity is strong – certainly not whether the putative hole actually has properties consistent with the Kerr metric. The stars closest to our Galactic Centre likewise lie so far out from the putative hole (their speeds are less than 1 percent that of light) that their orbits are essentially Newtonian.

We can infer from AGNs that 'gravitational pits' exist, which must be deep enough to allow several percent of the rest mass of infalling material to be radiated from a region compact enough to vary on timescales as short as an hour. But we still lack quantitative probes of the relativistic region. We believe in general relativity primarily because it has been resoundingly vindicated in the weak field limit (by high-precision observations in the Solar System, and of the binary pulsar) – not because we have evidence for black holes with the precise Kerr metric.

Relativists would seize eagerly on any relatively 'clean' probe of the strong-field domain. The emission from most accretion flows is concentrated towards the centre, where the potential well is deepest and the motions fastest. Such basic features of the phenomenon as the overall efficiency, the minimum variability timescale, and the possible extraction of energy from the hole itself all depend on inherently relativistic features of the metric – on whether the hole is spinning or not, how it is aligned, etc. But the data here are imprecise and 'messy'. We would occasionally expect to observe, even in quiescent nuclei, the tidal disruption of a star. Exactly how this happens would depend on distinctive precession effects around a Kerr metric, but the gas dynamics are so complex that even when a flare is detected it will not serve as a useful diagnostic of the metric in the strong-field domain. There are however several encouraging new possibilities.

3.2   X-Ray Spectroscopy of Accretion Flows

Optical spectroscopy tells us a great deal about the gas in AGNs. However, the optical spectrum originates quite far from the hole. This is because the innermost regions would be so hot that their thermal emission emerges as more energetic quanta. X-rays are a far more direct probe of the relativistic region. The appearance of the inner disc around a hole, taking doppler and gravitational shifts into account, along with light bending, was first calculated by Bardeen and Cunningham [14] and subsequently by several others (e.g. [15]). There is of course no hope (until X-ray interferometry is developed) of actually 'imaging' these inner discs. However, the large frequency-shifts could reveal themselves spectroscopically — substantial gravitational redshifts would be expected, as well as large doppler shifts [15]. Until recently, the energy resolution and sensitivity of X-ray detectors was inadequate to permit spectroscopy of extragalactic objects. The ASCA X-ray satellite was the first with the capability to measure emission line profiles in AGNs. There is already one convincing case [16] of a broad asymmetric emission line indicative of a relativistic disc, and others should soon follow. The value of (a/m) can in principle be constrained too, because the emission is concentrated closer in, and so displays larger shifts, if the hole is rapidly rotating, and there is some evidence that this must be the case in MCG -6-30-15 [17].

The Chandra and XMM X-ray satellites should be able to extend and refine these studies; they may offer enough sensitivity, in combination with time-resolution, to study flares, and even to follow a 'hot spot' on a plunging orbit.

The swing in the polarization vector of photon trajectories near a hole was long ago suggested [18] as another diagnostic; but this is still not feasible because X-ray polarimeters are far from capable of detecting the few percent polarization expected.

3.3   The Blandford-Znajek Process

Blandford and Znajek [19] showed that a magnetic field threading a hole (maintained by external currents in, for instance, a torus) could extract spin energy, converting it into directed Poynting flux and electron-positron pairs. Can we point to objects where this is definitively happening? The giant radio lobes from radio galaxies sometimes spread across millions of lightyears – 10^{10} times larger than the hole itself. If the Blandford-Znajek process is really going on, these huge structures may be the most direct manifestation of an inherently relativistic effect around a Kerr hole.

Jets in some AGNs definitely have Lorentz factors exceeding 10. Moreover, some are probably Poynting-dominated, and contain pair (rather than electron-ion) plasma. But there is still no compelling reason to believe that these jets are energised by the hole itself, rather than by winds and magnetic flux 'spun off' the surrounding torus. The case for the Blandford-Znajek mechanism would be strengthened if baryon-free jets were found with still higher Lorentz factors, or if the spin of the holes could be independently measured, and the properties of jets turned out to depend on (a/m).

The process cannot dominate unless either the field threading the hole is comparable with that in the orbiting material, or else the surrounding material radiates with low radiative efficiency. These requirements cannot be ruled out, though there has been recent controversy about how plausible they are. (It may be worth noting that the Blandford-Znajek effect could also be important in the still more extreme context of gamma-ray bursts, where a newly formed hole of a few solar masses could be threaded by a field exceeding 10^{15} G.)

3.4   What is the Expected Spin?

The spin of a hole affects the efficiency of 'classical' accretion processes; the value of a/m also determines how much energy is in principle extractable by the Blandford-Znajek effect. Moreover, the orientation of the spin axis may be important in relation to jet production, etc.

Spin-up is a natural consequence of prolonged disc-mode accretion: any hole that has (for instance) doubled its mass by capturing material that is all spinning the same way would end up with a/m being at least 0.5. A hole that is the outcome of a merger between two of comparable mass would also, generically, have a substantial spin. On the other hand, if it had gained its mass from capturing many low-mass objects (holes, or even stars) in randomly-oriented orbits, a/m would be small.

3.5   Precession and Alignment

Most of the literature on gas dynamics around Kerr holes assumes that the flow is axisymmetric. This assumption is motivated not just by simplicity, but by the expectation that Lense-Thirring precession would impose axisymmetry close in, even if the flow further out were oblique and/or on eccentric orbits. Plausible-seeming arguments, dating back to the pioneering 1975 paper by Bardeen and Petterson [20], suggested that the alignment would occur, and would extend out to a larger radius if the viscosity were low because there would be more time for Lense-Thirring precession to act on inward-spiralling gas. However, later studies, especially by Pringle, Ogilvie, and their associates, have shown that naive intuitions can go badly awry. The behaviour of the 'tilt' is much more subtle; the effective viscosity perpendicular to the disc plane can be much larger than in the plane. In a thin disc, the alignment effect is actually weaker when viscosity is low. What happens in a thick torus is still unclear, and will have to await 3-D gas-dynamical simulations.

The orientation of a hole's spin and the innermost flow patterns could have implications for jet alignment. An important paper by Pringle and Natarajan [21] shows that 'forced precession' effects due to torques on a disc can lead to swings in the rotation axis that are surprisingly fast (i.e. on timescales very much shorter than the timescale for changes in the hole's mass).

3.6   Stars in Relativistic Orbits?

Gas-dynamical phenomena are complicated because of viscosity, magnetic fields etc. It would be nice to have a 'cleaner' and more quantitative probe of the strong-field regime: for instance, a small star orbiting close to a supermassive hole. Such a star would behave like a test particle, and its precession would probe the metric in the 'strong field' domain. These interesting relativistic effects, have been computed in detail by Karas and Vokrouhlicky [22,23]. Would we expect to find a star in such an orbit?

An ordinary star certainly cannot get there by the kind of 'tidal capture' process that can create close binary star systems. This is because the binding energy of the final orbit (a circular orbit with the same angular momentum as an initially near-parabolic orbit with pericentre at the tidal-disruption radius) would have to be dissipated within the star, and that cannot happen without destroying it. Syer, Clarke and Rees [24] pointed out, however, that an orbit can be 'ground down' by successive impacts on a disc (or any other resisting medium) without being destroyed: the orbital energy then goes almost entirely into the material knocked out of the disc, rather than into the star itself. Other constraints on the survival of stars in the hostile environment around massive black holes – tidal dissipation when the orbit is eccentric, irradiation by ambient radiation, etc. – are explored by Podsiadlowski and Rees [25], and King and Done [26]. They can be thought of as close binary star systems with extreme mass ratios.

These stars would not be directly observable, except maybe in our own Galactic Centre. But they might have indirect effects: such a rapidly-orbiting star in an active galactic nucleus could signal its presence by quasiperiodically modulating the AGN emission.

3.7   Gravitational-Wave Capture of Compact Stars

Neutron stars or white dwarfs circling close to supermassive black holes would be impervious to tidal dissipation, and would have such a small geometrical cross section that the 'grinding down' process would be ineffective too. On the other hand, because they are small they can get into very tight orbits by straightforward stellar-dynamical processes. For ordinary stars, the 'point mass' approximation breaks down for encounter speeds above 1000 km/s – physical collisions are then more probable than large-angle deflections. But there is no reason why a 'cusp' of tightly bound compact stars should not extend much closer to the hole. Neutron stars or white dwarfs could exchange orbital energy by close encounters with each other until some got close enough that they either fell directly into the hole, or until gravitational radiation became the dominant energy loss. When stars get very close in, gravitational radiation losses become significant, and tend to circularise an elliptical orbit with small pericentre. Most such stars would be swallowed by the hole before circularisation, because the angular momentum of a highly eccentric orbit 'diffuses' faster than the energy does due to encounters with other stars, but some would get into close circular orbits [27,28].

A compact star is less likely than an ordinary star in similar orbit to 'modulate' the observed radiation in a detectable way. But the gravitational radiation (almost periodic because the dissipation timescale involves a factor $(M_{\rm hole}/m^*)$) would be detectable.

3.8   Scaling Laws and 'Microquasars'

Two galactic X-ray sources that are believed to involve black holes generate double radio structures that resemble miniature versions of the classical extragalactic strong radio sources. These are discussed in the papers by Mirabel and Fender. The jets have been found to display apparent superluminal motions across the sky, indicating that, like the extragalactic radio sources, they contain plasma that is moving relativistically.

There is no reason to be surprised by this analogy between phenomena on very different scales. Indeed, the physics is exactly the same, apart from very simple scaling laws. If we define $l = L/L_{\rm Ed}$ and $\dot{m} = \dot{M}/\dot{M}_{\rm crit}$, where $\dot{M}_{\rm crit} = L_{\rm Ed}/c^2$, then for a given value of $\dot{m}$, the flow pattern may be essentially independent of M. Linear scales and timescales, of course, are proportional to M, and densities in the flow scale as M^{-1}. The physics that amplifies and tangles any magnetic field may be scale-independent, and the field strength B scales as $M^{-1/2}$. So the bremsstrahlung or synchrotron cooling timescales go as M, implying that $t_{\rm cool}/t_{\rm dyn}$ is insensitive to M for a given $\dot{m}$. So also are ratios involving, for instance, coupling of electron and ions in thermal plasma. Therefore, the efficiencies and the value of l are insensitive to M, and depend only on $\dot{m}$. Moreover, the form of the spectrum, for given $\dot{m}$, depends on M only rather insensitively (and in a manner that is easily calculated).

The kinds of accretion flow inferred in, for instance, M87, giving rise to a compact radio and X-ray source, along with a relativistic jet, could operate just as well if the hole mass was lower by a hundred million, as in the galactic LMXB sources. So we can actually study the same processes involved in AGNs in microquasars close at hand within our own galaxy. And these miniature sources may allow us to observe a simulacrum of the entire evolution of a strong extragalactic radio source, its life-cycle speeded up by a similar factor.

3.9   Discoseismology

Discs or tori that are maintained by steady flow into a black hole can support vibrational modes [29-31]. The frequencies of these modes can, as in stars, serve as a probe for the structure of the inner disc or torus. The amplitude depends on the importance of pressure, and hence on disc thickness; how they are excited, and the amplitude they may reach, depends, as in the Sun, on interaction with convective cells and other macroscopic motions superimposed on the mean flow. But the frequencies of the modes can be calculated more reliably. In particular, the lowest g-mode frequency is close to the maximum value of the radial epicyclic frequency k. This epicyclic frequency is, in the Newtonian domain, equal to the orbital frequency. It drops to zero at the innermost stable orbit. It has a maximum at about $9GM/c^2$ for a Schwarzschild hole; for a Kerr hole, k peaks at a smaller radius (and a higher frequency for a given M). The frequency is 3.5 times higher for $(a/m) = 1$ than for the Schwarzschild case.

Novak and Wagoner [31] pointed out that these modes may cause an observable modulation in the X-ray emission from galactic black hole candidates. Just such effects have been seen in GRS 1915+105 [32]. The amplitude is a few percent (somewhat larger at harder X-ray energies) suggesting that the oscillations involve primarily the hotter inner part of the disc. The fluctuation spectrum shows a peak in Fourier space at around 67 Hz. This frequency does not change even when the X-ray luminosity doubles, suggesting that it relates to a particular radius in the disc. If this is indeed the lowest g-mode, and if the simple disc models are relevant, then the implied mass is $10.2 \, M_\odot$ for Schwarzschild, and $35 \, M_\odot$ for a 'maximal Kerr' hole (Nowak et al 1997). The mass of this system is not well known. However, this technique offers the exciting prospect of inferring (a/m) for holes whose masses are independently known.

GRS 1915+105 is one of the objects with superluminal radio jets. The simple scaling arguments in Section 3.8 imply that the AGNs which it resembles might equally well display oscillations with the same cause. However, the periods there would be measured in days, rather than fractions of a second.

4   Gravitational Radiation as a Probe

4.1   Gravitational Waves from Newly-Forming Massive Holes?

The gravitational radiation from black holes offers impressive tests of general relativity, involving no physics other than that of spacetime itself.

At first sight, the formation of a massive hole from a monolithic collapse might seem an obvious source of strong wave pulses. The wave emission would be maximally intense and efficient if the holes formed on a timescale as short as (r_g/c), where $r_g = (GM/c^2)$ – something that might happen if they built up via coalescence of smaller holes (cf. [12]).

If, on the other hand, supermassive black holes formed as suggested in Section 2 – directly from gas (some, albeit, already processed through stars), perhaps after a transient phase as a supermassive object – then the process may be too gradual to yield efficient gravitational radiation. That is because post-Newtonian instability is triggered at a radius $r_i \gg r_g$. Supermassive stars are fragile because of the dominance of radiation pressure: this renders the adiabatic index γ only slightly above 4/3 (by an amount of order $10(M/\mathrm{M}_\odot)^{-1/2}$). Since $\gamma = 4/3$ yields neutral stability in Newtonian theory, even the small post-Newtonian corrections then destabilize such 'superstars'. The characteristic collapse timescale when instability ensues is longer than r_g/c by the 3/2 power of the collapse factor.

The post-Newtonian instability is suppressed by rotation. A differentially rotating supermassive star could in principle support itself against post-Newtonian instability until it became very tightly bound. It could then perhaps develop a a bar-mode instability and collapse within a few dynamical times. To achieve this tightly-bound state without drastic mass loss, the object would need to have deflated over a long timescale, losing energy at no more than the Eddington rate.

The formation of a hole 'in one go' from a supermassive star is an unpromising source of gravitational waves. On the other hand, strong signals are expected when already-formed holes coalesce, as the aftermath of mergers of their host galaxies.

The gravitational waves associated with supermassive holes would be concentrated in a frequency range around a millihertz – too low to be accessible to ground-based detectors, which lose sensitivity below 100 Hz, owing to seismic and other background noise. Space-based detectors are needed. One such, proposed by the European Space Agency, is the Laser Interferometric Spacecraft (LISA) – three spacecraft on solar orbit, configured as a triangle, with sides of 5 million km long whose length is monitored by laser interferometry.

4.2   Gravitational Waves from Coalescing Supermassive Holes

The guaranteed sources of really intense gravitational waves in LISA's frequency range would be coalescing supermassive black holes. Many galaxies have experienced a merger since the epoch $z > 2$ when, according to 'quasar demography' arguments they acquired central holes. The holes in the two merging galaxies would spiral together, emitting, in their final coalescence, up to ~ 10 per cent of their rest mass as a burst of gravitational radiation in a timescale of only a few times r_g/c. These pulses would be so strong that LISA could detect them with high signal-to-noise even from large redshifts. Whether such events happen often enough to be interesting can to some extent be inferred from observations (we see many galaxies in the process of coalescing), and from simulations of the hierarchical clustering process whereby galaxies and other cosmic structures form. Haehnelt [33 and later references] has calculated the merger rate of the large galaxies believed to harbour supermassive holes: it is only about one event per century, even out to redshifts $z = 4$. However, big galaxies are probably the outcome of many successive mergers. As discussed in Section 2, we still have no direct evidence – nor firm theoretical clues – on whether these small galaxies harbour black holes (nor, if they do, of what the hole masses typically are). However it is certainly possible that enough holes of (say) $10^5 \, M_\odot$ lurk in small early-forming galaxies to yield, via subsequent mergers, more than one event per year detectable by LISA.

LISA is potentially so sensitive that it could detect the nearly-periodic waves from stellar-mass objects orbiting a $10^5 - 10^6 \, M_\odot$ hole, even at a range of a hundred Mpc, despite the m^*/M_{hole} factor whereby the amplitude is reduced compared with the coalescence of two objects of comparable mass M_{hole}. The stars in the observed 'cusps' around massive central holes in nearby galaxies are of course (unless almost exactly radial) on orbits that are far too large to display relativistic effects. Occasional captures into relativistic orbits can come about by dissipative processes – for instance, interaction with a massive disc [24,34]. But unless the hole mass were above $10^8 \, M_\odot$ (in which case the waves would be at too low a frequency for LISA to detect), solar-type stars would be tidally disrupted before getting into relativistic orbits. Interest therefore focuses on compact stars, for which dissipation due to tidal effects or drag is less effective. As described in Section 3.7, compact stars may get captured as a result of gravitational radiation, which can gradually 'grind down' an eccentric orbit with close pericenter passage into a nearly-circular relativistic orbit. The long quasi-periodic wave trains from such objects, modulated by orbital precession (cf. [22,23]) in principle carries detailed information about the metric.

The attraction of LISA as an 'observatory' is that even conservative assumptions lead to the prediction that a variety of phenomena will be detected. If there were many massive holes not associated with galactic centres (not to mention other speculative options such as cosmic strings), the event rate would be much enhanced. Even without factoring on an 'optimism factor' we can be confident that LISA will harvest a rich stream of data.

LISA is now being actively studied both in Europe and the US. If funded jointly by ESA and NASA, it could fly within ten years.

4.3   Gravitational-Wave Recoil

Is there any way of learning, before that date, something about gravitational radiation? The dynamics (and gravitational radiation) when two holes merge has so far been computed only for cases of special symmetry. The more general problem – coalescence of two Kerr holes with general orientations of their spin axes relative to the orbital angular momentum – is a 'grand challenge' computational project being tackled at the MPI in Potsdam, and at US

362   Martin J. Rees

centres. When this challenge has been met (and it will almost certainly not take all the time until LISA flies) we shall find out not only the characteristic wave form of the radiation, but the recoil that arises because there is a net emission of linear momentum.

There would be a recoil due to the non-zero net linear momentum carried away by gravitational waves in the coalescence. If the holes have unequal masses, a preferred longitude in the orbital plane is determined by the orbital phase at which the final plunge occurs. For spinning holes there may, additionally, be a rocket effect perpendicular to the orbital plane, since the spins break the mirror symmetry with respect to the orbital plane. [35]

The recoil is a strong-field gravitational effect which depends essentially on the lack of symmetry in the system. It can therefore only be properly calculated when fully 3-dimensional general relativistic calculations are feasible. The velocities arising from these processes would be astrophysically interesting if they were enough to dislodge the resultant hole from the centre of the merged galaxy, or even eject it into intergalactic space. This recoil could displace the hole from the centre of the merged galaxy – it might therefore be relevant to the low−z quasars that seem to be asymmetrically located in their hosts (and which may have been activated by a recent merger). Even galaxies that do not harbour a central hole may, therefore, once have done so in the past. The core of a galaxy that has experienced such an ejection event may retain some trace of it (perhaps, for instance, an unusual profile), because the energy transferred to stars via dynamical friction during the merger process (cf [36]).

The recoil might even be so violent that the merged hole breaks loose from its galaxy and goes hurtling through intergalactic space. This disconcerting thought should at least impress us with the reality and 'concreteness' of the extraordinary entities to whose discovery Riccardo Giacconi contributed so much.

5   Acknowledgements

I am grateful to several colleagues, especially Mitch Begelman, Andy Fabian and Martin Haehnelt for discussions and collaboration on topics mentioned here. I thank the Royal Society for support, and the organisers of this conference for the opportunity to participate in celebrating Riccardo Giacconi's extraordinary record of research and scientific leadership.

References

1. Watson, W.D. and Wallin, B.K. 1994 Astrophys. J. (Lett) 432, L35
2. Miyoshi, K. et al 1995 Nature 373, 127.
3. Lynden-Bell, D. and Rees, M.J. 1971 MNRAS 152, 461.

 4. Rees, M.J. 1982 in 'The Galactic Center' ed G. Riegler and R.D. Blandford (A.I.P) p166.
 5. Melia, F. 1994 Astrophys. J. 426, 577
 6. Narayan, R., Yi, I, and Mahadevan, R. 1995 Nature 374, 623.
 7. Williams, R. et al 1996 Astron. J. 112, 1335.
 8. Rees, M.J. 1978, Observatory 98, 210.
 9. Rees, M.J. 1993 Proc. Nat. Acad. Sci 90, 4840
10. Haehnelt M. and Rees, M.J. 1993 MNRAS 263, 168
11. Baumgarte, T.W. and Shapiro, S.L. 1999 Ap. J. 526, 941.
12. Quinlan, G.D. and Shapiro, S.L. 1990 Astrophys. J. 356, 483.
13. Haehnelt, M. and Kauffmann G. 1999 MNRAS in press
14. Bardeen, J. and Cunningham, J., 1972 Ap. J. 173, L137.
15. N.E. White et al 1989 MNRAS 238, 729.
16. Tanaka, Y. et al 1995 Nature 375, 659
17. Iwasawa, I. et al 1999, MNRAS 306, L191.
18. Connors, P.A., Piran, T. and and Stark, R.F. 1980 Ap. J 235, 224
19. Blandford, R.D. and Znajek, R.L. 1977 MNRAS 179, 433
20. Bardeen, J. and Petterson, J.A. 1975 Ap. J. 195, L65.
21. Natarajan, P. and Pringle, J.E. 1999 Ap. J. in press.
22. Karas, V., and Vokrouhlicky, D., 1993 MNRAS 265, 365
23. Karas, V., and Vokrouhlicky, D., 1994 Astrophys. J 422, 208
24. Syer, D., Clarke, C.J. and Rees, M.J. 1991 MNRAS 250, 505.
25. Podsiadlowski, P. and Rees, M.J. 1994 in 'Evolution of X-ray binaries' ed S Holt and C.Day (AIP) p403.
26. King, A.R. and Done. C., 1993 MNRAS 264, 388
27. Hils, D. and Bender, P.L. 1995 Astrophys. J. (Lett) 445, L7
28. Sigurdsson, S. and Rees M.J. 1997 MNRAS 284, 318.
29. Kato, S. and Fukui, J. 1980 PASJ 32, 377
30. Novak, M.A. and Wagoner, R.V. 1992 Astrophys. J. 393, 697
31. Novak, M.A. and Wagoner, R.V. 1993 Astrophys J. 418, 187
32. Morgan, E., Remillard, R. and Greiner, J. 1996 IAU Circular No. 6392.
33. Haehnelt M. 1994 MNRAS 269,199
34. Canizzo, J.K., Lee, H.M. and Goodman, J. 1990 Astrophys. J. 351, 38
35. Redmount, I. and Rees, M.J. 1989 Comm. Astrophys. Sp. Phys 14, 185.
36. Ebisuzaki, T., Makino, J., and Okumura, S.K. 1991 Nature 354, 212

The Formation and Evolution of Supermassive Black Holes and their Host Galaxies

Martin G. Haehnelt and Guinevere Kauffmann

Max-Planck-Institut für Astrophysik, Karl-Schwarzschild-Str. 1,
85740 Garching, Germany

Abstract. We discuss constraints on the assembly history of supermassive black holes from the observed remnant black holes in nearby galaxies and from the emission caused by accretion onto these black holes. We also summarize the results of a specific model for the evolution of galaxies and their central black holes which traces their hierarchical build-up in CDM-like cosmogonies. The model assumes (i) that black holes, ellipticals and starbursts form during major mergers of galaxies (ii) that the gas fraction in galaxies decreases with decreasing redshift (iii) that the optically bright phase of a QSO lasts for about 10^7 years. The model successfully reproduces the evolution of cold gas as traced by damped Lyα systems, the evolution of optically bright QSOs, the remnant black hole mass distribution and the host-galaxy luminosities of QSOs.

1 Introduction

The evidence for the existence of supermassive black holes has been steadily increasing over the last years. For the two most convincing cases our own galactic centre and NGC4258 [1–3], the evidence is now beyond reasonable doubt. The evidence that most nearby galaxies contain supermassive black holes is also compelling. Early suggestions of a linear relation between the black-hole mass and the bulge mass have been corroborated by larger samples [4,5]. It is generally believed that we observe a significant fraction if not all the material falling into supermassive black holes by the radiation emitted by active galactic nuclei. A supermassive black hole therefore seems to "know" in which galaxy it will end up at the present day. This and the fact that supermassive black holes contain as much as 0.2 to 0.6 percent of the baryonic mass of the galaxy [5,6], suggests that the formation of stars in the bulges of galaxies and the assembly of supermassive black holes at their centre are closely linked [7,8]. On the other hand, there is strong evidence that structures in the Universe form hierarchically, i.e. larger structures build up by merging of smaller structures. This is a generic feature of a wide class of structure formation scenarios, the so called cold dark matter (CDM) cosmogonies. Both galaxies and AGN activity have been successfully modelled within such hierarchical cosmogonies [9–15]. Here we first review observational constraints on the accretion history of supermassive black holes and discuss some clues for the formation mechanism. We then summarize the results of a specific model that describes the joint evolution of galaxies and supermassive black holes within a hierarchical cosmogony (see Kauffmann & Haehnelt [16] for more details). We assume here $\Omega_{\mathrm{mat}} = 0.3$, $\Omega_\Lambda = 0.7$, $h = 0.65$ and $\sigma_8 = 1$.

2   The Formation and Evolution of Supermassive Black Holes

2.1   Black-Hole Mass Densities

We can get some information on the assembly history of supermassive black holes in nearby galaxies by comparing the mass density of remnant black holes to that required to produce the radiation emitted during the accretion process [17]. The mass density in remnant black holes can be inferred from the total mass density in bulges and the mass ratio of black hole mass to bulge mass:

$$\rho_{\mathrm{BH}} = 7.2 \times 10^5 \left(\frac{M_{\mathrm{bh}}/M_{\mathrm{bulge}}}{0.002} \right) \left(\frac{\Omega_{\mathrm{bulge}}}{0.003} \right) M_\odot \, \mathrm{Mpc}^{-3}.$$

The total mass density in bulges has been estimated by Fugukita, Hogan and Peebles to be $0.001h^{-1} \leq \Omega_{\mathrm{bulge}} \leq 0.003h^{-1}$ [18]. The normalization of the bulge to black hole mass correlation is still a matter of debate. Magorrian et al. [5] claim a value of 0.6 percent while van der Marel [6] argues that a value of 0.2 percent is more realistic. We will adopt the latter value in the remainder of this paper.

The integrated emission by optically bright QSOs due to accretion onto supermassive black holes can also be used to infer the corresponding mass density in supermassive black holes if an efficiency for the transformation of accreted rest mass into optical $f_{\mathrm{B}}\epsilon$ light is assumed [19,20]:

$$\rho_{\mathrm{Opt}} = 1.4 \times 10^5 \left(\frac{f_{\mathrm{B}}\,\epsilon}{0.01} \right)^{-1} M_\odot \, \mathrm{Mpc}^{-3}.$$

Here ϵ is the overall efficiency of transforming accreted rest mass energy into radiation and f_{B} is the fraction emitted in the B-band. Similarly we can estimate the black hole mass density which results from the emission of hard X-rays [21]:

$$\rho_{\mathrm{X-ray}} = 3.8 \times 10^5 \left(\frac{f_{\mathrm{X-ray}}\,\epsilon}{0.01} \right)^{-1} M_\odot \, \mathrm{Mpc}^{-3},$$

where we have assumed a total hard X-ray flux of $140\,\mathrm{keV\,s}^{-1}\,\mathrm{cm}^{-2}\,\mathrm{sr}^{-1}$. The sources producing the hard X-ray background are generally assumed to be a class of AGN different to optically bright QSOs that has not yet been identified. We adopt an effective emission redshift $z_{\mathrm{em}} = 1.5$ for these unidentified sources.

Part of the IR-background should also be produced by AGN, although it has been argued that their contribution should not exceed 30 percent [22]. If 30 percent of the IR background were indeed emitted by AGN [23], then

$$\rho_{\mathrm{IR}} = 7.5 \times 10^5 \left(\frac{f_{\mathrm{IR}} \epsilon}{0.1} \right)^{-1} M_\odot \, \mathrm{Mpc}^{-3},$$

where we have assumed a total IR flux of $15\,\mathrm{nW\,s^{-1}\,cm^{-2}\,sr^{-1}}$ and again $z_{\mathrm{em}} = 1.5$. There has been some debate if the mass density inferred from the optical emission is large enough to explain the mass density in remnant black holes alone. Even for the low value of the black-hole to bulge mass ratio of 0.2 percent adopted here, there is still a discrepancy of a factor of about five. This suggests (i) that there is either a significant contribution to the black-hole mass density by accretion other than that traced by optically bright QSOs or (ii) that the efficiency for producing optical light during the accretion is lower than usually assumed or (iii) that the black hole to bulge mass ratio is lower than 0.2 percent [24,25]. The possible additional accretion may well explain the hard X-ray background and part of the infrared background. It may or may not trace the evolution of optically bright quasars.

2.2   Possible Assembly Histories of Supermassive Black Holes

There is still a rather wide range of possible assembly histories for the super-massive black holes in nearby galaxies (see Haehnelt, Natarajan & Rees [25] for a more detailed discussion). In Fig. 1 we sketch three possible options out of this range. In the *left panel* most of the mass is assembled in supermassive black holes during the epoch of optically bright QSOs around $z \sim 2.5$. In the *middle panel* only 20 % of the mass is assembled in supermassive black holes during the epoch of optically bright QSO. The rest is accreted at low redshift, possibly in the form of hot gas in an advection dominated accretion flow. In the *right panel* most of the mass is assembled into small supermassive

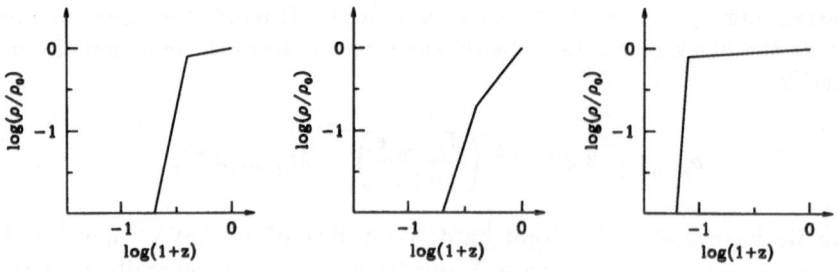

Fig. 1. A sketch of possible assembly histories of supermassive black holes in nearby galaxies. Plotted is the overall black-hole mass density relative to its present-day value. For a further description see text.

black holes at very high redshift. Present-day supermassive black holes form predominantly by merging of smaller black holes. Accretion of gas during the epoch of bright QSOs or at a later epoch does not change the mass density much.

2.3 Clues for the Formation Mechanism of the Typical Supermassive Black Hole

A variety of physical mechanisms for assembling mass into supermassive black holes have been suggested (see Rees [26] and in these proceedings for a review):

- the dynamical evolution of a dense cluster of stellar objects,
- the build-up of a supermassive black hole by merging of smaller black holes,
- the viscous evolution and/or merger-driven collapse of a self-gravitating gaseous object.

All of these processes certainly occur and can lead to the formation of supermassive black holes. The last option seems, however, the most attractive way of explaining the observed black-hole vs. bulge mass relation. The main reason is the high formation efficiency inferred from the large black-hole to bulge mass ratios. It is hard to see how as much as one percent of all available cold gas could end up in a supermassive black hole of $10^9 \, M_\odot$ if an intermediate state of a dense stellar cluster is involved. Initially the relaxation timescales in such a cluster would be long and a considerable fraction of stars would evaporate before the cluster becomes dense enough to evolve rapidly [27]. It is also problematic to build up supermassive black holes predominantly by merging of smaller black holes that have formed well before the epoch of optically bright QSOs. This would require large black-hole formation efficiencies in shallow potential wells with $v_c < 100 \, \mathrm{km \, s^{-1}}$. It seems more plausible that black holes should form less efficiently in smaller potential wells due to the feedback of the energy released both by accretion onto the (forming) supermassive black hole and due to supernovae. It is also unclear whether supermassive black holes in galaxies merge efficiently or whether sling-shot ejection plays a role [28,30]. We therefore consider the assembly history in the right panel of Fig. 1 to be rather improbable.

In the next section we will discuss a specific model for the evolution of galaxies and their supermassive black holes within a hierarchical cosmogony. This model assumes that the optically bright QSOs do trace the accretion history of supermassive black holes well (as in the left panel of Fig.1), but it could easily be altered to accommodate other accretion modes.

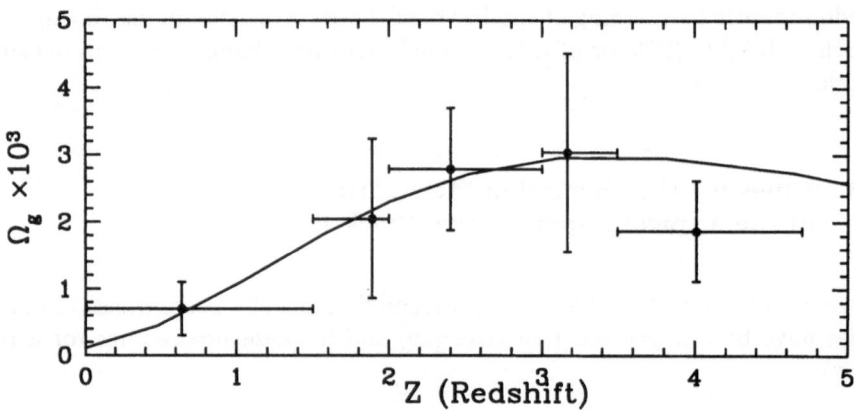

Fig. 2. The cosmological mass density in cold gas in galaxies as a function of redshift. The data are taken from Storrie-Lombardi [29].

3   Modelling the Assembly of Supermassive Black Holes and their Host Galaxies

3.1   Merging Galaxies, Starbursts and AGN

In CDM-like cosmogonies, galaxies build up by hierarchical merging. The formation and evolution of galaxies in such cosmologies has been studied extensively using Monte-Carlo realizations of the hierarchical build-up of galaxies which include simple prescriptions to describe gas cooling, star formation, supernova feedback and merging rates of galaxies. These models reproduce many observed properties of galaxies both at low and at high redshifts [11,13–15]. In the models, the quiescent accretion of gas from the halo results in the formation of a disk. If two galaxies of comparable mass merge, a spheroid is formed and the remaining gas undergoes a starburst. We assume here that such major mergers are also responsible for the growth and fuelling of black holes in galactic nuclei. If two galaxies of comparable mass merge, the central black holes of the progenitors coalesce and a few percent of the gas in the merger remnant is accreted by the new black hole. We have made the following assumptions in our model:

- The fraction of cold gas that forms stars over one dynamical timescale increases with decreasing redshift.
- A fraction $0.01/[1 + (280/v_c)^2]$ of the cold gas in the merging galaxies is accreted by the black hole, where v_c is the circular velocity of their combined dark matter halo.
- The accretion timescale of the gas scales with the dynamical timescale, $t_{\mathrm{acc}} = 2.5 \times 10^7 [0.7 + 0.3(1 + z)^3]^{-0.5}$ yr.

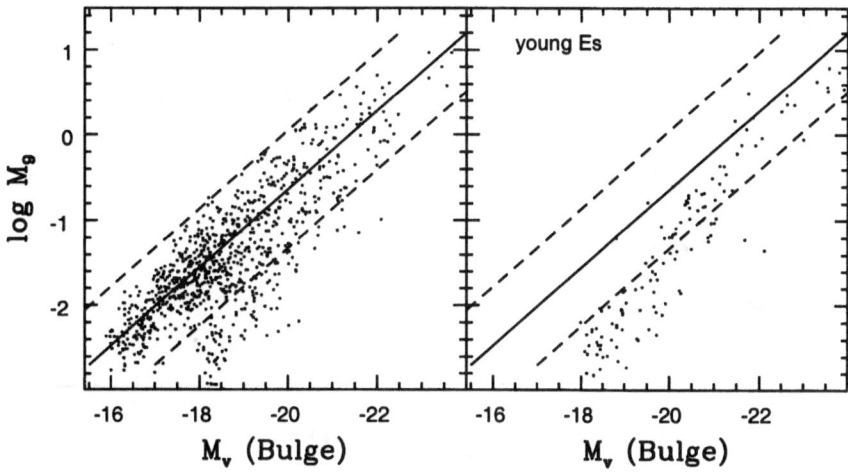

Fig. 3. The correlation between the logarithm of the mass of the central black hole expressed in units of 10^9 M$_\odot$ and the absolute V-band magnitude of the bulge. The dots are an absolute V-band magnitude limited sample of bulges in our model. The thick solid line is the M_V(bulge) vs. M$_{BH}$ relation obtained by Magorrian et al. [5] for nearby normal galaxies. The dashed lines give an indication of the 1σ scatter in the observations. The right panel is the prediction for young ellipticals.

- A fixed fraction of the accreted rest mass energy is radiated away in the optical. The luminosity cannot exceed the Eddington limit.

Our assumptions result in a strong decrease of the gas fraction in galaxies with redshift, from about 75 percent at $z = 3$ to 10 percent at $z = 0$. Our model also fits the strong decrease of the overall density of cool gas in the universe as inferred from the incidence rate of damped Lyα absorbers (Fig. 2). Because of this change in gas fraction with redshift, the gas fraction in major mergers that produce bulges is systematically higher for fainter bulges, which form on average at higher redshift. This might explain the systematic differences in the the slope of the stellar density distribution in the cores of high-luminosity and low-luminosity ellipticals [30].

3.2 Remnant Black Holes in Nearby Galaxies

Fig. 3 shows scatterplots of black hole mass versus bulge luminosity drawn from absolute magnitude-limited catalogues of bulges produced from our models. The thick solid line shows the relation derived by Magorrian et al. [5] and the dashed lines show the 1σ scatter of their observational data around this relation. Both the slope and the scatter predicted by our models agree

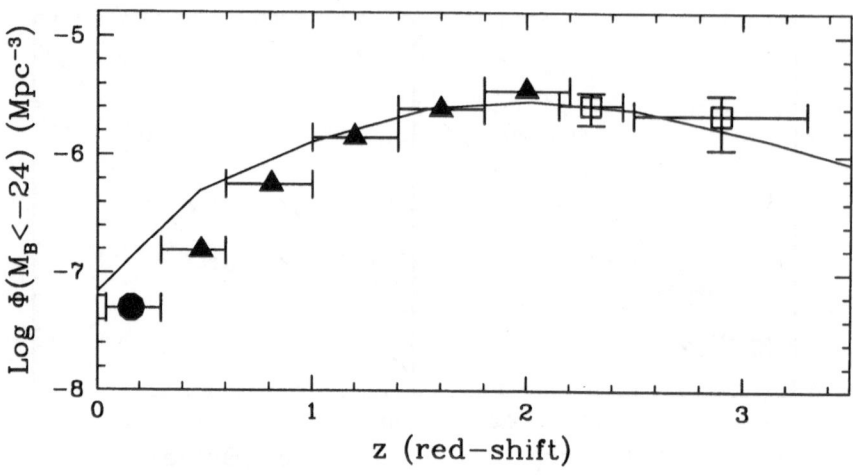

Fig. 4. The evolution of the space density of quasars with $M_B < -24$. The data points are a compilation by Grazian et al. [35]

reasonably well with the observed relation. The normalization is set by the assumed fraction of the cool gas in merger remnants which is accreted onto the black hole. Up to one percent of the cold gas has to find its way into the central supermassive black hole to reproduce a present-day black hole to bulge mass ratio of 0.2 percent. It is interesting to note that a central object with a mass fraction of about two percent can stabilize a bar instability in a surrounding gas disc which may be responsible for driving the gas to the centre [31–33]. One observationally testable prediction of our model is that elliptical galaxies that formed recently should harbour black holes with *smaller* masses than the spheroid population as a whole. This is illustrated in the second panel of Fig. 3, where we show the relation between bulge luminosity and black-hole mass for young ellipticals.

3.3   Evolution of Optically Bright QSOs

One of the striking features of the QSO population is their rapid evolution. The observed rapid decline of the space density at low redshift is not trivial to understand [34]. An important clue is probably the similar rapid drop of the overall amount of cool gas in the Universe inferred from the rate of incidence of damped Lyα systems. The solid curve in Fig. 4 shows the model prediction for the evolution of the QSOs with $M_B < -24$ compared to observational data compiled by Grazian et al. [35]. The agreement is reasonably good. In our model the strong decrease in quasar activity at low redshift results from a combination of three factors: i) a decrease in the merging rates of

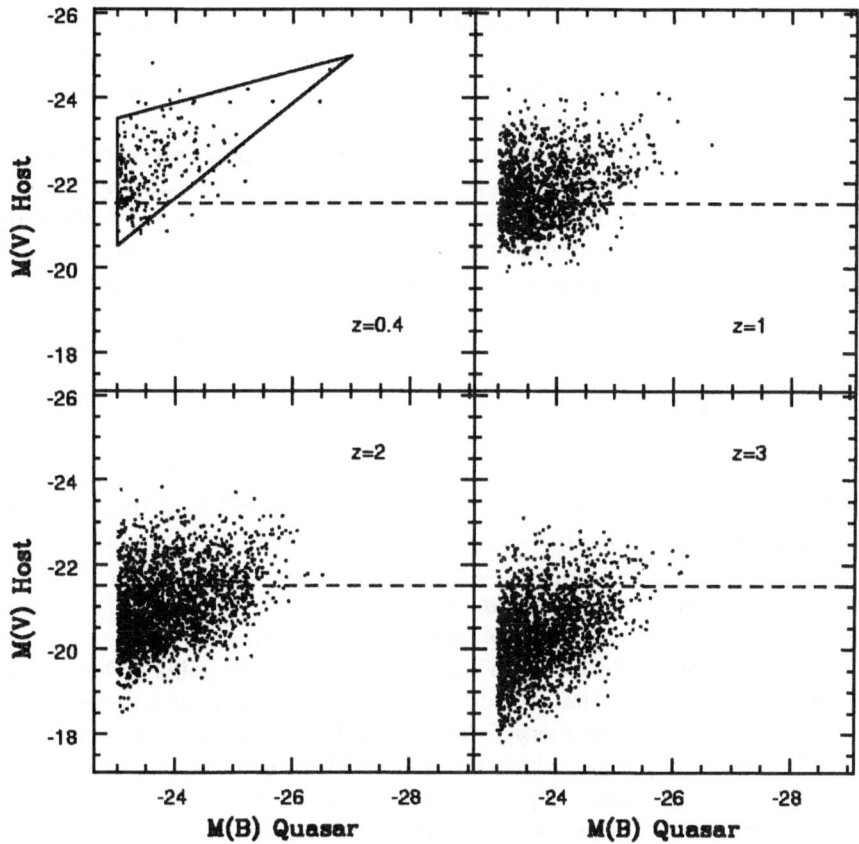

Fig. 5. Host galaxy versus quasar absolute magnitudes at a series of redshifts. The dashed line shows the present value of L_* for galaxies. The triangular box in the top-left panel shows the region spanned by the data set compiled by McLeod, Rieke & Storrie-Lombardi [36].

intermediate mass galaxies at late times, ii) a decrease in the gas fraction of galaxies with decreasing redshift iii) the assumption that black holes accrete gas at a lower rate at late times.

3.4   QSO Host Galaxies

In Fig. 5 we show scatterplots of host-galaxy luminosity versus quasar luminosity at a series of different redshifts. For reference, the horizontal line in each plot shows the present-day value of L_* for galaxies. At low redshift, quasars with magnitudes brighter than $M_B = -23$ reside mostly in galaxies more luminous than L_*. Our results at low redshift agree remarkably well with a compilation of ground-based and HST observations of quasar hosts by Mcleod, Rieke & Storrie-Lombardi [36]. The triangle in Fig. 5 marks the region spanned by their observational data points. At high redshift, our models predict that the quasars should be found in progressively *less luminous* host galaxies. This is not surprising because in hierarchical models, the massive galaxies that host luminous quasars at the present epoch are predicted to have assembled recently [13]. The luminosities of quasars hosted by galaxies at different epochs depends, however, on the redshift scaling of t_{acc}. Recently there have been a number of detections of hosts of high redshift QSOs [37–39]. Typically these seem to have $\sim L_*$ luminosity, suggesting that our assumed t_{acc} and its scaling with redshift is indeed correct.

4   Conclusions

There is agreement to within a factor of a few between the black-hole mass density inferred from black holes in nearby galaxies and that inferred from the radiation emitted by optically bright QSOs. This is consistent with the possibility that optically bright quasars trace the assembly history of supermassive black holes well, but significant accretion in a different accretion mode with a different redshift evolution is also viable.

The large black hole to bulge mass ratio in nearby galaxies argues for a formation mechanism that avoids the intermediate step of a dense stellar cluster. A scenario in which supermassive black holes are assembled by mergers of smaller black holes which formed well before the epoch of optically bright QSOs would require high formation efficiencies (about 10%) in shallow potential wells. The most plausible mechanism by which the mass in a typical present-day supermassive black hole is assembled, is the collapse and accretion of cold gas plus some additional accretion of hot gas and merging of black holes at late times.

It is possible to build a unified model for the evolution of galaxies their central black holes and AGN activity by assuming that black holes, and bulges of galaxies form together during the frequent (major) mergers predicted by hierarchical cosmogonies. Such a model can reproduce the observed rapid evolution of the space density of bright QSOs with redshift, the mass distribution of remnant black holes in nearby galaxies and the luminosity of QSO host galaxies.

Interesting implications of our model are the following. The typical duration of the optically bright QSO phase should be 10^7 yr. Young ellipticals should harbour black holes with smaller masses than the spheroid population as a whole. QSO hosts are typically brighter than L_* at low redshift and should become fainter with increasing redshift. Important for the rapid decline of the space density of bright QSOs and the cosmological density of cold gas, is that the gas fraction in galaxies decreases with decreasing redshift. As a consequence, fainter ellipticals have formed in more gas-rich mergers than bright ellipticals. Supermassive binaries and merging of supermassive binaries should occur frequently in hierarchical cosmogonies. The latter is good news for space-borne gravitational wave experiments like LISA [40,41].

We acknowledge helpful discussions with Andrea Cattaneo, Stefano Cristiani, David Merrit, Prija Natarajan, Joel Primack, Martin Rees, Hans-Walter Rix and Simon White.

References

1. Genzel R., Eckart A., Ott T., Eisenhauer F., 1997, MNRAS, 201, 219
2. Watson W.D., Wallin B.K., 1994, ApJ, 432, L35
3. Miyoshi M., Moran M., Hernstein J., Greenhill L., Nakai N., Diamond P., Inoue N., 1995, Nature, 373, 127
4. Kormendy J., Richstone D., 1995, ARAA, 33, 581
5. Magorrian J., et al., 1998, AJ, 115, 2285
6. van der Marel R. P., 1998, in Sanders D.B.,Barnes J., eds, IAU Symposium 186 Kyoto 1997. Kluwer
7. Richstone D., Ajhar, E.A., Bender, R., Bower, G., Dressler, A., Faber, S.M., Filippenko, A.V., Gebhardt, K. et al., 1998, Nat. Suppl., 395, 14
8. Cattaneo,A., Haehnelt, M.G. & Rees, M.J., 1999, 308, 77
9. Efstathiou G. P., Rees M. J., 1988, MNRAS, 230, 5p
10. Haehnelt M.G., Rees M.J., 1993, MNRAS, 263, 168
11. Kauffmann, G., 1996, MNRAS, 281, 487
12. Haiman Z., Loeb A., 1998, ApJ, 503, 505
13. Kauffmann, G. & Charlot, S., 1998, MNRAS, 297, 23
14. Baugh, C.M., Cole, S., Frenk, C.G. & Lacey, C.G., 1998, ApJ, 498, 504
15. Somerville, R.S., Primack, J.R. & Faber, S.M., 1999, submitted, astro-ph/9806228
16. Kauffmann G., Haehnelt M., 1999, MNRAS, in press
17. Salucci P., Szuszkievicz E., Monaco, P., Danese, L., 1999, MNRAS, in press, astro-ph/9811102
18. Fugukita M., Hogan C.J., Peebles P.J.E., 1998, ApJ, 503, 518
19. Soltan A., 1982, MNRAS, 200, 115.
20. Chokshi A., Turner E. L., 1992, MNRAS, 259, 421
21. Di Matteo T., Fabian A. C., 1997, MNRAS, 286, 393
22. Almaini O., Lawrence A., Boyle B.J., 1999, MNRAS, 305, L59
23. Puget J.L., Abergel A., Bernard J.P., Boulanger F., Burton W.B., Desert F.X., Hartmann D., 1996, A&A, 308, L5
24. Phinney E. S., 1997, talk presented at the IAU Symposium 186 Kyoto 1997
25. Haehnelt M., Natarajan P., Rees M.J., 1998, MNRAS, 300, 817
26. Rees M. J., 1984, ARAA, 22, 471
27. Quinlan G.D., Shapiro S.L., 1990, ApJ, 356, 483
28. Begelman M.C., Blandford R.D., Rees M.J., 1980, Nature, 287, 307
29. Storrie-Lombardi, L.J., MacMahon, R.G. & Irwin, M.J., 1996, MNRAS, 283, L79

30. Meritt D., 1999, in: "Galaxy Dynamics: From the Early Universe to the Present", Paris 1999, eds. F. Combes, G. Mamon, V. Charmandaris
31. Shlosman I., Begelman M.C., Julian F., 1990, Nat., 345, 679
32. Norman C., Sellwood J.A., Hasan H., 1996, ApJ, 462, 114
33. Sellwood J.A., Moore E.M., 1999, ApJ, 510, 125
34. Cavaliere A., Vittorini V., 1997, in Müller V. et al., 1997, Proc. 12th Potsdam cosmology workshop., astro-ph/9712295
35. Grazian, A., Cristiani, S., D'Odorico, V., Omizzolo, A., & Pizella, A., 2000, preprint
36. McLeod, K.K., Rieke, G.H. & Storrie-Lombardi, L.J.,1999, ApJ, 511, L67
37. Aretxaga I., Terlevich R.J., Boyle B.J., 1997, MNRAS,
38. Rix H.-W., Falco E., Impey C., Kochanek C., Lehar J., McLeod B., Munoz J., Peng C., 1999, in "Gravitational Lensing: Recent Progress and Future Goals", eds. T. Brained and C. Kochanek, astroph/9910190
39. Ridgway S., Heckman T., Calzetti D., Lehnert M., 1999, in: " Lifecycles of Radio Galaxies", eds. J. Biretta et al, New Astronomy Reviews, astroph/9911049
40. Haehnelt M.G., 1994, MNRAS, 269, 199
41. Haehnelt M.G., 1999 in: "The second international LISA symposium", Pasadena 1998, Ed. Folkner W., AIP, p. 45

Black Hole Formation in Dark Matter Halos

Juan José Gracia Calvo and Max Camenzind

Landessternwarte Königstuhl, D-69117 Heidelberg, Germany

1   Introduction

The formation of supermassive black holes is not fully understood. The proposed mechanisms predict the formation of these at late times. We study a new scenario where black holes form in a CDM cosmology. While the baryons are smoothly distributed until after the recombination era, cold dark matter decouples early from background expansion, allowing the dark matter inhomogeneities to grow long before recombination. After recombination baryonic matter decouples from photons and falls into the deep gravitational potential wells of the dark component.

2   Model and Results

The evolution of a spherical symmetric system of collisionless DM can be studied analytically [3] or with N-body simulations [1,4]. A stationary solution is reached within a few Jeans timescales $t_j = (4\pi G \rho_0)^{-1/2}$. The resulting density profile is a power law $\rho \sim r^\alpha$. The DM gravitational potential is determined by this density profile.

After recombination baryonic gas starts falling into the dark matter potential wells. Cosmology predicts in principal all necessary initial conditions after specifying a redshift z. We studied the spherical accretion of $10^7\,M_\odot$ baryonic gas onto the potential well of $10^8\,M_\odot$ dark matter, taking into account optically thin Compton-cooling.

Initially the gravitational potential is dominated exclusively by dark matter. A density fluctuation with $1000\,pc$ diameter contracts within a few Jeans timescales. During the contraction phase the potential due to baryon self-gravity builds up. Pressure support begins to hold at $10 - 50\,pc$. The dark component is gravitationally negligible by then. The baryon cloud contracts further, driven by its selfgravity, until infalling mass shells shock at the inner pressure supported, optically thick region. This shock propagates outwards and leaves behind a heated core ($T \approx 10^7\,K$). The last mass shell finally bounces off the core and matter is ejected into the surrounding medium.

The baryonic gas in the outer regions is effectively cooled by Compton-scattering. Pressure cannot prevent baryonic gas from falling into the very center. The innermost region is very compact. Fowler [2] showed that a self-gravitating system suffers from relativistic gravitational instability whenever

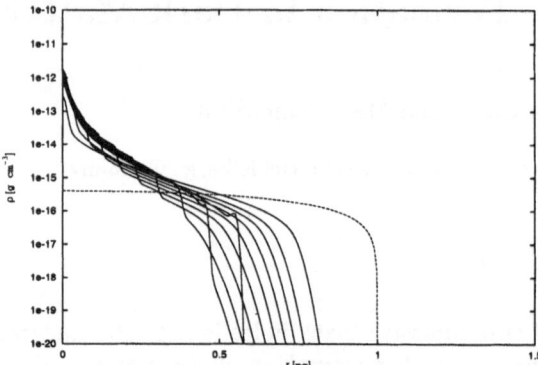

Fig. 1. Time-sequence $0 \leq t \leq 2.6\, t_j$ of the density profile. *Dashed lines* are initial values (top hat density profile). The baryons collapse within a few Jeans timescales.

the density exceeds a critical value. This can be expressed in terms of a critical compactness parameter $(R_S/R)_{crit} = 1.26 \times 10^{-3}\, M_6^{-1/2}$. Considering additional cooling mechanisms or radiation transport might allow us to reach this critical value.

3   Discussion

We numerically investigate the spherical accretion of dense baryonic matter into the potential well of dark matter, including baryon self-gravity and optically thin Compton-cooling. Though these are only preliminary results, our simulations suggest that massive cores with masses of $10^7\, M_\odot - 10^8\, M_\odot$ might form as early as $z = 50 - 25$. If these cores stay optically thick, they will cool, contract further, eventually become gravitationally unstable and collapse to form a massive black hole. If angular momentum of the baryonic gas was sufficiently high, a self-gravitating disk is expected to form. As angular momentum is distributed broadly among the initial density fluctuations, the present model would correspond to high density fluctuations with only moderate angular momentum. Angular momentum transport in such configurations might be acquired through the radiation drag mechanism as proposed in [5].

References

1. H. Couchman, C.S. Frenck, et al. MNRAS, 296:1061-1071, 1998.
2. W.A. Fowler. *Rev. mod. Phys.*, 36:545, 1964.
3. A.V. Gurevich and K.P. Zybin. *Sov. Phys. JETP*, 67:1-12, 1988.
4. Navarro, J. F., Frenk, C. S. & White, S. D. M. 1997, ApJ, 490, 493
5. Umemura, M. , Loeb, A. & Turner, E. L. 1993, ApJ, 419, 459

Author Index

ESO ASTROPHYSICS SYMPOSIA
European Southern Observatory

Series Editor: Jacqueline Bergeron

G. Meylan (Ed.), **QSO Absorption Lines**
Proceedings, 1994. XXIII, 471 pages. 1995.

D. Minniti, H.-W. Rix (Eds.), **Spiral Galaxies in the Near-IR**
Proceedings, 1995. X, 350 pages. 1996.

H. U. Käufl, R. Siebenmorgen (Eds.), **The Role of Dust in the Formation of Stars**
Proceedings, 1995. XXII, 461 pages. 1996.

P. A. Shaver (Ed.), **Science with Large Millimetre Arrays**
Proceedings, 1995. XVII, 408 pages. 1996.

J. Bergeron (Ed.), **The Early Universe with the VLT**
Proceedings, 1996. XXII, 438 pages. 1997.

F. Paresce (Ed.), **Science with the VLT Interferometer**
Proceedings, 1996. XXII, 406 pages. 1997.

D. L. Clements, I. Pérez-Fournon (Eds.), **Quasar Hosts**
Proceedings, 1996. XVII, 336 pages. 1997.

L. N. da Costa, A. Renzini (Eds.), **Galaxy Scaling Relations: Origins, Evolution and Applications**
Proceedings, 1996. XX, 404 pages. 1997.

L. Kaper, A. W. Fullerton (Eds.), **Cyclical Variability in Stellar Winds**
Proceedings, 1997. XXII, 415 pages. 1998.

R. Morganti, W. J. Couch (Eds.), **Looking Deep in the Southern Sky**
Proceedings, 1997. XXIII, 336 pages. 1999.

J. R. Walsh, M. R. Rosa (Eds.), **Chemical Evolution from Zero to High Redshift**
Proceedings, 1998. XVIII, 312 pages. 1999.

J. Bergeron, A. Renzini (Eds.), **From Extrasolar Planets to Cosmology:**
The VLT Opening Symposium
Proceedings, 1999. XXVIII, 575 pages. 2000.

A. Weiss, T. G. Abel, V. Hill (Eds.), **The First Stars**
Proceedings, 1999. XIII, 355 pages. 2000.

A. Fitzsimmons, D. Jewitt, R. M. West (Eds.), **Minor Bodies in the Outer Solar System**
Proceedings, 1998. XV, 192 pages. 2000.

L. Kaper, E. P. J. van den Heuvel, P. A. Woudt (Eds.), **Black Holes in Binaries and Galactic Nuclei:**
Diagnostics, Demography and Formation
Proceedings, 1999. XXIII, 378 pages. 2001.

Series homepage – http://www.springer.de/phys/books/eso/